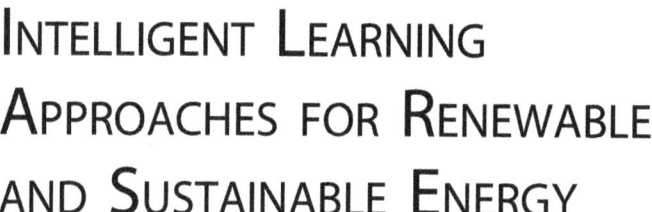

INTELLIGENT LEARNING
APPROACHES FOR RENEWABLE
AND SUSTAINABLE ENERGY

INTELLIGENT LEARNING APPROACHES FOR RENEWABLE AND SUSTAINABLE ENERGY

Edited by

JOSEP M. GUERRERO
*AAU Energy, Center for Research on Microgrids (CROM),
Aalborg University, Aalborg, Denmark*

PANKAJ GUPTA
*Indira Gandhi Delhi Technical University for Women,
Delhi, India*

RITU KANDARI
*Indira Gandhi Delhi Technical University for Women,
Delhi, India*

ALEXANDER MICALLEF
*Department of Electrical Engineering,
University of Malta, Msida, Malta*

ELSEVIER

Elsevier
Radarweg 29, PO Box 211, 1000 AE Amsterdam, Netherlands
125 London Wall, London EC2Y 5AS, United Kingdom
50 Hampshire Street, 5th Floor, Cambridge, MA 02139, United States

ISBN: 978-0-443-15806-3

For Information on all Elsevier publications
visit our website at https://www.elsevier.com/books-and-journals

Publisher: Megan Ball
Acquisitions Editor: Edward Payne
Editorial Project Manager: Joshua Mearns
Production Project Manager: Kamesh R.
Cover Designer: Miles Hitchen

Typeset by MPS Limited, Chennai, India

Working together
to grow libraries in
developing countries

www.elsevier.com • www.bookaid.org

Contents

Section I Introduction to intelligent learning approaches for renewable and sustainable energy

Section II Applications of intelligence learning approaches for renewable and sustainable energy

Section III Intelligent learning methods for optimizing integrated energy systems

9. Intelligent power quality disturbance detection methods in virtual power plants: state-of-the-art   267

Gökay Bayrak and Alper Yilmaz

List of contributors

Mohammad Ali
K.A. CARE Energy Research and Innovation Center, King Fahd University of Petroleum and Minerals, Dhahran, Saudi Arabia

Deivis Avila
Higher Polytechnic School of Engineering (EPSI), University of La Laguna, Santa Cruz de Tenerife, Spain

Gökay Bayrak
Smart Grid Laboratory, Department of Electrical and Electronics Engineering, Bursa Technical University, Bursa, Turkey

Víctor Caínzos López
Department of Industrial Engineering, University of A Coruña, CTC, CITIC Research, Rúa Mendizábal, Ferrol, A Coruna, Spain

Jose Luis Calvo-Rolle
Department of Industrial Engineering, University of A Coruña, CTC, CITIC Research, Rúa Mendizábal, Ferrol, A Coruna, Spain

Yanelys Cuba
Centre for Advanced and Sustainable Manufacturing Studies, University of Matanzas, Matanzas, Cuba

Jean Laurent Duchaud
Sciences for Environment Laboratory, University of Corsica Pasquale Paoli, Ajaccio, France

Ghjuvan Antone Faggianelli
Sciences for Environment Laboratory, University of Corsica Pasquale Paoli, Ajaccio, France

Zhang Ge
Key Laboratory of Power System Optimization and Energy Saving Technology, Guangxi University, Nanning, P.R. China

Muhammad Majid Gulzar
Control and Instrumentation Engineering Department, Interdisciplinary Research Center for Renewable Energy and Power Systems (IRC-REPS), King Fahd University of Petroleum and Minerals, Dhahran, Saudi Arabia

Esteban Jove
Department of Industrial Engineering, University of A Coruña, CTC, CITIC Research, Rúa Mendizábal, Ferrol, A Coruna, Spain

Muhammad Khalid
Electrical Engineering Department, Interdisciplinary Research Center for Renewable Energy and Power Systems (IRC-REPS), King Fahd University of Petroleum and Minerals, Dhahran, Saudi Arabia

Zheng Liqin
Key Laboratory of Power System Optimization and Energy Saving Technology, Guangxi University, Nanning, P.R. China; State Grid Xiamen Electric Power Supply Company, Xiamen, P.R. China

Zhengxuan Liu
Faculty of Architecture and the Built Environment, Delft University of Technology, Delft, The Netherlands

Graciliano N. Marichal
Higher Polytechnic School of Engineering (EPSI), University of La Laguna, Santa Cruz de Tenerife, Spain

Hasan Meral
Smart Grid Laboratory, Department of Electrical and Electronics Engineering, Bursa Technical University, Bursa, Turkey; Sibernetik Machinery & Automation R&D Center, Bursa, Turkey

Álvaro Michelena
Department of Industrial Engineering, University of A Coruña, CTC, CITIC Research, Rúa Mendizábal, Ferrol, A Coruna, Spain

Gilles Notton
Sciences for Environment Laboratory, University of Corsica Pasquale Paoli, Ajaccio, France

Sarah Ouédraogo
Sciences for Environment Laboratory, University of Corsica Pasquale Paoli, Ajaccio, France

Wang Puming
Key Laboratory of Power System Optimization and Energy Saving Technology, Guangxi University, Nanning, P.R. China

Hector Quintian
Department of Industrial Engineering, University of A Coruña, CTC, CITIC Research, Rúa Mendizábal, Ferrol, A Coruna, Spain

Ramón Quiza
Centre for Advanced and Sustainable Manufacturing Studies, University of Matanzas, Matanzas, Cuba

Zhu Songyang
Key Laboratory of Power System Optimization and Energy Saving Technology, Guangxi University, Nanning, P.R. China

Miriam Timiraos
Department of Industrial Engineering, University of A Coruña, CTC, CITIC Research, Rúa Mendizábal, Ferrol, A Coruna, Spain

Cyril Voyant
Sciences for Environment Laboratory, University of Corsica Pasquale Paoli, Ajaccio, France

Shaojun Wang
China Construction Fourth Engineering Division, Corp. Ltd, Guangdong, P.R. China

Bai Xiaoqing
Key Laboratory of Power System Optimization and Energy Saving Technology, Guangxi University, Nanning, P.R. China

Shi Xiaoqing
Key Laboratory of Power System Optimization and Energy Saving Technology, Guangxi University, Nanning, P.R. China

Alper Yilmaz
Smart Grid Laboratory, Department of Electrical and Electronics Engineering, Bursa Technical University, Bursa, Turkey

Linfeng Zhang
Department of Underground Engineering, School of Transportation, Southeast University, Nanjing, P.R. China

Preface

The union of intelligent learning approaches with cutting-edge technologies in the field of renewable and sustainable energy has triggered a revolution. This book, "Intelligent Learning Approaches for Renewable and Sustainable Energy," serves as a comprehensive guide to understanding and leveraging the transformative power of artificial intelligence (AI) and machine learning (ML) in the pursuit of a more sustainable energy future.

The world is at a critical juncture, facing challenges of climate change, depleting natural resources, and an increasing demand for energy. It is against this backdrop that this book navigates the intricate landscape of intelligent learning approaches, unravelling their potential applications and contributions to renewable and sustainable energy systems. The chapters within this volume traverse the spectrum from theoretical foundations to practical implementations, offering insights, methodologies, and real-world case studies. This book is a collaborative effort of esteemed contributors who are at the forefront of research and application in the field of renewable and sustainable energy. As we navigate the intricate landscape of intelligent learning approaches, we invite readers, researchers, practitioners, and policymakers to embark on a journey that transcends theoretical boundaries and unlocks the potential of intelligent technologies in shaping a sustainable energy future.

Section I: Introduction to intelligent learning approaches for renewable and sustainable energy

The journey begins with an in-depth exploration of the fundamental shifts in the energy landscape. Chapter 1 introduces readers to the fusion of AI, ML, renewable energy sources (RES), energy storage, electric vehicles (EVs), and prosumers. The authors investigate the status of RES and storage systems, showcasing a case study that explains the application of AI in power electronics-driven RES. In Chapter 2, the focus shifts to the crucial aspect of managing the demand side in the context of EV charging stations. The proposed method presents an innovative approach to optimizing charging parameters, addressing the evolving needs of an electric mobility future. The integration of stochastic processes in renewable energy systems is explored in Chapter 3. By combining the Monte Carlo method and mixture density networks, the authors explain the path to reliable predictions in the face of the inherent variability of RES. Chapter 4 presents a comprehensive overview on the improvement of efficiency and profitability of photovoltaic microgrids through advanced forecasting tools. The chapter also explains various control strategies, demonstrating the pivotal role of solar production forecasting in energy management.

Section II: Applications of intelligence learning approaches for renewable and sustainable energy

The second section goes into practical applications of intelligent learning models in renewable energy forecasting. Chapter 5 provides an in-depth look into various case studies, forecasting/modeling techniques, and their impact on real-world scenarios. Chapter 6 extends the exploration to the intricate challenge of predicting multienergy loads. The authors present an AI method for predicting load characteristics, offering a comprehensive analysis and outlook for this critical side of energy planning. Chapter 7 addresses the significant role of demand-side controllers in the context of building-integrated photovoltaics. The chapter explores challenges, reviews existing literature, and presents a case study, shedding light on the potential of intelligent learning in optimizing energy consumption in buildings.

Section III: Intelligent learning methods for optimizing integrated energy systems

The final section explores the complexities of optimizing integrated energy systems. Chapter 8 introduces advanced approaches to optimization, considering the inherent uncertainties in energy systems. The authors present a case study that emphasizes the practicality and effectiveness of the proposed methods. Finally, Chapter 9 explores the critical domain of power quality in virtual power plants. The chapter evaluates integration criteria, planning parameters, and physical evaluations, providing a state-of-the-art analysis of power quality disturbance detection.

Introduction to intelligent learning approaches for renewable and sustainable energy

CHAPTER ONE

Transforming the grid: AI, ML, renewable, storage, EVs, and prosumers

Mohammad Ali[1], Muhammad Khalid[2] and
Muhammad Majid Gulzar[3]

[1]K.A. CARE Energy Research and Innovation Center, King Fahd University of Petroleum and Minerals, Dhahran, Saudi Arabia
[2]Electrical Engineering Department, Interdisciplinary Research Center for Renewable Energy and Power Systems (IRC-REPS), King Fahd University of Petroleum and Minerals, Dhahran, Saudi Arabia
[3]Control and Instrumentation Engineering Department, Interdisciplinary Research Center for Renewable Energy and Power Systems (IRC-REPS), King Fahd University of Petroleum and Minerals, Dhahran, Saudi Arabia

1.1 Introduction

The depletion of fossil fuels coupled with a rising global energy demand necessitates the human race's transition to natural and renewable energy sources. It is debatable whether current technology will enable these intermittent sources to serve as the primary fulfiller of the energy demand; however, the current scenario suggests that these sources will become significant contenders to the existing fossil fuel-based utilities. Modern power systems with large-scale renewable energy source (RES) penetrations are evolving into mature and stabilized systems through continuous monitoring and forecasting. Artificial intelligence and machine learning are playing, and will play, an essential role in establishing resilient systems that respond more effectively to the renewable energy sources' intermittent behavior. The following sections discuss the advancement in the technologies that have led to the transformation of the modern grid. Artificial intelligence, machine learning, renewable energy integration, electric vehicles, and energy storage elements such as batteries and supercapacitors make a blend that has paved the way for the modernization and continuous transformation of the grid. In Table 1.1, the features of the conventional grid and the modern grid are shown and compared. It can be concluded that the modern grid behaves

Intelligent Learning Approaches for Renewable and Sustainable Energy
DOI: https://doi.org/10.1016/B978-0-443-15806-3.00001-2

Table 1.1 Features of the conventional grid and the modern grid.

Features	Normal grid	Modern grid
Generation type	Centralized due to large fossil fuel-based power plants	Distributed due to the employment of renewable energy systems
Maintenance method	Mostly manual equipment check	Remote monitoring
Fault location determination and monitoring	Difficult and manual	Remote determination and predictable
Technology	Noncommunicable electromechanical devices	Remotely accessed and controlled digital devices
In case of failure of equipment	Postfailure may lead to blackouts	Adaptive to failure and can be automatically isolated and reconnected
Communication flow	From consumer to utility	Two-way communication is possible for prosumers
Power flow	From the grid to the consumer	Prosumer-based two-way power flow

smartly due to better communication paradigms and thus leads to a more secure and stable grid with all the renewable energy sources and their intermittent behaviors.

1.2 Artificial intelligence and machine learning in the modern grid

The modern electrical grid faces a myriad of scalability, autonomous control, and reliability challenges arising from the industrial revolution and population growth (Sun et al., 2019). The smart grid (SG) paradigm seems to be the most promising solution to these challenges. The concept of SG has revolutionized the existing grid with better reliability and transparency (You et al., 2020). The distributed RES market is booming and is estimated to surpass the traditional sources market (Ullah et al., 2020). Thus, robust and intelligent platforms are required to play an essential role in penetrating the Distributed Energy Resources (DERs) into the existing power system. For their smooth and mature penetration, the modern distribution network requires a robust and intelligent coordination platform between the different elements of power systems. Sophisticated problems can be effectively addressed through the fusion of AI and SG, leveraging AI's learning, reasoning, and decision-making capabilities to streamline operations (Vinuesa et al., 2020). As modern power systems embark toward an AI-based transition, the technologies under consideration span from the component level, such as the power electronic converters, to larger system operations addressing economic power dispatch (Ahmad et al., 2020; Ali & Choi, 2020; Ali, Tariq, Lodi, et al., 2021; Brockway et al., 2019; Ourahou et al., 2020; Ullah et al., 2020; Yin et al., 2020; Zhao, & Blaabjerg, & Wang, 2021). The main applications can be visualized in Fig. 1.1, while the applications with some inference can be seen in Table 1.2.

1.2.1 AI-based load forecasting

Load forecasting involves the analysis of data pertaining to previous load consumption, the environmental impact on load behavior, optimal operational costs, power quality, amongst others. The prediction of the future is essential for ensuring the stable operation of the overall electrical system. These data must be analyzed over short, medium, and long durations (Refaat et al., 2021). Table 1.3 highlights AI-based techniques applied in load forecasting across different timeframes. Long-term forecasting requires data ranging from one year to 5 years (Kaytez, 2020; Nalcaci et al., 2019). Similarly, medium-term and short-term forecasting require data ranging from a few days to one month (Boroojeni et al., 2017; Shah

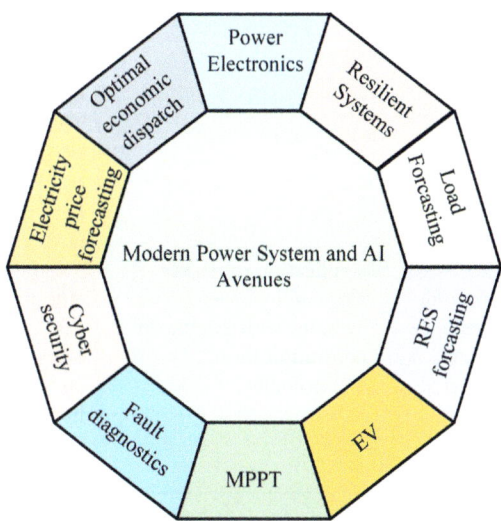

Figure 1.1 Applications of AI in the modern grid.

Table 1.2 Various applications of AI in modern power system scenarios.

Ref.	Application	Inference
Ahmad et al. (2020); Bou-Rabee et al. (n.d.); Hossain et al. (2023); Massaoudi et al. (2021)	Renewable Energy forecasting using ANN and ML Models for modern power systems	Investigating the short-term forecasting of solar and wind for power system applications is focused on comparison to long or medium-term forecasting.
Ourahou et al. (2020)	Smart Grid reliability analysis using AI with RES in picture	Grid's operational reliability is a significant concern that includes frequency and voltage deviations, overloading, voltage collapse, and control measures.
Yin et al. (2020)	Smart power dispatch using AI methods and parallel learning	AI and ML models are necessary for smart dispatch and generation control with regular improvement in robustness, reliability, and accuracy. Further, big data can play an important role in the grid technology transformation.
Ali, Tariq, Lodi, et al. (2021); Lodi et al. (2020); Upadhyay et al. (2022); Zhao and Chen (2021)	AI applications for power electronics	The application of AI in power electronics encompasses from its initial design phase to the final maintenance phase. The practical implementation of continuous improvising systems is to be researched.
"Achieving Net Zero Emissions with Machine Learning: The Challenge Ahead," (2022); Heo et al. (2022); Ullah et al. (2020)	AI application in smart cities and net-zero avenues with intelligent communication technologies	5 G and big data will empower the development of smart cities. For this, the SG communication infrastructure development is most important.

Table 1.3 AI-based load forecasting of various durations.

Ref.	Horizon	AI model	Accuracy
Long-term forecasting			
Kaytez (2020)	5 years	Least square, autoregressive integrated moving average.	MAPE = 1.02%
Nalcaci et al. (2019)	2 years	Multivariate adaptive regression splines, artificial neural network, linear regression.	MAPE = 6%
Medium-term forecasting			
Shah et al. (2020)	One month	Autoregressive, nonparametric autoregressive, nonparametric autoregressive, autoregressive integrated moving average.	MAPE = 4.85%
Boroojeni et al. (2017)	One week	Autoregressive, moving average.	MAPE = 4.46%
Short-term forecasting			
Aly (2020)	One day	Kalman filtering, wavelet neural networks, artificial neural networks.	MAPE = 2.15%
Liu et al. (2020)	5 min	Asynchronous advantage actor-critic, deep deterministic policy gradient.	R^2 = 98%

MAPE, mean absolute percentage error; R^2, coefficient of determination.

et al., 2020) and from one second to a few hours (Aly, 2020; Liu et al., 2020), respectively. From the table, one can observe that various AI-based techniques have been used in literature. The accuracy of predictions derived from historical data directly influences the cost-effectiveness and stability facilitation of the evolving grid structure.

1.2.2 AI-based renewable energy forecasting

The energy generation from the RES is intermittent primarily due to the unpredictable changes in weather conditions. Solar energy is generated from a fluctuating irradiance in short duration during the day, varying average irradiance over the day in medium duration, and experiencing monthly and yearly variations in irradiance for long-term forecasting (Zhang et al., n.d.). Similarly, the intermittent nature of wind energy comes from the variations in the wind speed in short, medium, and long time durations throughout the year (Azad et al., 2014; Bou–Rabee et al.,

2020; Han & Tong, 2020; Hossain et al., 2023). Forecasting these energy sources in terms of their capacity (solar irradiance and wind speed) and the amount generation capability will enable prediction and react early to the threats of grid instability and unsatisfactory injection of power to the transforming grid. Various AI-based techniques applied in the literature for the prediction of wind and solar energy are summarized in Table 1.4. Continuous research on forecasting models is required to enhance the accuracy and reduce the computation time (Refaat et al., 2021).

1.2.3 EVs operation, AI, and modern grid integration

The demand of the time is resilient and reliable electric vehicle (EV) operation, given the steep rise in the electric vehicle wave. Currently, the market

Table 1.4 AI-based duration forecasting of RES, focusing on wind and solar energy.

Ref.	Horizon	AI model	Accuracy	RES
Long-term forecasting				
Azad et al. (2014)	1 year	Nonlinear autoregressive exogenous.	MAPE = 0.8%	Wind speed
Medium-term forecasting				
Yousuf et al. (2019)	One month	Gaussian kernel regression, random feature extraction.	RMSE = 11.37 MW	Wind power and speed
Bou-Rabee et al. (2020)	One month	Particle swarm optimization, artificial neural networks.	MAPE = 3%−6%	Wind speed
Short-term forecasting				
Han & Tong (2020)	10 min	Wavelet packet decomposition, improved gray wolf optimization.	$R^2 = 99.27\%$	Wind power
Chen & Liu (2020)	15 min	Convolutional long short-term memory, back–propagation neural networks, modified whale optimization.	MAPE = 2.62%	Wind speed
Zhang et al. (n.d.)	1 hour	Generative adversarial networks.	MAE = 0.10 MW	Solar PV

MAE, mean absolute error; *RMSE*, root mean square error.

is valued at approximately 205 billion USD and is projected to grow eight-fold, reaching 1716 billion USD in the next ten years (Electric Vehicle Market - Global Industry Analysis, Size, Share, Growth, Trends\n, Regional Outlook, & Forecast 2023–2032, 2023). The attention drawn by EVs is attributed to its pollution-free characteristics, contributing to a reduction in carbon footprint. Additionally, it is also recognized for its social and economic impact on users and stakeholders, particularly through community microgrids (Ahmad et al., 2019). Lithium–ion (Li-ion) batteries are primarily used to power the electric motors of EVs. Exploration of V2G (vehicle-to-grid) aims to leverage the contribution of vehicles to the grid, marking the initiation of another pollution-free distributed energy source in the modern grid and EV integration. However, the implementation is challenging, demanding extensive prediction and estimation on both the load and source sides. Estimation of the State-of-Charge (SoC) is a critical necessity to assess the performance of electric vehicles (EVs) both on and off-road. This was demonstrated by How et al. (2019) for Li-ion batteries by adopting a predictive method with a neural network configured based on empirical heuristics. Recently, deep reinforcement learning has been employed Qiu et al. (2020) to solve the pricing problem of EVs integrated with the grid.

1.2.4 AI in modern grid fault diagnostics

Modern grids face vulnerabilities from traditional electrical faults, and the integration of RES introduces new fault scenarios. Repairing faults incurs high costs, especially when a delay occurs due to manual diagnosis and lack of engineering expertise. Correct fault diagnosis is essential to operate the DER-based modern grids safely. AI can effectively contribute to predicting and detecting faults. Detailed discussions on SG faults are presented by Labrador Rivas & Abrão (2020), suggesting the superior performance of AI-based techniques compared to analytical fault detection methods.

1.3 Status of RES and storage systems in the modern grid

The structure of the modern power system network is shown in Fig. 1.2, depicting the integration of nonconventional renewable energy sources and energy storage systems. Fig. 1.3 provides an overview of the

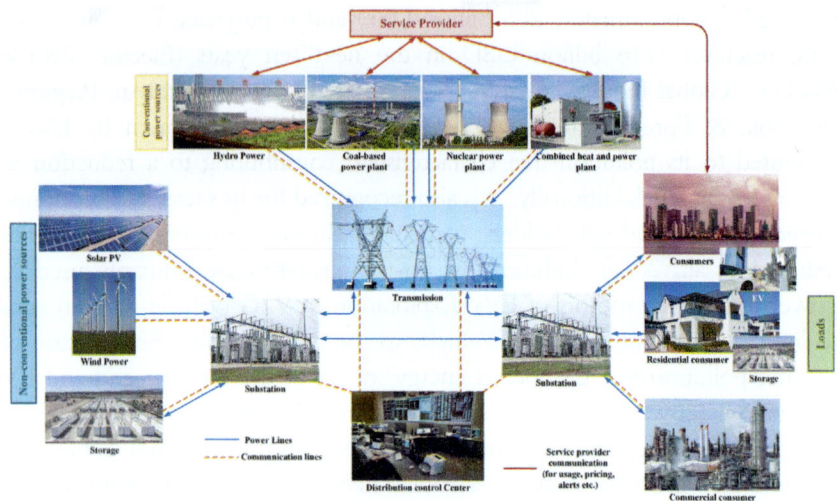

Figure 1.2 Schematic showing the modern power system and its components.

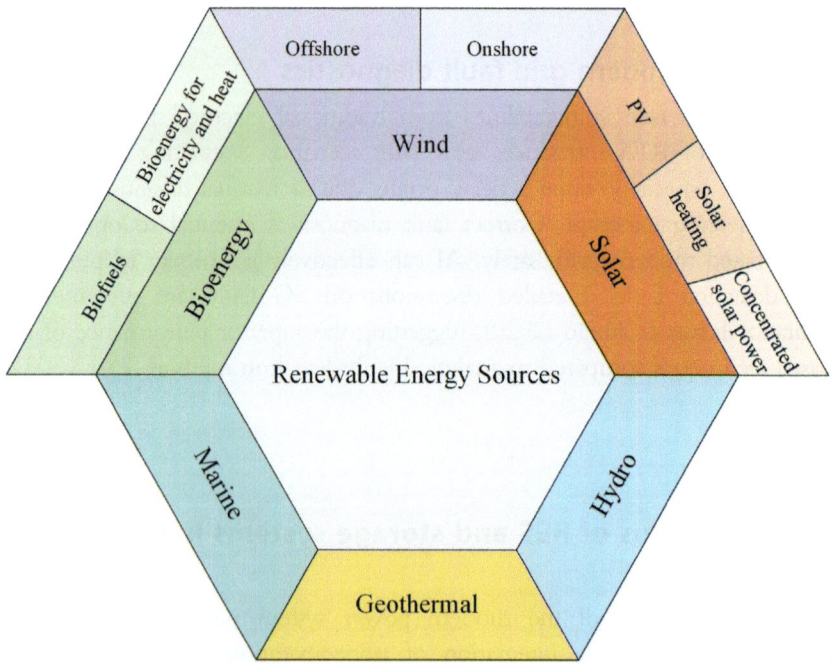

Figure 1.3 Common renewable energy sources.

various RES technologies currently being explored, categorizing them into six main groups. The rising penetration of RES into the existing system is imperative to meet the ever-increasing global energy demand (CanmetENERGY, 2020; ExxonMobil, 2019). This rise in RES integration has led to a reduction in capital costs, and projections suggest that the penetration level is expected to surpass its current state by 2050 (U.S. Energy Info. Admin., Annual Energy Outlook 2022 (with projections to 2050), 2022). The stable operation of these RES is crucial to ensure the satisfactory performance of the modern grid.

1.3.1 Status of RES in the modern grid

1.3.1.1 Solar energy

Solar energy is deployed in the form of solar PV, solar thermal heating and cooling, and concentrated solar power (CSP). In solar PV systems, the sun's insolation is transformed into electrical energy. Semiconductor devices form the photo voltaic (PV) cells, which are connected to form a PV module. Multiple PV modules are connected together in series and parallel to form an array. These solar PV arrays are deployed in large numbers to form solar farms with capacities in the MW range. The PV systems are supported by additional components, including power electronic converters, energy storage, and mounting systems to enhance its efficiency and functionality. Solar PV technology is often considered to be superior to CSP due to its ability to harness the diffused component of sunlight thus making PV systems capable of partially generating electricity even in cloudy conditions (Alam et al., 2020). As the penetration of electrical energy from solar PV is increasing, the modern grid must advance its capabilities to accommodate intermittent generation conditions. Further more, it must be capable of quickly mitigating the counter-effects, thereby ensuring a stable and robust system (Refaat et al., 2021).

1.3.1.2 Wind energy

Wind's kinetic energy is converted to electrical energy by employing electrical generators connected through wind turbines. According to the Betz's limit, the wind turbine can theoretically capture a maximum of 59.3% of wind power. Practically, the conversion is limited to 35%–45% (Ali et al., 2022; Refaat et al., 2021). As the captured power is directly proportional to the blade sweep area, the objective is to design wind turbines with the largest possible blade sweep area. This necessitates keeping the nacelle at elevated heights, consequently incurring higher installation costs. To mitigate expenses, there is a desire for fewer components, as well as enhanced system and component-level operation. The installation of offshore wind farms has experienced a significant increase compared to onshore installations, particularly in the Scandinavian region.

Advanced turbines use variable speed drives (VSDs) to align with the requirements of the modern grid. The two main arrangements commonly used by the industry include synchronous generators connected to the grid using full power converters and doubly fed induction generators connected to the grid with slip-power rating power electronic converters. VSDs play a crucial role in decoupling the impact of the wind's intermittent profile from the grid, allowing for the production of controlled quality waveforms. In terms of power conversion, either. double-stage voltage source converters can be employed (Ali et al., 2022), or direct AC-AC three-phase or multiphase converters may also be employed (Ali et al., 2017, 2018). AI-based optimization tools, such as differential evolution, can also be applied to control these power converters (Ali et al., 2019; Ali & Khalid, 2022.).

1.3.1.3 Other renewable energy sources

Renewable energy sources other than solar and wind are limited due to geographical locations. Hydropower is one of the largest and most reliable RES but presents a geographical limitation, particularly in arid regions like the Middle East where water resources are scarce. Despite these limitations, countries such as China, Canada, Brazil, and the USA are using their water resources effectively, leveraging hydropower to meet a significant portion of their electricity demand. Tidal and ocean energy are primarily accessible near the shore; moreover, deploying installations for generating bulk electrical power remains a challenge that necessitates further research. Projections suggest that the maximum global installation by 2050 may reach around 300 to 350 GW. Meanwhile, notable strides have been made in exploring geothermal energy avenues in countries such as Turkey, Kenya, Indonesia, Mexico, and Japan. A target of 140 GW from geothermal energy sources is anticipated by the year 2050 (Refaat et al., 2021).

1.3.2 Status of storage systems in the modern grid

The need to store surplus energy has always been desired due to fluctuations in daily requirements and seasonal reasons. Thus it is required to have ample capacity storage to meet the average demand rather than the peak demand (Al-Humaid et al., 2021). Various energy storage systems can be seen in Fig. 1.4. Further, the penetration of RES with intermittent nature into the conventional grid hits the grid's reliability. The ESS can be implemented to

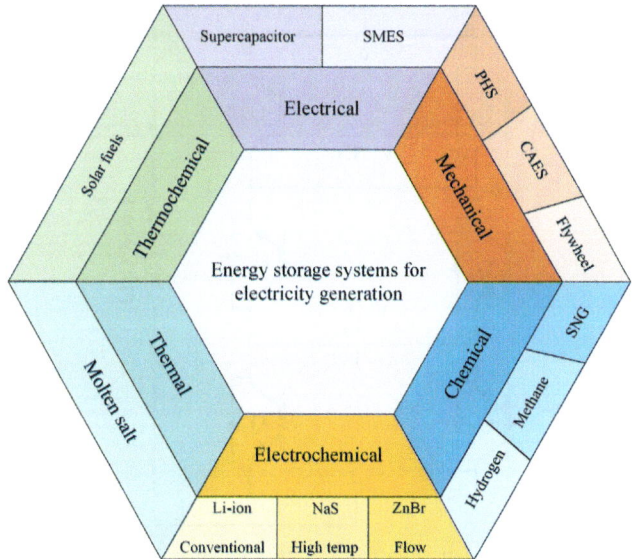

Figure 1.4 Various energy storage systems for electricity generation.

improve the RES operation under disturbances, thus leading to efficient power generation (Rauf et al., 2022). The ESS exhibited in Fig. 1.4 are regularly explored regarding cost, point of applications, and mitigation of source intermittencies. ESS and its various forms will undoubtedly support the modern grid infrastructure, where their capacity and control can be judged effectively (Alam et al., 2019). According to (Renewables, 2019), pumped storage hydropower is a significant player in ESS, which makes the total ES capacity 5.5% of the total installed generation capacity. Considerable increase in battery storage, thermal, and electromechanical storage in recent years. But, a large amount of ESS is anticipated to be employed in the future and will stabilize the future grid performance (Alam et al., 2023; Khalid, n.d.).

1.4 Case study: application of AI in power electronics driven RES

In Ali, Tariq, Lodi, et al. (2021), an artificial neural network-based controller was used to operate the modified Packed U-Cell (MPUC-5) multilevel inverter by Vahedi et al. (2018) in a solar PV application subjected to variable solar insolation. The MPUC-5 represents a multilevel

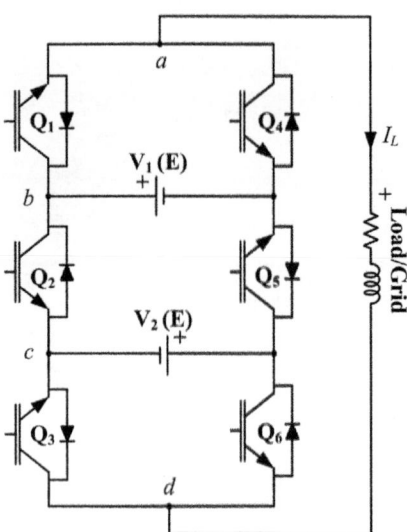

Figure 1.5 Schematic of a modified PUC-5 multilevel inverter. Source: *From M. Ali et al., Robust ANN-Based Control of Modified PUC-5 Inverter for Solar PV Applications, IEEE Trans. Ind. Appl., 57(4) (2021), 3863–3876.*

inverter capable of connecting two sources of equal magnitude, as shown in Fig. 1.5. The variation in solar isolation results in a variable boost converter voltage output, aimed at maintaining the load line at the maximum power point (MPP). When employing a standard stepped output voltage, the fundamental voltage component of the MPUC-5 output changes. Thus, optimum angles must be determined to obtain an output voltage with a constant fundamental, minimum total harmonic distortion (THD) and minimal DC-link voltage changes. Genetic Algorithm (GA) can help to generate the required dataset for training the ANN-based controller. This, in turn, facilitates the operation of the MPUC-5 under variable DC-link voltages. GA, along with other metaheuristic techniques, has been successfully employed in various Power Electronics applications (Ali et al., 2019, 2021; Ali, Tariq, Lin, et al., 2021; Iqbal et al., 2019; Lodi et al., 2020). Fig. 1.6 shows the two waveforms with zero and one notch, 2-angle and 4-angle operation. The states of operation of the MPUC are shown in Fig. 1.7 and summarized in Table 1.5.

1.4.1 Problem formulation

Increasing the switching angles enhances the harmonics elimination capability but at the cost of complexity. The 5-level waveforms for

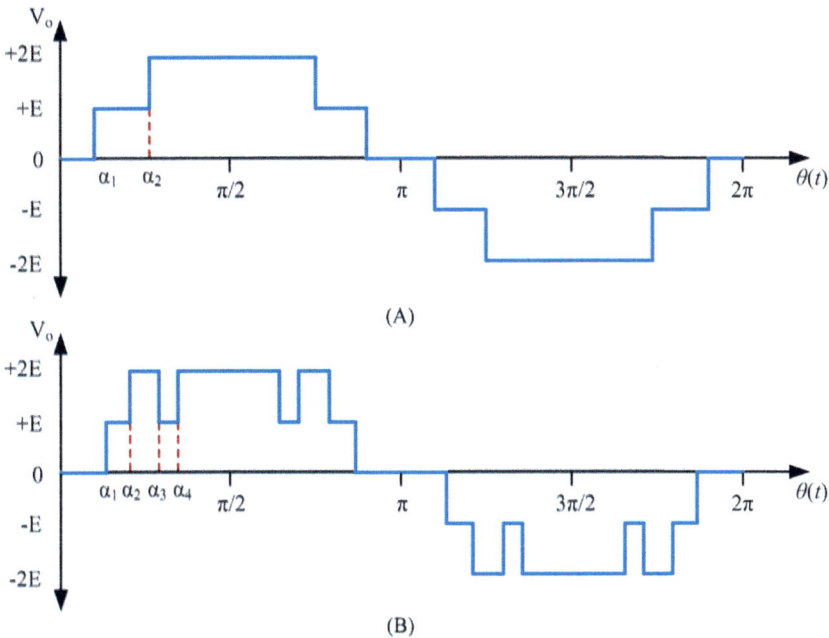

Figure 1.6 (A) 2-angle and (B) 4-angle 5-level waveforms. *Source: From M. Ali et al., Robust ANN-Based Control of Modified PUC-5 Inverter for Solar PV Applications, IEEE Trans. Ind. Appl., 57(4) (2021), 3863–3876.*

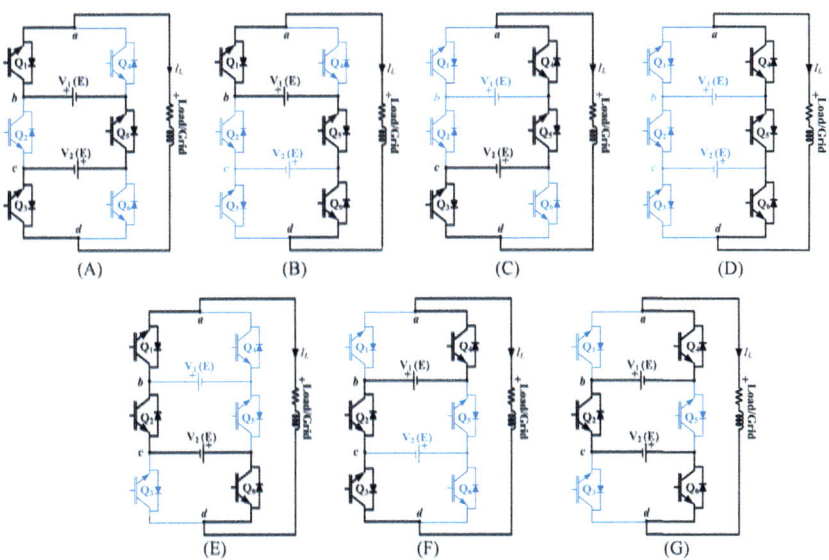

Figure 1.7 Possible operating states of MPUC-5 with following levels: (A) +2E, (B) + E, (C) + E, (D) 0, (E) − E, (F) − E, and (G) −2E levels. *Source: From M. Ali et al., Robust ANN-Based Control of Modified PUC-5 Inverter for Solar PV Applications, IEEE Trans. Ind. Appl., 57(4) (2021), 3863–3876.*

Table 1.5 Switching table for modified PUC-5 inverter.

State	Voltage	Q_1	Q_2	Q_3	Q_4	Q_5	Q_6	V_o
0	V_a	1	0	1	0	1	0	+2E
1	$V_a - V_c$	1	0	0	0	1	1	+E
2	V_c	0	0	1	1	1	0	+E
3	0	0	0	0	1	1	1	0
4	0	1	1	1	0	0	0	0
5	$-V_a$	1	1	0	0	0	1	-E
6	$V_c - V_a$	0	1	1	1	0	1	-E
7	$-V_c$	0	1	0	1	0	1	-2E

Source: From M. Ali et al., Robust ANN-Based Control of Modified PUC-5 Inverter for Solar PV Applications, IEEE Trans. Ind. Appl., 57(4) (2021), 3863−3876.

MPUC-5 operation are shown in Fig. 1.6, which is expressed mathematically as:

$$V_o = E[u\left(\omega t - \alpha_1\right) + \sum_{m=2}^{4}(-1)^m u(\omega t - \alpha_m) - u\left(\omega t - \left(\pi - \alpha_1\right)\right)$$

$$- \sum_{m=2}^{4}(-1)^m u(\omega t - (\pi - \alpha_m)) - u\left(\omega t - \left(\pi + \alpha_1\right)\right)$$

$$- \sum_{m=2}^{4}(-1)^m u(\omega t - (\pi + \alpha_m)) + u\left(\omega t - \left(2\pi - \alpha_1\right)\right)$$

$$- \sum_{m=2}^{4}(-1)^m u(\omega t - (2\pi - \alpha_m))]$$

The waveforms in Fig. 1.6 possess half and quarter-wave symmetries. Thus, the Fourier expansion will contain the odd harmonic number:

$$b_n = \frac{2}{T_o}\int_{T_o} V_o(t)\sin(n\omega_o t)dt \quad \forall n = 1, 3, 5, \cdots \qquad (1.2)$$

Solving further results in:

$$b_n = \frac{2E}{n\omega_o T_o}[4\cos(n\omega_o \alpha_1) + 4\cos(n\omega_o \alpha_2) - 4\cos(n\omega_o \alpha_3) + 4\cos(n\omega_o \alpha_4)]$$

$$(1.3)$$

The time period T_o is 2π. This results in $\omega_o = 1$ and thus:

$$b_n = \frac{4E}{n\pi}[\cos(n\alpha_1) + \cos(n\alpha_2) - \cos(n\alpha_3) + \cos(n\alpha_4)] \qquad (1.4)$$

The THD for the 4-angle operation can be expressed as:

$$f_1^{4\ ang}(\alpha) = \frac{\sqrt{\sum_{n=3,\ 5,\cdots}^{\infty} \frac{4E}{n\pi}[\cos(n\alpha_1) + \cos(n\alpha_2) - \cos(n\alpha_3) + \cos(n\alpha_4)]}}{\frac{4E}{n\pi}[\cos(\alpha_1) + \cos(\alpha_2) - \cos(\alpha_3) + \cos(\alpha_4)]}$$

(1.5)

To eliminate the third harmonic content and odd harmonics, Eq. 1.3 can be modified as follows:

$$f_2^{4ang}(\alpha) = b_{3n} = \frac{4E}{3n\pi}[\cos(3n\alpha_1) + \cos(3n\alpha_2) - \cos(3n\alpha_3) + \cos(3n\alpha_4)] = 0$$

(1.6)

Further, the following three conditions behave as constraints:

$$\left.\begin{array}{l}\cos(3n\alpha_1) + \cos(3n\alpha_2) = 0 \\ \cos(3n\alpha_4) - \cos(3n\alpha_3) = 0\end{array}\right\} A\, or \left.\begin{array}{l}\cos(3n\alpha_1) - \cos(3n\alpha_3) = 0 \\ \cos(3n\alpha_4) + \cos(3n\alpha_2) = 0\end{array}\right\} Bor \left.\begin{array}{l}\cos(3n\alpha_1) + \cos(3n\alpha_4) = 0 \\ \cos(3n\alpha_2) - \cos(3n\alpha_3) = 0\end{array}\right\} C$$

(1.7)

The above combinations are used for third harmonic elimination by using them with:

$$\left.\begin{array}{l}\alpha_1 + \alpha_2 = \pi/3 \\ \alpha_3 + \alpha_4 = 2\pi/3\end{array}\right\} A\, or \left.\begin{array}{l}\alpha_1 + \alpha_3 = 2\pi/3 \\ \alpha_2 + \alpha_4 = \pi/3\end{array}\right\} Bor \left.\begin{array}{l}\alpha_1 + \alpha_4 = \pi/3 \\ \alpha_2 + \alpha_3 = 2\pi/3\end{array}\right\} C$$
(1.8)

As shown in Ali, Tariq, Lodi, et al. (2021), only condition A can be used along with the following constraint:

$$0^o < \alpha_1 < \alpha_2 < \alpha_3 < \alpha_4 < 90^o$$
(1.9)

Thus, the functions $f_1(\alpha)$ and $f_2(\alpha)$ as derived in (1.5) and (1.6), respectively, are subject to the constrained minimization under conditions (1.8 A) and (1.9). In addition, the minimization problem is further constrained by the following:

$$E \times [\cos\alpha_1 + \cos\alpha_2 - \cos\alpha_3 + \cos\alpha_4] = \frac{V_{max}\pi}{4}$$
(1.10)

to produce a constant output voltage.

1.4.2 System under investigation

Two same-rating solar-PV arrays are considered for power delivery to an RL load. The system under examination is illustrated in Fig. 1.8, and the 5-level boosted operation is implemented using the configuration outlined

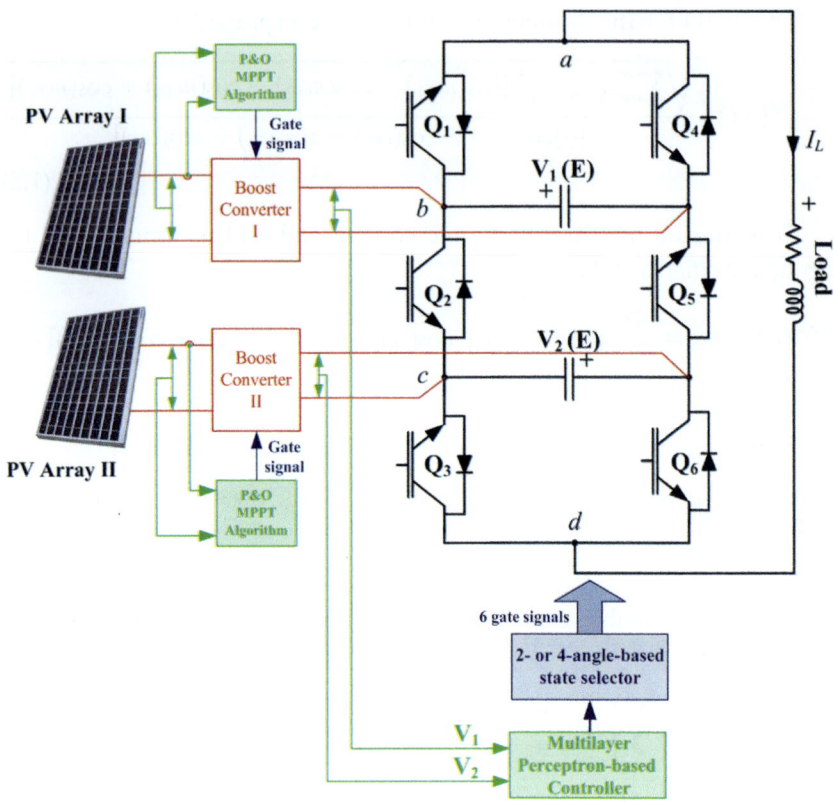

Figure 1.8 Schematic of the solar-PV integration under investigation. Source: *From M. Ali et al., Robust ANN-Based Control of Modified PUC-5 Inverter for Solar PV Applications, IEEE Trans. Ind. Appl., 57(4) (2021), 3863–3876.*

in Table 1.5. The PV panel's DC output is controlled by employing DC-DC boost converters controlled by perturb and observe (P&O) MPP tracking algorithm. This keeps the load line along the MPP locus on the Power-Voltage and Current-Voltage curves. Further more, an MLP-based ANN is employed to operate the MPUC-5 inverter. Following the training of the ANN-based controller, which operates on the input-output mapping, the controller exhibits an overall robust performance. The input values supplied to the controller consist of measured DC-link voltage values. As an output, the controller generates switching angles, and the inverter state is selected based on the output of the controller.. Extensive data, gathered over various epochs, is obtained by employing

Table 1.6 PV panel specifications.

Parameter	Rating
Panel type	Trina Solar TSM-250PA
Maximum Power per panel (W)	249.66
Open circuit voltage V_{oc} (V)	37.6
Cells per module (N_{cell})	60
Voltage at maximum power point V_{mp} (V)	31
Current at maximum power point I_{mp} (A)	8.06
Short-circuit current I_{sc} (A)	8.55
Shunt resistance R_{sh} (ohms)	301.8149
Temperature coefficient of I_{sc} (%/deg.C)	0.06
Series resistance R_s (ohms)	0.247

Source: From M. Ali et al., Robust ANN-Based Control of Modified PUC-5 Inverter for Solar PV Applications, IEEE Trans. Ind. Appl., 57(4) (2021), 3863–3876.

GA. The procedure and its application are discussed further in Section 1.4.3. For comprehensive insights, the specifications of the PV panels are given in Table 1.6.

1.4.3 Genetic algorithm for data generation

The genetic algorithm has been applied in this context for data generation by solving the above-mentioned equations. The steps employed in this algorithm are shown in Fig. 1.9. The primary stages involve the genetic evolution of the population using selection, crossover, and mutation to obtain mature data, which is then adapted as the final solution. For further insights, refer to the work by Ali, Tariq, Lodi, et al. (2021a). In this study, the population of the design variables consists of binary strings of eight bits.

1.4.3.1 Objective function

The design of the objective function adheres to the following conditions:
1. A minimum THD output voltage waveform (given by (1.5)),
2. A minimum triplen harmonic output waveform (provided by (1.6)), and
3. A constant fundamental-output voltage irrespective of DC-link voltage change (expression (1.10)).

For operation with four angles, the first objective function with minimum THD, and the second one with minimum THD and zero triplen harmonics, are defined as follows:

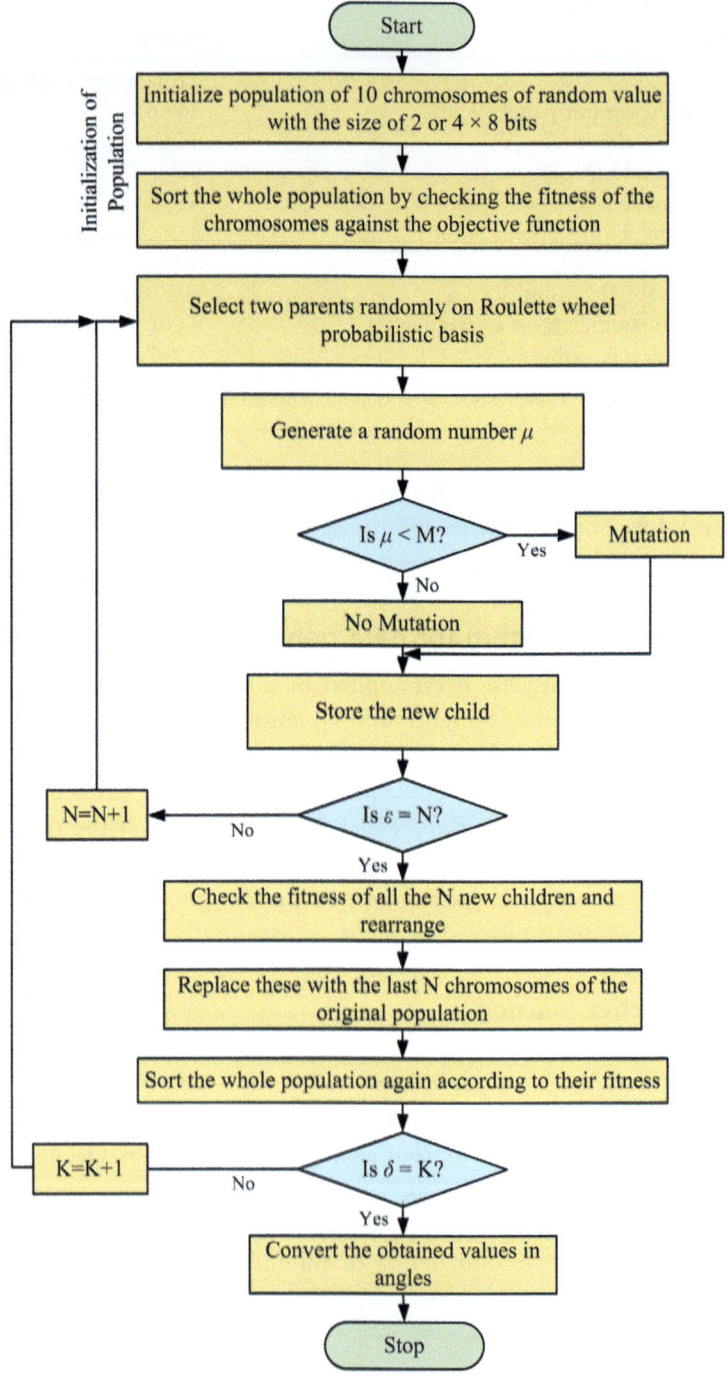

Figure 1.9 Flowchart of GA. Source: *From M. Ali et al., Robust ANN-Based Control of Modified PUC-5 Inverter for Solar PV Applications, IEEE Trans. Ind. Appl., 57(4) (2021), 3863–3876.*

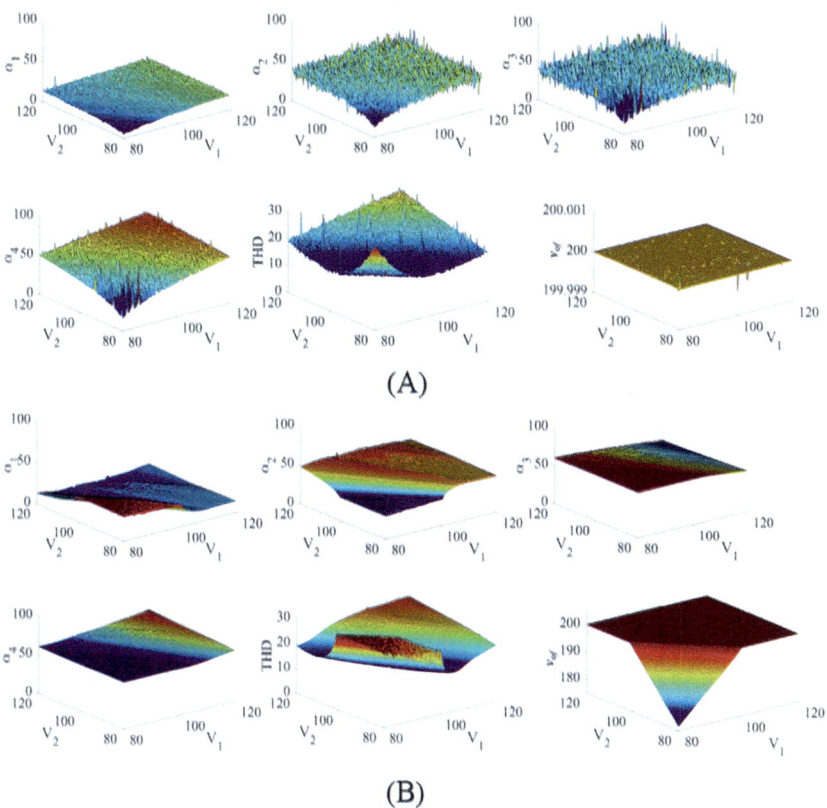

Figure 1.10 4-angle operation: Behavior of α_1, α_2, α_3, α_4, total harmonic distortion (THD), and the fundamental component's magnitude with objective functions as (A) $f_1(\alpha)$; and (B) combination of $f_1(\alpha)$ and $f_2(\alpha)$. Source: *From M. Ali et al., Robust ANN-Based Control of Modified PUC-5 Inverter for Solar PV Applications, IEEE Trans. Ind. Appl., 57(4) (2021), 3863–3876.*

$$f_{obj1}^{4ang}(\alpha) = f_1^{4ang}(\alpha) + \epsilon_1 \left(E[\cos\alpha_1 + \cos\alpha_2 - \cos\alpha_3 + \cos\alpha_4] - \frac{V_{max}}{4}\pi \right)$$

$$f_{obj2}^{4ang}(\alpha) = f_2^{4ang}(\alpha) + \epsilon_1 \left(E[\cos\alpha_1 + \cos\alpha_2 - \cos\alpha_3 + \cos\alpha_4] - \frac{V_{max}}{4}\pi \right)$$

$$+ \epsilon_2 \left(\alpha_1 + \alpha_2 - \frac{\pi}{3} \right) + \epsilon_3 \left(\alpha_3 + \alpha_4 - \frac{2\pi}{3} \right)$$

where, ϵ_1, ϵ_2 and ϵ_3 are constants of large value. To modify the angles from binary form to decimal, the following can be employed:

$$\alpha_x^{rad} = \alpha_{x-1}^{rad} + \frac{\pi/2 - \alpha_{x-1}^{rad}}{2^7 - 1} \sum_{k=0}^{7} 2^k b_k^x \qquad (1.12)$$

where, the kth bit of the xth angle is b_k^x, and α_o is assumed as equal to 0. This expression is valid for 4-angle and 2-angle scenarios. The data obtained from the GA is shown in Fig. 1.10.

1.4.4 ANN-based controller

ANN-based controllers have the capability to function as robust controllers, emulating the human learning processes that are inherently complex and draw inspiration from the various methods discussed in the literature (Haykin, 1999; Lodi et al., 2020). Here a multilayer perceptron (MLP) model trained by the backpropagation technique is employed, as shown in (Fig. 1.11B). This is an MLP structure with hidden layer/s between the output and input neuron layers. In the backpropagation training method, the error propagates from the output layer toward the input layer, adjusting the weights in successive iterations until the validation of input-output relation is achieved. In this process, the measured DC-bus voltages are fed into the input layer. Correspondingly, the network must generate the optimal angles (2 or 4), as shown in Fig. 1.11C. After the ANN is trained for a wide range of data in offline mode, it is then implemented as shown in Fig. 1.11C. The network hyperparameters are outlined in Table 1.7. The offline training process, as discussed in the previous subsection, involves training the network and then utilizing it as shown in Fig. 1.11C. Finally, the ANN is integrated in the actual controller as shown in Fig. 1.8. The training of the ANN includes validation and testing phases that are performed in the MATLAB environment as shown in Fig. 1.11D. The dataset (generated in Section 1.4.3) and assisted by the GA is randomly divided for training, validation, and testing purposes. Out of the generated data, 70% is used to train the network, followed by validation using 15% of the data, and the remaining 15% of data is allocated for

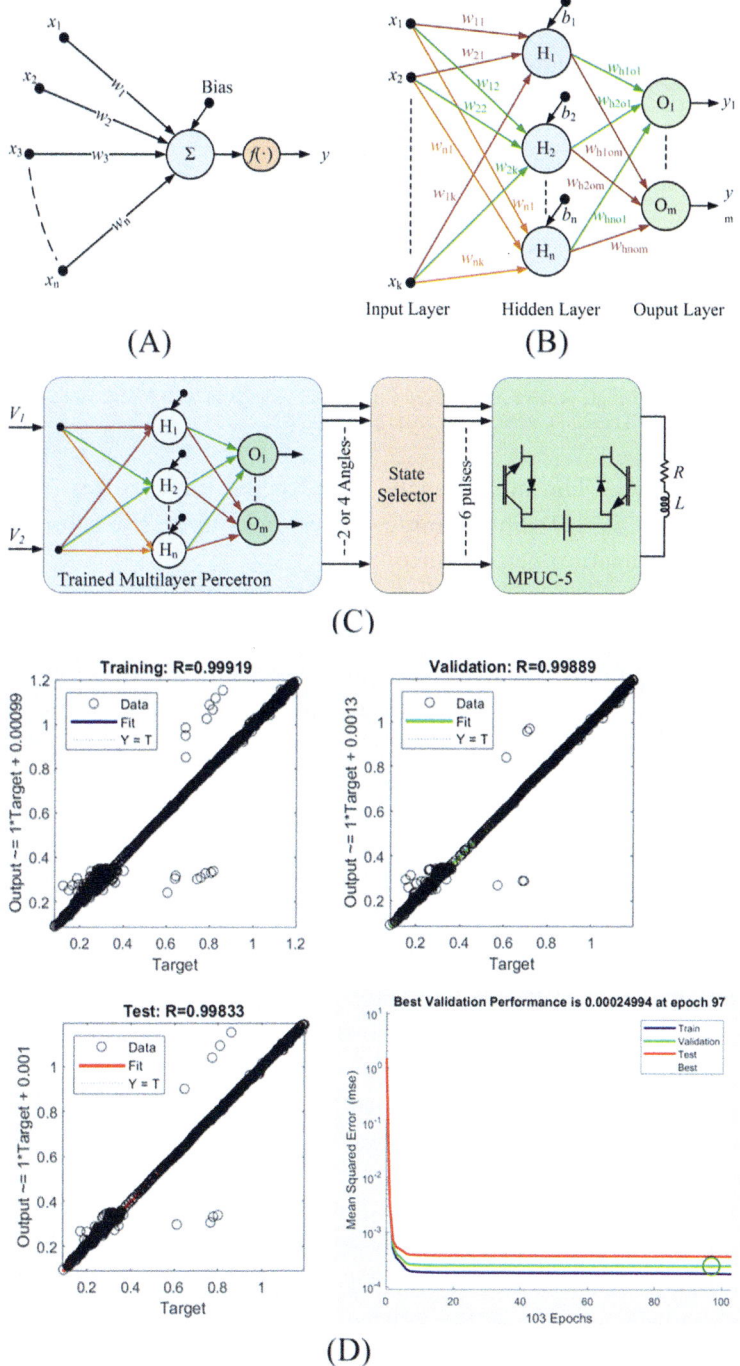

Figure 1.11 The ANN training: (A) a basic artificial neuron, (B) structure of an MLP, (C) MLP's implementation as controller of MPUC-5, and (D) Results of ANN training i^{th} case. Source: *From M. Ali et al., Robust ANN-Based Control of Modified PUC-5 Inverter for Solar PV Applications, IEEE Trans. Ind. Appl., 57(4) (2021), 3863–3876.*

Table 1.7 Training hyperparameters of the ANN.

S. No.	Parameter	Value
1	Input neurons	2
2	Hidden layers	1
3	Hidden neurons	20
4	Output neurons	2 or 4
5	Iterations	103
6	Method	Back Propagation
7	μ	10^{-6}
8	σ	0.1

Source: From M. Ali et al., Robust ANN-Based Control of Modified PUC-5 Inverter for Solar PV Applications, IEEE Trans. Ind. Appl., 57(4) (2021), 3863–3876.

testing purposes. The mean-square error for the three processes is presented in Fig. 1.11D. The training process halts upon the achievement of the best validation performance.

1.4.5 Results

After training, the ANN controller was tested with different voltage conditions at the converter's main and auxiliary DC link. The results for the 2- and 4-angle operation are presented in Fig. 1.12. Fig. 1.12A and 1.12B exhibit the deliberate change in voltage magnitude at the DC links (V_1 and V_2) in MATLAB/Simulink. Four distinct DC-link voltages are defined in a duration of 0 to 0.35 sec. The resulting output voltage and load current of the MPUC-5 inverter are presented in Fig. 1.12C. The resulting output is a stepped waveform with a constant fundamental component, which was extracted in Fig. 1.12D. The THD of all four zones of inverter operation is below the IEEE prescribed value, and the triplen harmonics are mitigated, which supports the effectiveness of the controller (Fig. 1.12E).

To further validate the performance of the controller, real-time irradiance data spanning 6 seconds from 11:59:55 AM to 12:00:01 PM (NRcan & ca, n.d.) was obtained and applied to the power conversion system. The results are shown in Fig. 1.13, showcasing the constant fundamental value of the output voltage despite variable DC-link voltage (Fig. 1.13F). The active power delivered at both DC links is the same (Fig. 1.13D and E). A zoomed-out view of the voltage and the THD at various instances is presented in Fig. 1.13G. The peak of the fundamental components at every instant of operation is approximately 200 V, providing evidence of the controller's robust performance.

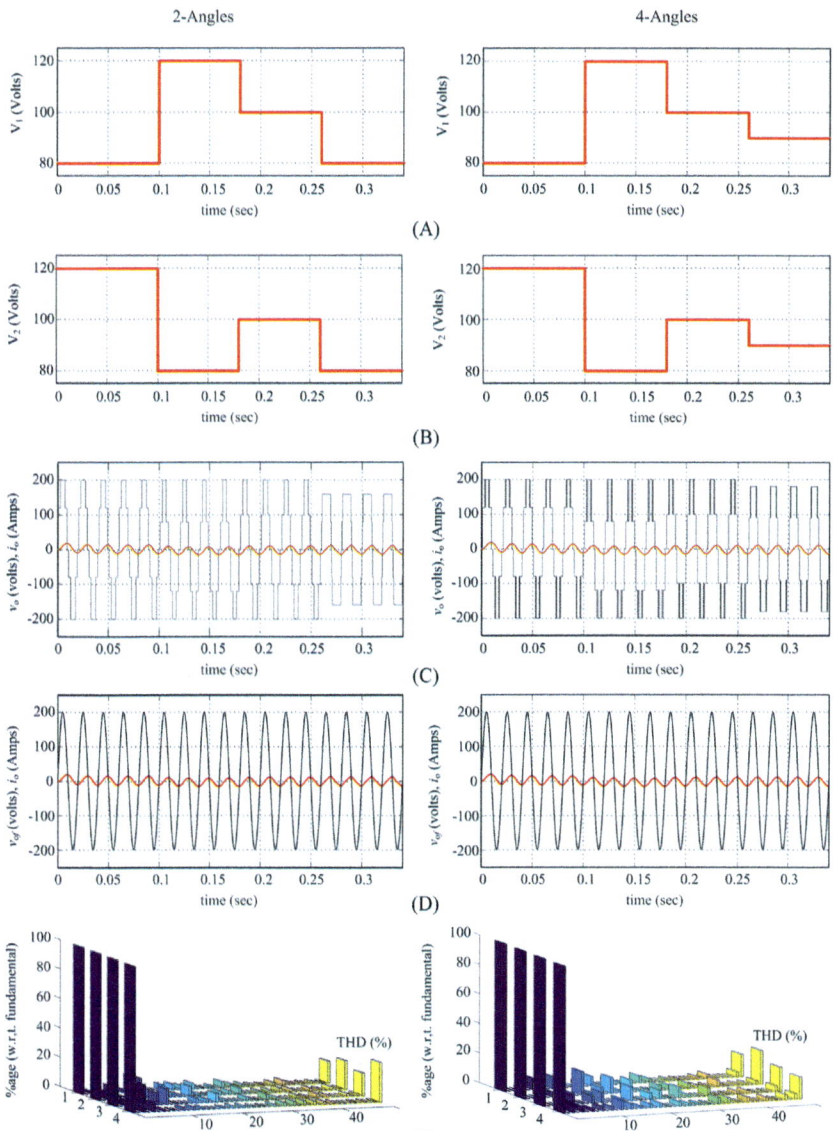

Figure 1.12 (A) DC-link voltage V $_1$, (B) auxiliary DC-link voltage V $_2$; Results for operation with 2-angle and 4-angle waveforms with (C) actual output voltage and current, (D) fundamental component of the output voltage and load current waveform, and, (E) THD profile in four DC-link voltage conditions. Source: *From M. Ali et al., Robust ANN-Based Control of Modified PUC-5 Inverter for Solar PV Applications, IEEE Trans. Ind. Appl., 57(4) (2021), 3863–3876.*

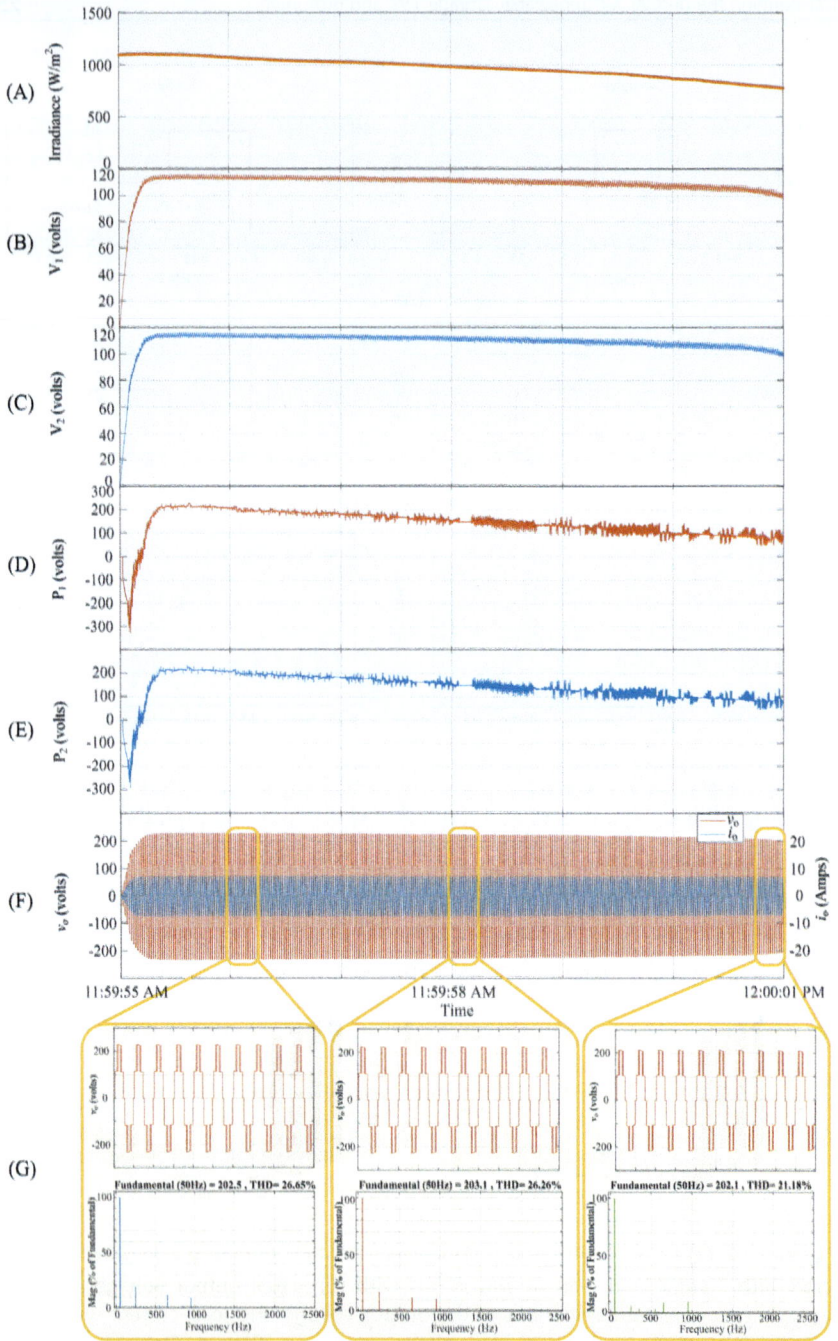

Figure 1.13 Real-time data simulation results: (A) variation in solar irradiance; output DC voltages of (B) the first boost converter and (C) the second boost converter; power available at (D) first DC-bus, and (E) the second DC-bus of MPUC-5; (F) output voltage and load current, and (G) zoomed output voltage and its THD profile at various operational instances. *Source: From M. Ali et al., Robust ANN-Based Control of Modified PUC-5 Inverter for Solar PV Applications, IEEE Trans. Ind. Appl., 57(4) (2021), 3863—3876.*

References

Achieving net zero emissions with machine learning: The challenge ahead. *Nature Machine Intelligence*, 4, 661−662 (2022). https://doi.org/10.1038/s42256-022-00529-w.

Ahmad, F., Alam, M. S., Shariff, S. M., & Krishnamurthy, M. (2019). A cost-efficient approach to EV charging station integrated community microgrid: A case study of indian power market. *IEEE Transactions on Transportation Electrification*, 5(1), 200−214. Available from https://doi.org/10.1109/TTE.2019.2893766.

Ahmad, T., Zhang, H., & Yan, B. (2020). A review on renewable energy and electricity requirement forecasting models for smart grid and buildings. *Sustainable Cities and Society*, 55102052. Available from https://doi.org/10.1016/j.scs.2020.102052.

Alam, M. S., Al-Ismail, F. S., & Abido, M. A. (2019). Power management and state of charge restoration\nof direct current microgrid with improved voltage-shifting controller. *Journal of Energy Storage*, 44. Available from https://doi.org/10.1016/j.est.2021.103253.

Alam, M. S., Al-Ismail, F. S., Al-Sulaiman, F. A., & Abido, M. A. (2023). Energy management in DC microgrid with an efficient voltage compensation mechanism. *Electric Power Systems Research* (214). Available from https://doi.org/10.1016/j.epsr.2022.108842.

Alam, Md. S., Al-Ismail, F. S., Salem, A., & Abido, M. A. (2020). High-level penetration of renewable energy sources into grid utility: Challenges and solutions. *IEEE Access*, 8, 190277−190299. Available from https://doi.org/10.1109/access.2020.3031481.

Ali M., Khalid M. (2022). Multiphase matrix converter modulation for wind energy systems using genetic algorithm. In *IEEE-IES Saudi Arabia Smart Grid Conference (SASG 2022)*, pp. 1−5, Jeddah, Saudi Arabia.

Ali, S. S., & Choi, B. J. (2020). State-of-the-art artificial intelligence techniques for distributed smart grids: A review. *Electronics (Switzerland)*, 9(6), 1−28. Available from https://doi.org/10.3390/electronics9061030, https://www.mdpi.com/2079-9292/9/6/1030/pdf.

Al-Humaid, Y. M., Khan, K. A., Abdulgalil, M. A., & Khalid, M. (2021). Two-stage stochastic optimization of sodium-sulfur energy storage technology in hybrid renewable power systems. *IEEE Access*, 9, 162962−162972. Available from https://doi.org/10.1109/ACCESS.2021.3133261.

Ali M. Tariq M. Lodi K.A. Chakrabortty R.K. Ryan M.J. Alamri B. Bharatiraja C. 2021 Robust ANN-based control of modified PUC-5 inverter for solar PV applications. *IEEE Transactions on Industry Applications Institute of Electrical and Electronics Engineers Inc. India* 57 (4) 3863−3876. Retrieved July 1, 2021, Available from https://doi.org/10.1109/TIA.2021.3076032.

Ali, M., Amrr, S. M., & Khalid, M. (2022). Speed control of a wind turbine−driven doubly fed induction generator using sliding mode technique with practical finite-time stability. *Frontiers in Energy Research* (10). Available from https://doi.org/10.3389/fenrg.2022.970755.

Ali, M., Iqbal, A., Anees, M. A., Khan, M. R., Rahman, K., & Ayyub, M. (2019). Differential evolution-based pulse-width modulation technique for multiphase MC. *IET Power Electronics*, 12(9), 2224−2235. Available from https://doi.org/10.1049/iet-pel.2018.5862.

Ali, M., Iqbal, A., & Khan, M. R. (2018). *AC-AC converters* (pp. 417−456). Elsevier BV. Available from 10.1016/b978-0-12-811407-0.00014-3.

Ali, M., Iqbal, A., Khan, M. R., Ayyub, M., & Anees, M. A. (2017). Generalized theory and analysis of scalar modulation techniques for a m × n matrix converter. *IEEE Transactions on Power Electronics*, 32(6), 4864−4877. Available from https://doi.org/10.1109/TPEL.2016.2600034.

Ali, M., Tariq, M., Lin, C. H., Chakrobortty, R. K., Alamri, B., Alahmadi, A., & Ryan, M. J. (2021). Operation of a UXE-type 11-level inverter with voltage-balance modulation using NLC and ACO-based SHE. *Sustainability (Switzerland)*, *13*(16). Available from https://doi.org/10.3390/su13169035.

Ali, M., Tariq, M., Chakrabortty, R. K., Ryan, M. J., Alamri, B., & Bou-Rabee, M. A. (2021). 11-level operation with voltage-balance control of WE-type inverter using conventional and DE-SHE techniques. *IEEE Access*, *9*, 64317–64330. Available from https://doi.org/10.1109/ACCESS.2021.3072905.

Aly, Hamed H. H. (2020). A proposed intelligent short-term load forecasting hybrid models of ANN, WNN and KF based on clustering techniques for smart grid. *Electric Power Systems Research*, *182*106191. Available from https://doi.org/10.1016/j.epsr.2019.106191.

Azad, H. B., Mekhilef, S., & Ganapathy, V. G. (2014). Long-term wind speed forecasting and general pattern recognition using neural networks. *IEEE Transactions on Sustainable Energy*, *5*(2), 546–553. Available from https://doi.org/10.1109/TSTE.2014.2300150.

Boroojeni, K. G., Amini, M. H., Bahrami, S., Iyengar, S. S., Sarwat, A. I., & Karabasoglu, O. (2017). A novel multi-time-scale modeling for electric power demand forecasting: From short-term to medium-term horizon. *Electric Power Systems Research*, *142*, 58–73. Available from https://doi.org/10.1016/j.epsr.2016.08.031.

Bou-Rabee, M., Lodi, K. A., Ali, M., Ansari, M. F., Tariq, M., & Sulaiman, S. A. (2020). One-month-ahead wind speed forecasting using hybrid AI model for coastal locations. *IEEE Access*, *8*, 198482–198493. Available from https://doi.org/10.1109/ACCESS.2020.3028259.

Brockway, P. E., Owen, A., Brand-Correa, L. I., & Hardt, L. (2019). Estimation of global final-stage energy-return-on-investment for fossil fuels with comparison to renewable energy sources. *Nature Energy*, *4*(7), 612–621. Available from https://doi.org/10.1038/s41560-019-0425-z.

CanmetENERGY. (2020). High-resolution solar radiation datasets. Government of Canada. https://natural-resources.canada.ca/energy/renewable-electricity/solar-photovoltaic/18409.

Chen, C., & Liu, H. (2020). Medium-term wind power forecasting based on multi-resolution multi-learner ensemble and adaptive model selection. *Energy Conversion and Management*, *206*, 112492 Available from https://doi.org/10.1016/j.enconman.2020.112492.

Electric Vehicle Market - Global Industry Analysis, Size, Share, Growth, Trends\n, Regional Outlook, and Forecast 2023–2032. 2023.

ExxonMobil. (2019). ExxonMobil 2019 outlook for energy: A perspective to 2040. https://corporate.exxonmobil.com/Energy-and-environment/Looking-forward/Outlook-for-Energy.

Han, Y., & Tong, X. (2020). Multi-step short-term wind power prediction based on three-level decomposition and improved grey wolf optimization. *IEEE Access*, *8*, 67124–67136. Available from https://doi.org/10.1109/ACCESS.2020.2984851.

Haykin, S. S. (1999). *Neural networks: A comprehensive foundation*. Prentice Hall.

Heo, S. K., Ko, J., Kim, S. Y., Jeong, C., Hwangbo, S., & Yoo, C. K. (2022). Explainable AI-driven net-zero carbon roadmap for petrochemical industry considering stochastic scenarios of remotely sensed offshore wind energy. *Journal of Cleaner Production* (379). Available from https://doi.org/10.1016/j.jclepro.2022.134793.

Hossain, M. A., Gray, E., Lu, J., Islam, M. R., Alam, M. S., Chakrabortty, R., & Pota, H. R. (2023). Optimized forecasting model to improve the accuracy of very short-term wind power prediction. *IEEE Transactions on Industrial Informatics*, 1–13. Available from https://doi.org/10.1109/TII.2022.3230726.

How D.N.T. Hannan M.A. Lipu M.S.H. Sahari K.S.M. Ker P.J. Muttaqi K.M. 2019 *State-of-charge estimation of Li-ion battery in electric vehicles: A deep neural network approach*. IEEE Industry Applications Society Annual Meeting, IAS 2019 Institute of Electrical and Electronics Engineers Inc. 9 1 2019/09/01 2019. Available from https://doi.org/10.1109/IAS.2019.8912003.

Iqbal, A., Meraj, M., Tariq, M., Lodi, K. A., Maswood, A. I., & Rahman, S. (2019). Experimental investigation and comparative evaluation of standard level shifted multi-carrier modulation schemes with a constraint GA based SHE techniques for a seven-level PUC inverter. *IEEE Access, 7*, 100605−100617. Available from https://doi.org/10.1109/ACCESS.2019.2928693, http://ieeexplore.ieee.org/xpl/RecentIssue.jsp?punumber = 6287639.

Kaytez, F. (2020). A hybrid approach based on autoregressive integrated moving average and least-square support vector machine for long-term forecasting of net electricity consumption. *Energy, 197*, 117200

Khalid, M. A. (2019). Review on the selected applications of battery-supercapacitor hybrid energy storage systems for microgrids. Energies, 12, 4559. Available from https://doi.org/10.3390/en12234559

Labrador Rivas, A. E., & Abrão, T. (2020). Faults in smart grid systems: Monitoring, detection and classification. *Electric Power Systems Research, 189*, 106602 Available from https://doi.org/10.1016/j.epsr.2020.106602.

Liu, H., Yu, C., Wu, H., Duan, Z., & Yan, G. (2020). A new hybrid ensemble deep rein-forcement learning model for wind speed short term forecasting. *Energy, 202*, 117794 Available from https://doi.org/10.1016/j.energy.2020.117794.

Lodi, K. A., Ali, M., Tariq, M., Meraj, M., Iqbal, A., Chakrabortty, R. K., & Ryan, M. J. (2020). Modulation with metaheuristic approach for cascaded-MPUC49 asymmetrical inverter with boosted output. *IEEE Access, 8*, 96867−96877. Available from https://doi.org/10.1109/ACCESS.2020.2995782.

Massaoudi, M., Refaat, S. S., Chihi, I., Trabelsi, M., Oueslati, F. S., & Abu-Rub, H. (2021). A novel stacked generalization ensemble-based hybrid LGBM-XGB-MLP model for Short-Term Load Forecasting. *Energy* (214). Available from https://doi.org/10.1016/j.energy.2020.118874.

Nalcaci, G., Özmen, A., & Weber, G. W. (2019). Long-term load forecasting: Models based on MARS, ANN and LR methods. *Central European Journal of Operations Research, 27* (4), 1033−1049. Available from https://doi.org/10.1007/s10100-018-0531-1.

Ourahou, M., Ayrir, W., Hassouni, B. E. L., & Haddi, A. (2020). Review on smart grid control and reliability in presence of renewable energies: Challenges and prospects. *Mathematics and Computers in Simulation, 167*, 19−31. Available from https://doi.org/10.1016/j.matcom.2018.11.009.

Qiu D., Ye Y., Papadaskalopoulos D., & Strbac G. (2020). A deep reinforcement learning method for pricing electric vehicles with discrete charging levels. IEEE Transactions on Industry Applications, 56(5), 5901−5912. Available from https://doi.org/10.1109/TIA.2020.2984614.

Rauf, A., Kassas, M., & Khalid, M. (2022). Data-driven optimal battery storage sizing for grid-connected hybrid distributed generations considering solar and wind uncertainty. *Sustainability, 14*(17), 11002. Available from https://doi.org/10.3390/su141711002.

Refaat, S. S., Ellabban, O., Bayhan, S., Abu-Rub, H., Blaabjerg, F., & Begovic, M. M. (2021). *Smart grid and enabling technologies. Smart Grid and Enabling Technologies* (pp. 1−512). Qatar: Wiley Blackwell. Available from https://doi.org/10.1002/9781119422464.

Renewables, Market analysis and forecast from 2019 to 2024. 2019.

Refaat, S. S., Ellabban, O., Bayhan, S., Abu-Rub, H., Blaabjerg, F., & Begovic, M. M. (2021). Smart Grid and Enabling Technologies. Wiley.

Shah, I., Iftikhar, H., & Ali, S. (2020). Modeling and forecasting medium-term electricity consumption using component estimation technique. *Forecasting, 2*(2), 163−179. Available from https://doi.org/10.3390/forecast2020009.

Sun, Q., Zhang, N., You, S., & Wang, J. (2019). The dual control with consideration of security operation and economic efficiency for energy hub. *IEEE Transactions on Smart Grid, 10*(6), 5930−5941. Available from https://doi.org/10.1109/TSG.2019.2893285.

Ullah, Z., Al-Turjman, F., Mostarda, L., & Gagliardi, R. (2020). Applications of artificial intelligence and machine learning in smart cities. *Computer Communications, 154*, 313−323. Available from https://doi.org/10.1016/j.comcom.2020.02.069.

Upadhyay, D., Ali, M., Tariq, M., Khan, S. A., Alamri, B., & Alahmadi, A. (2022). Thirteen-level UXE-type inverter with 12-band hysteresis current control employing PSO based PI controller. *IEEE Access, 10*, 29890−29902. Available from https://doi.org/10.1109/ACCESS.2022.3146355.

U.S. Energy Info. Admin., Annual Energy Outlook 2022 (with projections to 2050). (2022). https://www.eia.gov/outlooks/aeo/pdf/AEO2022_Narrative.pdf.https://www.eia.gov/outlooks/aeo/.

Vahedi, H., Shojaei, A. A., Dessaint, L. A., & Al-Haddad, K. (2018). Reduced DC-link voltage active power filter using modified PUC5 converter. *IEEE Transactions on Power Electronics, 33*(2), 943−947. Available from https://doi.org/10.1109/TPEL.2017.2727325.

Vinuesa, R., Azizpour, H., Leite, I., Balaam, M., Dignum, V., Domisch, S., Felländer, A., Langhans, S. D., Tegmark, M., & Fuso Nerini, F. (2020). The role of artificial intelligence in achieving the sustainable development goals. *Nature Communications, 11*(1). Available from https://doi.org/10.1038/s41467-019-14108-y.

Yin, L., Gao, Q., Zhao, L., Zhang, B., Wang, T., Li, S., & Liu, H. (2020). A review of machine learning for new generation smart dispatch in power systems. *Engineering Applications of Artificial Intelligence, 88*, 103372 Available from https://doi.org/10.1016/j.engappai.2019.103372.

You, M., Zhang, X., Zheng, G., Jiang, J., & Sun, H. (2020). A versatile software defined smart grid testbed: Artificial Intelligence enhanced real-time co-evaluation of ICT systems and power systems. *IEEE Access, 8*, 88651−88663. Available from https://doi.org/10.1109/access.2020.2992906.

Yousuf, M. U., Al-Bahadly, I., & Avci, E. (2019). Current perspective on the accuracy of deterministic wind speed and power forecasting. *IEEE Access, 7*, 159547−159564. Available from https://doi.org/10.1109/ACCESS.2019.2951153.

Zhang C., Li R., Shi H., Li F., Deep learning for day-ahead electricity price forecasting.

Zhao, S., Blaabjerg, F., & Wang, H. (2021). An overview of artificial intelligence applications for power electronics. *IEEE Transactions on Power Electronics, 36*(4), 4633−4658. Available from https://doi.org/10.1109/TPEL.2020.3024914.

Zhao, T., & Chen, D. (2021). A power adaptive control strategy for further extending the operation range of single-phase cascaded H-bridge multilevel PV inverter. *IEEE Transactions on Industrial Electronics*. Available from https://doi.org/10.1109/TIE.2021.3060646.

CHAPTER TWO

A new artificial intelligence-based demand side management method for EV charging stations

Gökay Bayrak[1] and Hasan Meral[1,2]

[1]Smart Grid Laboratory, Department of Electrical and Electronics Engineering, Bursa Technical University, Bursa, Turkey
[2]Sibernetik Machinery & Automation R&D Center, Bursa, Turkey

2.1 Introduction

Today, in parallel with the target of limiting the use of fossil fuels, electric vehicles (EVs) are replacing vehicles using fossil fuels. The use of EVs in transportation will inevitably accelerate and increase in the coming years. The rapid adoption of EVs and, consequently, the establishment of electric vehicle charging stations, will pose challenges for the grid due to the increase in energy demand. The rising energy demands have increased the importance of concepts such as smart grid, smart energy management, and demand side management (DSM) within the current grid conditions.

The smart grid is designed as an electrical grid model that can transmit electricity from generation points to active consumers in a controlled, intelligent manner. Smart Grid, smart energy management, and DSM approaches can reduce the investment cost required to upgrade grid capacity by enhancing the reliability of the power system and reducing peak demand over the long term (Siano, 2014). The electrical power system is undergoing a transformation with the increasing need for a more sustainable energy supply, more efficient use of the existing grid, and accordingly, the commissioning of smart grid infrastructures. Fig. 2.1 shows the impact of increased EV sales on the US electricity market (How will growing EV sales affect US power markets?, 2017).

Intelligent Learning Approaches for Renewable and Sustainable Energy
DOI: https://doi.org/10.1016/B978-0-443-15806-3.00002-4

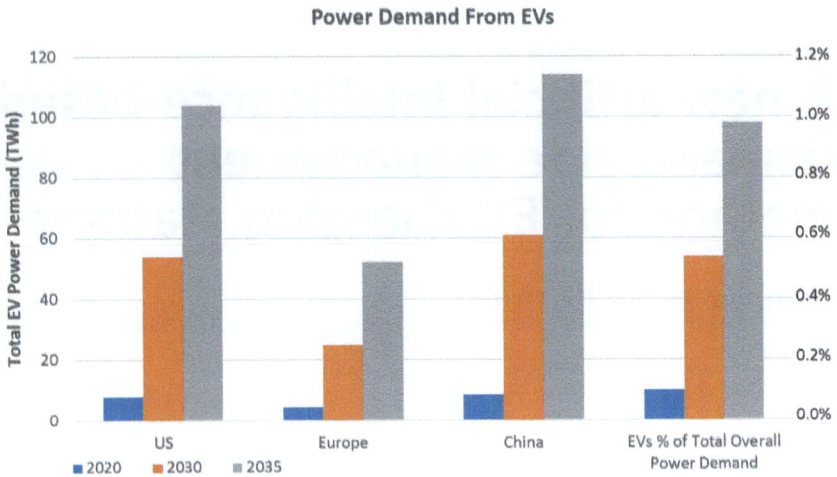

Figure 2.1 *The effects of increased EV sales on the US electricity market.*

DSM plays a vital role within the smart grid, offering users the opportunity to optimize their load usage patterns thereby balancing the energy supply and demand leading to a reduction in the peak-average ratio (PAR) (Nasir et al., 2021). DSM can be achieved through various applications such as trimming the peak load demand, increasing the load demand in low-load demand situations, rescheduling of load activation times, strategically increasing/decreasing the demand, and flexible load shaping. Fig. 2.2 shows various conventional DSM techniques (Jabir et al., 2018).

In this context, the boundaries of electricity grids are being redefined with the emergence of consumers contributing to the increase in energy efficiency. The need for advanced load control methods is on the rise, driven by the necessity to effectively manage supply and demand (Goy & Sancho-Tomás, 2019). This study focuses on the real-time monitoring of consumption within a microgrid. Fig. 2.3 illustrates a scenario of the proposed intelligent DSM system for charging stations with multiple charging units. In systems with more than one charging station, a smart load controller assesses information received from consumers along with instantaneous consumption values in the microgrid, using a machine learning-based method. The system then decides whether to apply the appropriate charging type and direct load control methods, such as load shifting.

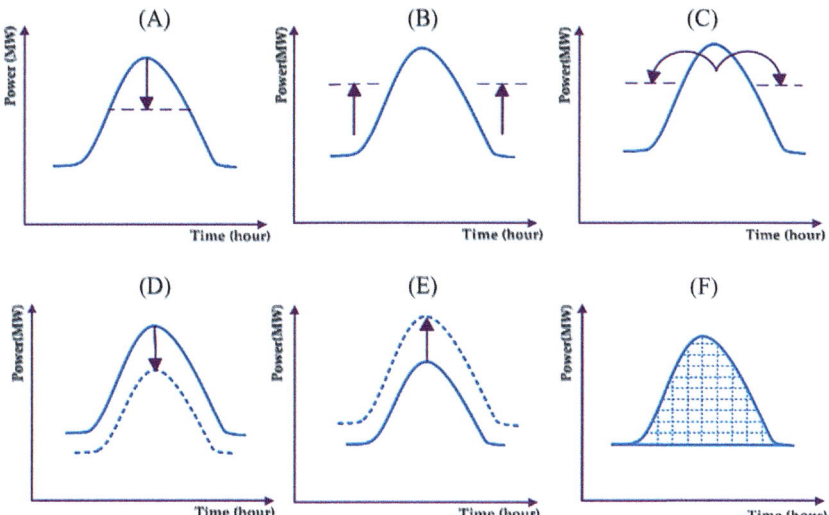

Figure 2.2 *Various DSM techniques.* (A) Peak clipping, (B) Valley filling, (C) Load shifting, (D) Energy efficiency, (E) Strategic load growth, and (f) Flexible load shape. *DSM*, demand side management.

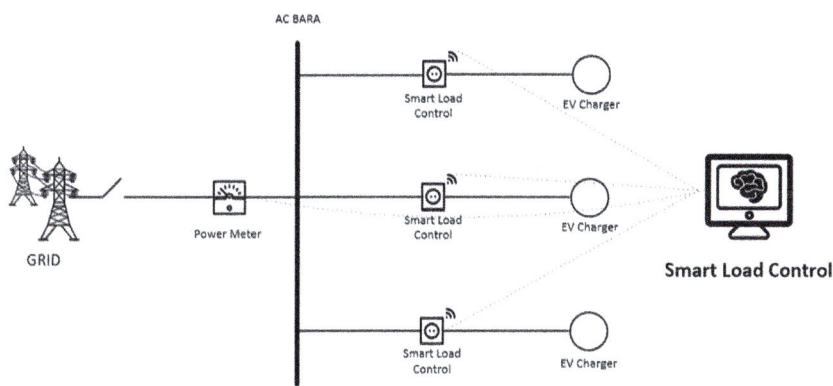

Figure 2.3 *Proposed structure for intelligent DSM studies for EV charging station.* *DSM*, demand side management; *EV*, electric vehicle.

2.1.1 Direct load control

Direct load control is among the techniques that can be employed to achieve the desired flat load profile. Direct load control play a significant role in the integration processes of EVs. The dispatchable demand side flexibility provided by direct load control can increase the grid's reliability

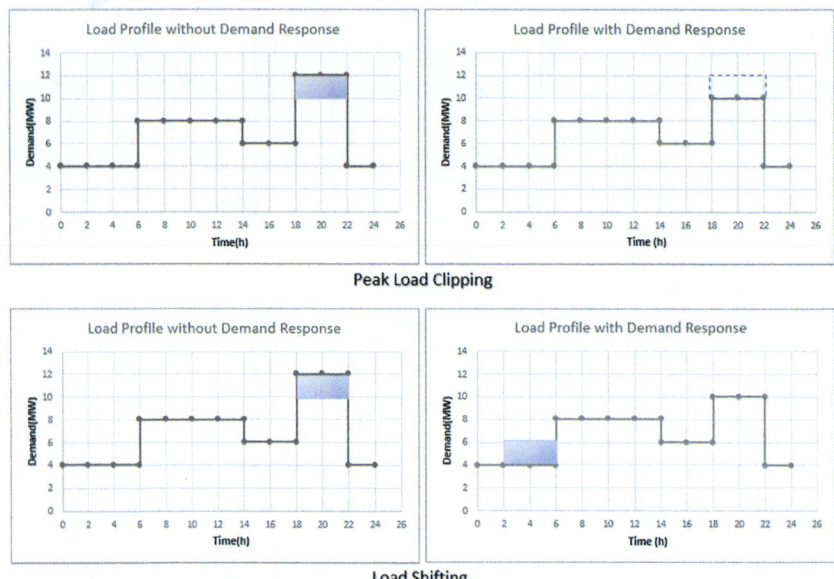

Figure 2.4 *Peak load trimming and load shift graphs.*

and can facilitate meaningful carbon reductions through by accelerating the low-cost integration of EVs (Hesser & Succar, 2012).

Load shifting, load scheduling, and load disabling are direct load control methods commonly used in DSM processes (Li et al., 2017). Fig. 2.4 shows examples for peak load clipping and load shift graphs over 24-hour periods. Methods such as load shifting and load deactivation can be adapted for domestic users, serving applications such as shaving peak loads, addressing power quality problems, and reducing usage costs (Bayrak, 2018; Yılmaz et al., 2022a, 2022b).

An energy management plan that uses direct load control should only impact devices whose supply may be interrupted without causing unacceptable levels of inconvenience to the consumer (Bayrak, Küçüker, et al., 2023; Bayrak, Yılmaz, et al., 2023; Yılmaz & Bayrak, 2019). This requirement is crucial in securing client acceptance for such an energy management scheme. Home appliances suitable for this form of control include cooling devices, storage/space heaters, and water heaters (McDonald et al., 1990).

Electric vehicle charging stations are set to be among the critical loads directly subject to load control applications in the near future. Using the administrative records of vehicle inspection stations obtained from the Ministry of Transport and Infrastructure, the Turkish Statistical Institute calculated vehicle-kilometer data according to vehicle types. The average kilometers traveled by various vehicles in 2016 were as follows: trucks 45,735 km, buses 44,491 km, minibusses 26,396 km, pickup trucks 17,845 km, cars 13,117 km, and motorcycles 3766 km (Türkiye İstatistik Kurumu Taşıt-kilometre İstatistikleri, 2018). The average daily travel distance of a personal vehicle was then estimated to be 36 km. A study conducted in Singapore, revealed that the average number of trips per day with a personal vehicle is three, with an average journey length of 17.3 km. The length of each journey is limited to a minimum of 1 km, and more than 90% of the journeys show a normal distribution not exceeding 37.5 km (Huber et al., 2012). The fact that EVs are generally not in use for long periods makes the charging processes of EVs important for smart demand management systems. With the widespread use of technologies such as V2G and V2H, scheduling the load and deactivating the load by evaluating the periods when the vehicle is not in use, as well as controlling the EV charging processes, can play a key role in preventing the formation of peak loads.

2.2 Problem description

Considering the current grid conditions in Turkey, the integration of EVs and EV charging stations is expected to result in increased demand, inevitably leading to peak loads. The rising demand highlights the importance of DSM applications. These applications can play a crucial role in achieving a flat load profile by preventing the occurrence of peak loads, particularity by using demand response (DR) techniques in combination with storage elements and appropriate switching methods (Dharmaraj & Natarajan, 2022).

EVs used for personal transportation often remain inactive for a significant portion of the day. In Turkey, where the average daily travel distance is predicted to be 36 km, users can charge their vehicles at their homes, on-site, or at their workplaces. Vehicles that can remain connected to the charging station for a longer period can be charged at different times, helping to alleviate peak load issues.

In charging stations equipped with multiple EV charging units, the speed of the charging station is also important, in addition to assessing the vehicle's suitability for direct load control. This allows users with ample time to reach the required charging rate to opt for AC charging stations instead of DC fast charging stations, helping to avoid the formation of peak loads.

Case 1. – Choosing the Appropriate Charger in Charging Stations with Multiple Charge Units

In this application type, the formation of peak charging load is prevented by evaluating information received from the user using machine learning-based methods. Unnecessary usage of high-speed charging stations is avoided to prevent the formation of peak load.

It was reported in the literature that low-speed charging in different charging strategies and targets leads to higher EV penetration into the grid, lower unit costs, and reduced peak load when compared to fast charging stations (Mehta et al., 2018). In order to select the appropriate charging speed mechanism in charging stations with multiple units, users are prompted for information on battery capacity, the current state of charge, the time that the vehicle can stay connected, and the subsequent trip distance. In addition to the data received from the user, the current instantaneous consumption values of the microgrid are collected. The data obtained for the Case 1 study is evaluated with smart methods, allowing the user to determine the most appropriate charging speed. A dataset that simulates 1000 different situations was created with data randomly generated within the limits determined by MATLAB® and calculations were derived from this data.

Case 2. – Evaluation of Suitability for Direct Load Control in Electric Vehicle Charging Stations for Personal Use.

In this application type, machine learning-based methods are employed to evaluate the information received from the user. The system utilizes these methods to prevent the formation of peak loads through direct load control. For example, in a microgrid designed for residential use which includes one charging station. If the EV charging station is used during peak hours of domestic energy consumption, it can significantly amplify the instantaneous load, thereby contributing to the formation of peak loads. In these scenarios, the inferences drawn from the information obtained from the EV user could be evaluated by a machine learning-based system. The system then makes the most appropriate decision by evaluating the suitability of these inferences for direct load control applications, such as load shifting.

2.3 Proposed method

In this study, a machine learning-based system was proposed to enhance the sustainability of the grid and prevent peak load formation in two different case studies. For both case studies, a dataset containing various information such as user inputs, instantaneous consumption details from the microgrid, and the power rating of the utilized charging station, was generated with random data within the predefined limits. As depicted in Table 2.1, this dataset was used to simulate 2000 cases, comprising 1000 instances of Case-1 and 1000 instances of Case-2.

To train the machine learning algorithm, a decision algorithm for charger-type selection was developed, as illustrated in Fig. 2.5.

2.3.1 RUS Boost tree ensemble classifiers

The ensemble classification structure combines the strengths of different classifiers to produce a classifier with higher accuracy. Ensemble classifiers are classifiers that operate on the basis that the model was created using multiple classifiers, instead of relying on just one classifier. This approach results in more accurate decision-making through a voting process (Polikar, 2006). It helps to assess the strengths of each classifier, mitigating the weaknesses of individual models, ultimately achieving high-accuracy classification through combined predictions (Alpaydin, 2007). Fig. 2.6 shows a comparison of the ensemble learning method performance with the decision tree learning method.

Table 2.1 A section of the data set created for the Case-2 study.

	EV Charger type	Instant consumption	Battery capacity kWh	State of charge	Time required for full charge	Time the vehicle is plugged in	Subsequent trip distance	Minimum energy requirement for the next trip kW	Load management unit (0 = start, 1 = load shifting, 2 = will not charge)
1	22,00	9,00	80	0,81	0,71	2	200,00	36	0
2	22,00	10,04	80	0,35	2,36	5	200,00	36	1
3	22,00	14,06	80	0,19	2,94	4	200,00	36	1
4	22,00	14,12	52	0,80	0,48	3	100,00	18	1
5	22,00	7,28	52	0,30	1,67	2	100,00	18	0
6	22,00	15,08	52	0,38	1,46	2	200,00	36	0
7	22,00	13,00	84	0,27	2,80	4	200,00	36	0
8	22,00	10,85	84	0,28	2,75	2	200,00	36	0
9	22,00	8,01	84	0,57	1,64	4	200,00	36	0
10	22,00	14,09	60	0,50	1,37	3	100,00	18	1
11	22,00	8,16	60	0,54	1,25	5	100,00	18	0
12	22,00	4,73	60	0,38	1,70	1	200,00	36	0
13	22,00	12,30	80	0,16	3,06	3	100,00	18	0
14	22,00	13,49	80	0,39	2,21	1	100,00	18	0
15	22,00	7,79	80	0,75	0,90	4	200,00	36	0

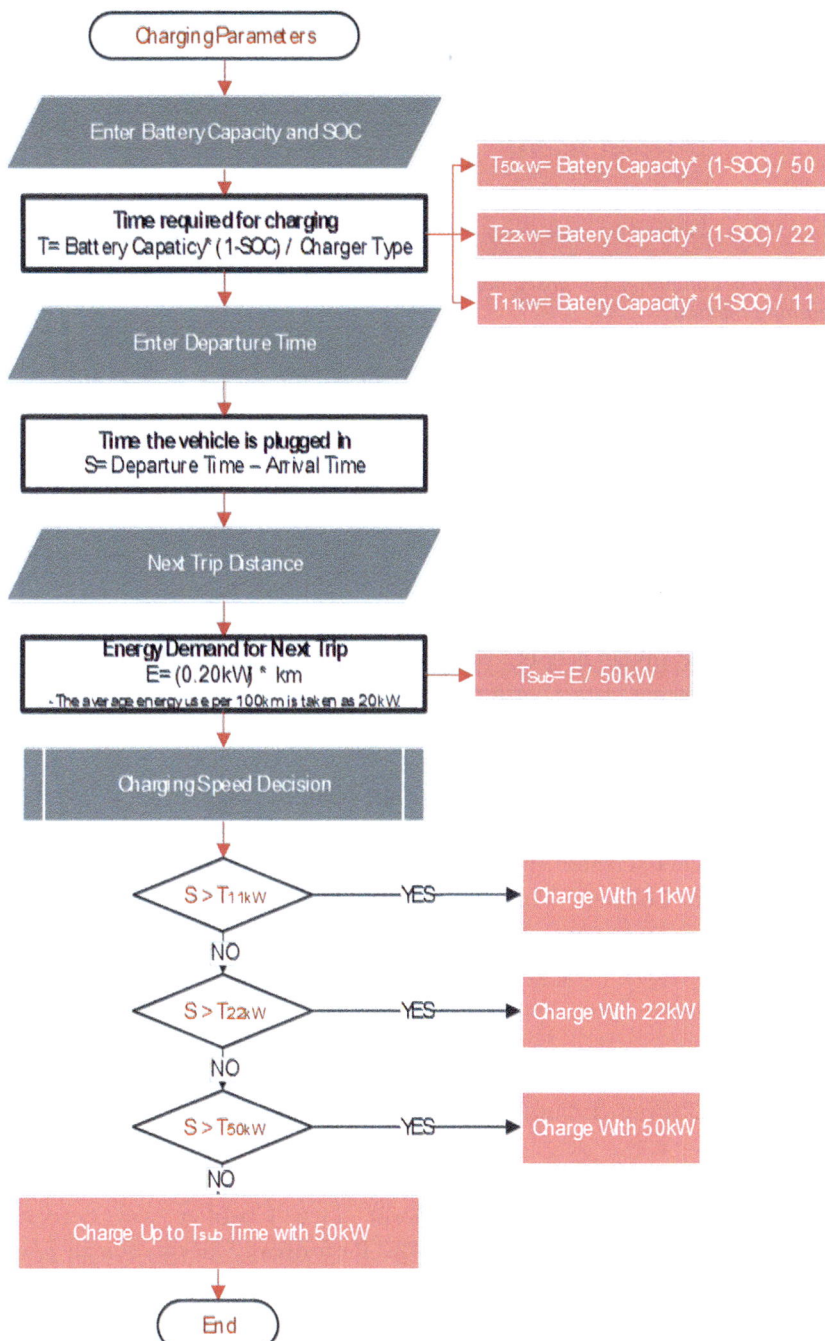

Figure 2.5 *Charger type selection decision algorithm, which is used in the training of machine learning algorithm.*

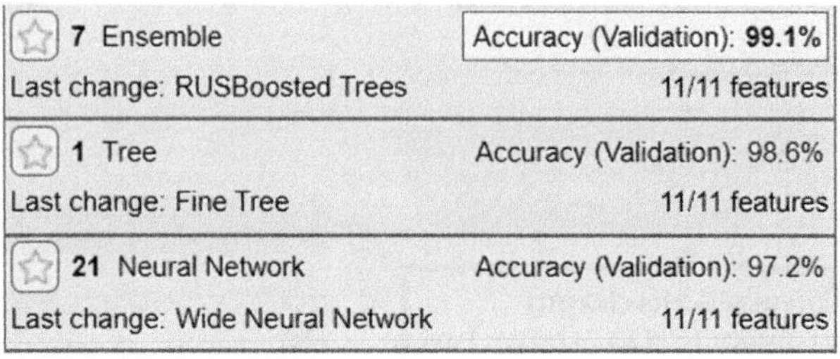

☆ **7** Ensemble	Accuracy (Validation): **99.1%**
Last change: RUSBoosted Trees	11/11 features
☆ **1** Tree	Accuracy (Validation): 98.6%
Last change: Fine Tree	11/11 features
☆ **21** Neural Network	Accuracy (Validation): 97.2%
Last change: Wide Neural Network	11/11 features

Figure 2.6 *Comparison of ensemble learning method performance with decision tree learning method.*

RUSBoost is another algorithm based on AdaBoost, and much like SMOTEBoost, it exhibits increased effectiveness when dealing with unstable data. Instead of using SMOTE to create a balance between classes, RUSBoost uses random subsampling (RUS) to achieve this equilibrium (Seiffert et al., 2010). The study utilized the RUSBoost Tree Ensemble classifier, an ensemble algorithm boasting the highest accuracy rate with the given dataset. RUSBoost proves particularly useful when handling unstable data (Forslund, 2022). Fig. 2.7 shows the RUSBoost algorithm flowchart.

Case-1. RUSBoost Ensemble Algorithm Results.

The dataset created for Case-1 was tested using the MATLAB Classification Learner. The RUSBoost Tree Classifier was executed, resulting in an impressive accuracy rate of 99.1%. Performance details for all tested classifiers were given in Tables 2.2 and 2.3. Fig. 2.8 also shows the RUSBoost Tree confusion matrix and ROC curve for Case-1 Dataset.

Case-2. RUSBoost Ensemble Algorithm Results.

The dataset created for Case-2 was tested using the MATLAB Classification Learner. The RUSBoost Tree Classifier was executed, resulting in an accuracy rate of 98.4%. Performance details for all tested classifiers were given in Figs. 2.9 and 2.10.

Inputs:
X: Original dataset features (shape: $[n_{samples}, n_{features}]$)
y: Original dataset labels (shape: $[n_{samples}]$)
T: Number of weak classifiers
base_learner: Examples of weak classifier to use (Decision Tree, SVM, Naïve Bayes, etc.)
max_iterations: Maximum number of iterations for AdaBoost
Output:
Final_ensemble_classifier: Final ensemble classifier
Pseudo code:
1. Initialize weights for each sample $w_i = 1/n\_samples$, where $n_{samples}$ is the number of samples in the dataset.
2. Create an empty list structure to hold the base learners.
3. Repeat for $t = 1$ to T:
   a. Randomly under-sample the majority class to create a balanced training set.
   b. Train the base learner on the under-sampled training set using the sample weights w_i.
   c. Make predictions on the entire dataset and calculate the error E_t of the base learner:
   $E_t = \text{sum}(w_i)$ for misclassified samples / $\text{sum}(w_i)$ for all samples.
   d. Calculate the base learner's weight in the ensemble structure:
   $alpha_t = 0{,}5 * \ln((1 - E_t)/E_t)$
   e. Update the sample weights:
   For $i = 1$ to $n_{samples}$, if the i_{th} sample is correctly classified by the base learner:
   $w_i = w_i * e^{-alpha_t}$, else $w_i = w_i * e^{alpha_t}$
   Normalize the sample weights that sum up to 1.
   f. Save the base learner and its weight $alpha_t$ in the list of base learners.
   g. If the error E_t is 0 or the max_iterations are reached, break the loop.
4. Create an empty list for the ensemble algorithms predictions.
5. Repeat for each sample in the dataset:
   a. Initialize ensemble algorithms prediction score = 0.
   b. For each base learner and its weight $(alpha_t)$:
   i. Make the base learner predict the sample's label.
   ii. Add the prediction multiplied by $alpha_t$ to the ensemble prediction score.
   c. If the ensemble prediction score is greater than or equal to 0.5, classify the sample as the positive class; otherwise, classify it as the negative class.
   d. Append the ensemble algorithms prediction to the list of ensemble predictions.
6. Return the final_ensemble_classifier, which can be used to predict labels for new unseen data.

Figure 2.7 *RUSBoost algorithm.*

Table 2.2 Performance comparison of different classifier models for Case-1.

Classifier	Accuracy	Prediction speed	Training time (s)
RUS boosted trees	%99.1	15,000obs/s	1.5504
Bagged trees	%98.7	10,000obs/s	2.6921
Fine tree	%98.6	61,000obs/s	0.7275
Wide neural network	%97.2	48,000obs/s	1.5294

Table 2.3 Performance comparison of different classifier models for Case-2.

Classifier	Accuracy	Prediction speed	Training time (s)
RUS boosted trees	%98.4	11,000 obs/s	2.003
Wide neural network	%98.0	44,000 obs/s	1.4844
Bagged trees	%97.8	9,400 obs/s	2.6858
Quadratic SVM	%97.5	45,000 obs/s	1.4044

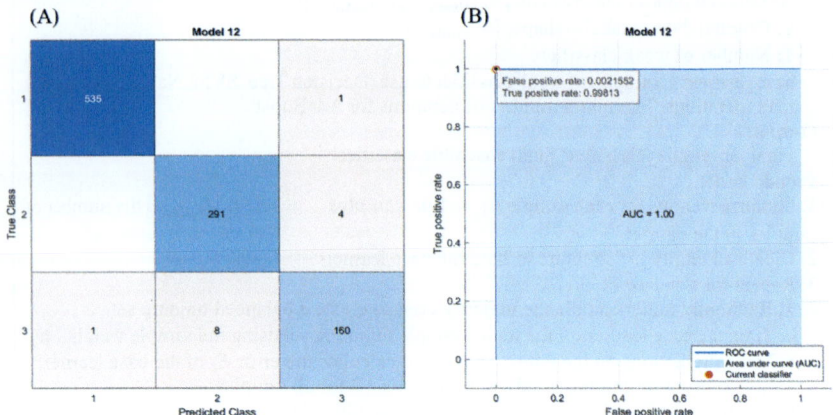

Figure 2.8 *(A) Case-1 Dataset RUSBoost trees confusion matrix, (B) ROC curve.*

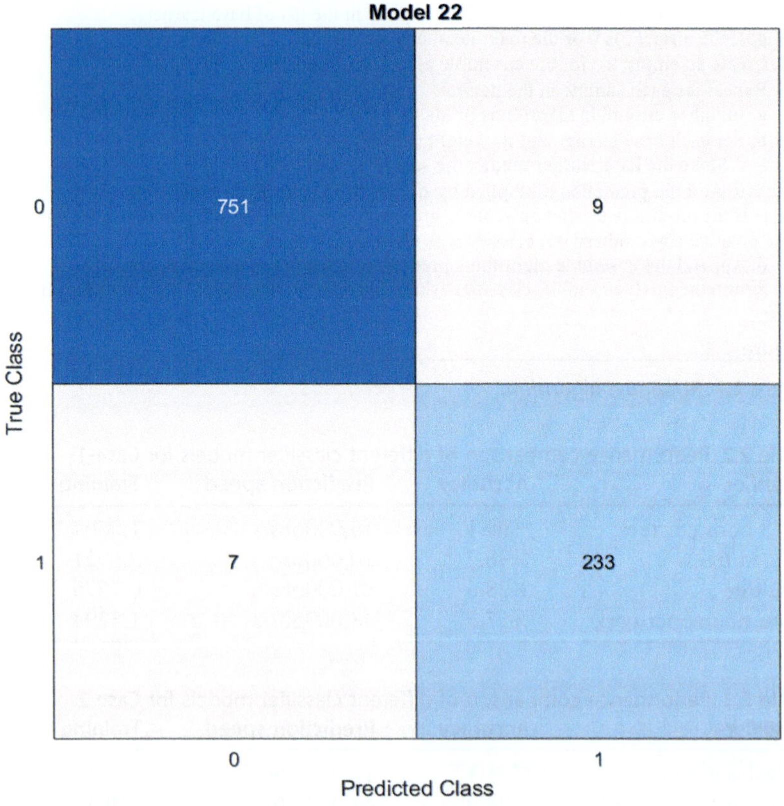

Model 22

751 | 9

7 | 233

True Class

Predicted Class

Figure 2.9 *Case-2 dataset RUSBoost trees confusion matrix.*

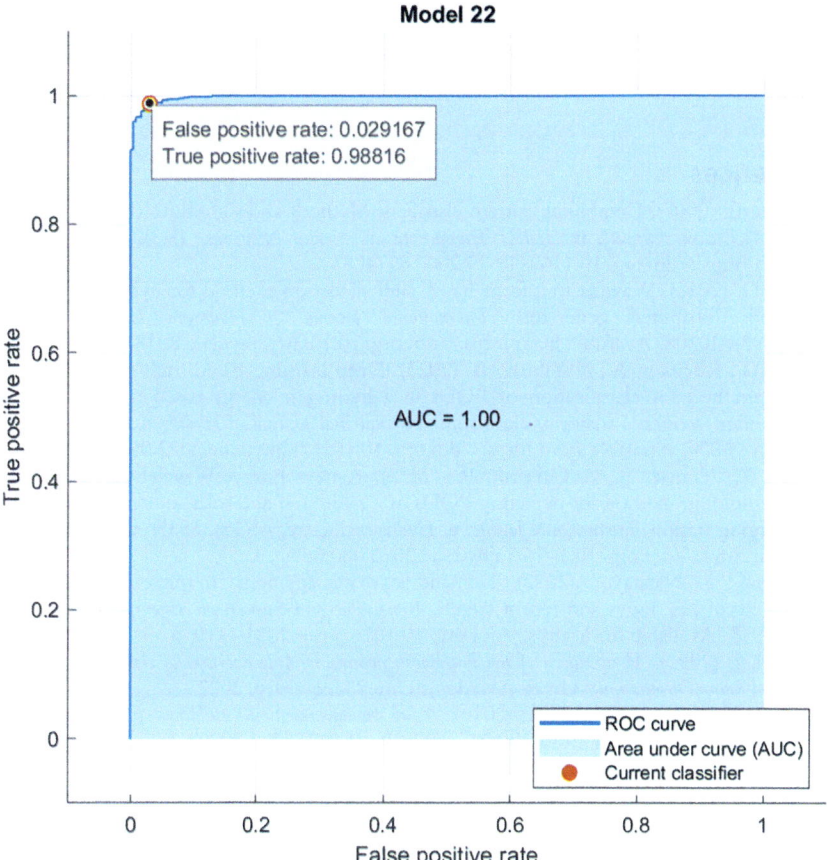

Figure 2.10 *Case-2 Dataset RUSBoost trees confusion matrix and ROC curve.*

2.4 Conclusion

In this study, two different scenarios were considered. Datasets were created for each case, to maximize classification performance using machine learning methods. The RUSBoost classifier exhibited the highest accuracy, achieving 99.1% for Case-1 and 98.4% for Case-2. The results indicate that intelligent load control with high accuracy can be achieved using RUSBoost tree ensemble classifiers, offering a potential solution for preventing the formation of peak loads. The observed performance of Ensemble Learning classifier methods suggests the potential for further enhancement in the future. These methods can be leveraged for direct load management applications, especially when dealing with complex datasets centered on increased utilization of renewable energy.

References

Alpaydin, E. (2007). Combining pattern classifiers: Methods and algorithms (Kuncheva, L.I.; 2004) [book review], in. *IEEE Transactions on Neural Networks, 18*(3), 964. Available from https://doi.org/10.1109/TNN.2007.897478.

Bayrak, G. (2018). Wavelet transform-based fault detection method for hydrogen energy-based distributed generators. *International Journal of Hydrogen Energy, 43*(44), 20293−20308. Available from https://doi.org/10.1016/j.ijhydene.2018.060.183.

Bayrak, G., Küçüker, A., & Yılmaz, A. (2023). Deep learning-based multi-model ensemble method for classification of PQDs in a hydrogen energy-based microgrid using modified weighted majority algorithm. *International Journal of Hydrogen Energy, 48*(18), 6824−6836. Available from https://doi.org/10.1016/j.ijhydene.2022.050.137.

Bayrak, G., Yılmaz, A., & Çakmak, R. (2023). A new Fuzzy&Wavelet-based adaptive thresholding method for detecting PQDs in a hydrogen and solar-energy powered EV charging station. *International Journal of Hydrogen Energy, 48*(18), 6855−6870. Available from https://doi.org/10.1016/j.ijhydene.2022.080.067.

Dharmaraj., & Natarajan. (2022). Demand side management in microgrid: A critical review of key issues and recent trends. *Renewable and Sustainable Energy Reviews, 156,* 111915. Available from https://doi.org/10.1016/j.rser.2021.111915.

Forslund, I. (2022). *Modification of the RusBoost algorithm: A comparison of classifiers on imbalanced data (Dissertation)* .Umeå (Sweden), Umeå University, 2022.

Goy, S., & Sancho-Tomás, A. (2019). *Load management in buildings* (pp. 137−179). Elsevier BV. Available from 10.1016/b978-0-12-811553-4.00004-4.

Hesser, T., & Succar, S. (2012). *Renewables integration through direct load control and demand response* (pp. 209−233). Elsevier BV. Available from https://doi.org/10.1016/b978-0-12-386452-9.00009-7.

How will growing EV sales affect US power markets? (2017). https://www.woodmac.com/our-expertise/focus/power-renewables/how-will-growing-ev-sales-affect-us-power-markets/.

Huber, T., Kuhn., & Hamacher, T. (2012). Effects of large-scale EV and PV integration on power supply systems in the context of Singapore. *Innovative Smart Grid Technologies (ISGT Europe), 2012 3rd IEEE PES International Conference and Exhibition,* 1−8. Available from https://doi.org/10.1109/ISGTEurope.2012.6465831.

Jabir, H. J., Teh, J., Ishak, D., & Abunima, H. (2018). Impact of demand-side management on the reliability of generation systems. *Energies, 11*(8), 2155. Available from https://doi.org/10.3390/en11082155.

Li, D., Chiu, W. Y., & Sun, H. (2017). *Demand side management in microgrid control systems microgrid: Advanced control methods and renewable energy system integration* (pp. 203−230). United Kingdom: Elsevier Inc. Available from http://www.sciencedirect.com/science/book/9780081017531.

McDonald, J. R., Lo, K. L., & Maulana, H. (1990). *Operational constraints on the introduction of direct load control in electric power systems* (pp. 391−397). Elsevier BV. Available from https://doi.org/10.1016/b978-0-08-037539-7.50066-1.

Mehta, R., Srinivasan, D., Khambadkone, A. M., Yang, J., & Trivedi, A. (2018). Smart charging strategies for optimal integration of plug-in electric vehicles within existing distribution system infrastructure. *IEEE Transactions on Smart Grid, 9*(1), 299−312. Available from https://doi.org/10.1109/TSG.2016.2550559.

Nasir, T., Bukhari, S. S. H., Raza, S., Munir, H. M., Abrar, M., Muqeet, H. A. U., Bhatti, K. L., Ro, J.-S., Masroor, R., & Sun, Q. (2021). Recent challenges and methodologies in smart grid demand side management: State-of-the-art literature review. *Mathematical Problems in Engineering, 2021*, 1−16. Available from https://doi.org/10.1155/2021/5821301.

Polikar, R. (2006). Ensemble based systems in decision making. *IEEE Circuits and Systems Magazine, 6*(3), 21−45. Available from https://doi.org/10.1109/mcas.2006.1688199.

Seiffert, C., Khoshgoftaar, T. M., Van Hulse, J., & Napolitano, A. (2010). RUSBoost: A hybrid approach to alleviating class imbalance. *IEEE Transactions on Systems, Man, and Cybernetics - Part A: Systems and Humans, 40*(1), 185−197. Available from https://doi.org/10.1109/tsmca.2009.2029559.

Siano, P. (2014). Demand response and smart grids—A survey. *Renewable and Sustainable Energy Reviews, 30*, 461−478. Available from https://doi.org/10.1016/j.rser.2013.100.022.

Türkiye İstatistik Kurumu Taşıt-kilometre İstatistikleri. (2018). https://data.tuik.gov.tr/Bulten/Index?p = Vehicle-kilometer-Statistics-2016-30846.

Yılmaz, A., Küçüker, A., & Bayrak, G. (2022a). Automated classification of power quality disturbances in a SOFC&PV-based distributed generator using a hybrid machine learning method with high noise immunity. *International Journal of Hydrogen Energy, 47* (45), 19797−19809. Available from https://doi.org/10.1016/j.ijhydene.2022.020.033.

Yılmaz, A., & Bayrak, G. (2019). A real-time UWT-based intelligent fault detection method for PV-based microgrids. *Electric Power Systems Research, 177*, 105984. Available from https://doi.org/10.1016/j.epsr.2019.105984.

Yılmaz, A., Küçüker, A., Bayrak, G., Ertekin, D., Shafie-Khah, M., & Guerrero, J. M. (2022b). An improved automated PQD classification method for distributed generators with hybrid SVM-based approach using un-decimated wavelet transform. *International Journal of Electrical Power & Energy Systems, 136*, 107763. Available from https://doi.org/10.1016/j.ijepes.2021.107763.

CHAPTER THREE

Modeling stochastic renewable energy processes by combining the Monte Carlo method and mixture density networks

Deivis Avila[1], Yanelys Cuba[2], Graciliano N. Marichal[1] and Ramón Quiza[2]

[1]Higher Polytechnic School of Engineering (EPSI), University of La Laguna, Santa Cruz de Tenerife, Spain
[2]Centre for Advanced and Sustainable Manufacturing Studies, University of Matanzas, Matanzas, Cuba

3.1 Introduction to stochastic phenomena in renewable energies

Human civilization has been shaped by discoveries and advancements in harnessing new energy sources and optimizing their efficiency. One might say that there are two truly important periods in humanity's pursuit for sources of energy: the discovery of fire in the primitive era and the use of machines in the modern era, such as the steam engine and the combustion engine in the 18th and 19th centuries, both of which played key roles in the Industrial Revolution (Sánchez, 2009).

All these technologies traditionally relied on burning fuels such as coal, gasoline, diesel, etc., to generate energy to move the machinery. This results in the release of high levels of polluting gases, notably CO_2, the main cause of the greenhouse effect.

Although renewable energies have been used by humanity for thousands of years, such as wind for navigation and water to generate mechanical energy, it was not until the last few decades that many scientists, engineers, institutions, and governments focused their efforts on using renewable energies to decrease the reliance on fossil fuels and mitigate the emissions from greenhouse gases.

Renewable energy sources (RES) originate in nature and are continually replenished. This kind of energy can be defined as "Energy obtained

Intelligent Learning Approaches for Renewable and Sustainable Energy
DOI: https://doi.org/10.1016/B978-0-443-15806-3.00003-6

from continuous or repetitive currents recurring in the natural environment." RES are inexhaustible and safe sources of energy with similar properties, including low energy density, intermittent supply, and in many cases, stochastic behavior (Twidell & Weir, 2006).

Fig. 3.1 shows various RES and their resultant energies. These RES can be classified based on their origin: for example, directly from the Sun (e.g. photovoltaic (PV) and solar thermal) and indirectly from the Sun (e.g. wind energy, hydraulic energy, biomass and some ocean energies like wave energies, thermal gradient). Other RES include geothermal, derived from the internal energy of the Earth, and tidal energy, a consequence of the gravitational interaction between the Moon, Sun, and Earth (ADIRA, 2008; Twidell & Weir, 2006).

The use of RES has many benefits, such as reducing the use of fossil fuels, contributing to a cleaner environment, and improving social well-being. However, the intermittent nature of some of the most abundant and environment-friendly RES, such as solar, wind, and wave energy, poses many problems for its integration into electrical systems.

The stochastic behavior of renewable energy from the Sun and its resultant energies, such as wind and sea waves, is one of the central challenges for integrating RES into electrical grids. It is crucial to note that these energies depend on uncontrollable aspects, such as meteorological conditions. Weather forecasts can help reduce the uncertainty involving the generation of renewable energies (Ahmed & Khalid, 2019; Corizzo et al., 2021; Nottona et al., 2018). It is possible to establish synergies between different RES and create a hybrid system, which can be isolated or integrated into the grid. By taking into account the stochastic behavior of different renewable energies such as solar, wind, or wave, a hybrid

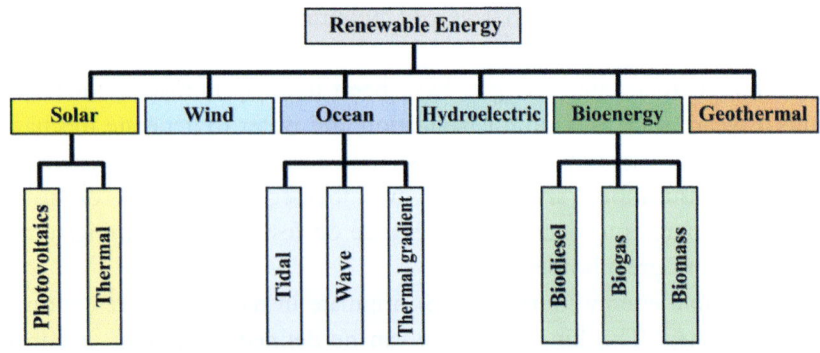

Figure 3.1 Renewable energy sources.

system with backup offers greater consistency in the electrical system (Avila, Marichal, Hernández, et al., 2021; Talaria et al., 2018). Many stochastic dependent and independent models have been used in the study of RES, as evident in the works of authors such as Talaria et al. (2018) and Su (2005). These authors discuss the importance of stochastic analyses in determining RES production and their integration into the electricity grid. This chapter explains the different types of methods that renewable energy power system operators can apply to address the uncertainties of different types of RES. Among the stochastic simulation methods discussed, the Monte Carlo Method (MCM) is considered one of the most precise and direct; however, it requires a significant computational effort.

The scientific community has worked for a long time with stochastic methods to predict the behaviors of various meteorological conditions such as radiation, wind, and so on. Numerous studies on this subject reflect the ongoing efforts to understand and forecast these conditions. As the demand for clean energy production from renewable resources continues to grow, it becomes increasingly evident that predicting these resources using stochastic methods is essential. Several recent studies have focused on the application of stochastic methods in predicting the behaviors of wind, solar energy, and hybrid systems (solar, wind, other) (Chideana et al., 2018; Corizzo et al., 2021; Eghbali et al., 2022; Liu & Wu, 2021; Nam et al., 2020; Sakki et al., 2022; Silva et al., 2022; Yadava et al., 2018; Yan et al., 2023; Zhang et al., 2022).

Ocean renewables are the only RES that were not previously modeled in hybrid systems. However, waves are the most appealing and abundant energy source for the world's coastal regions. The possible inclusion of wave energy in the energy mix would be a significant advancement for electricity networks in coastal regions, ensuring a diversified energy supply. However, the development of wave energy converters (WEC) faces challenges. Currently, operational marketable prototypes of WEC do not exist anywhere in the world; there are only experimental wave energy projects. The wave energy industry aspires to become a noteworthy sector in the future, playing a crucial role in coastal electricity networks. Achieving this potential will require substantial support from institutional and government organizations as well as private investors. (Avila et al., 2022; Avila, Marichal, Quiza, et al., 2021; Padrón et al., 2022; Sheng, 2019).

Currently, many research teams continue to work on experimental prototypes. These teams require robust and efficient tools to assess the energy transformation capabilities (energy matrix) under various sea conditions worldwide. In an effort to achieve consistent forecasts of energy conversion capabilities, models that define the behavior of stochastic sea waves are required. Several studies in this domain have been conducted, including those by Avila, Marichal, Quiza, et al. (2021), Avila et al. (2022), Ahn et al. (2020, 2021), Sandvik et al. (2019) and Azharul et al. (2020).

Taking into account all of the above, the main goal of this chapter is to develop an approach for predicting the potential transformation of the energy of sea wave energy. This approach relies on a specific type of artificial neural network, namely mixture density networks (MDN), and the MCM, while taking into account the power matrix of different WEC.

3.2 Monte Carlo method (MCM)

3.2.1 Foundations

The term "Monte Carlo method" (MCM) refers to a method for solving complex problems by using random numbers. Despite its name, it is not a conventional method, characterized by a set of ordered steps like a recipe or algorithm; on the contrary, it is a simulation philosophy.

The name "Monte Carlo" was introduced in the mid-20th century by Ulam, von Neumann, and Metropolis, inspired by the random behavior of the roulette wheels in the famous gambling casinos in the city of Monte Carlo (Stevens, 2023). Despite this, the principles of the MCM were in use much earlier.

The MCM can be applied to solve two main types of problems: those where data constructed by using random numbers are used to generate purely deterministic output variables and those where they are used to generate probability distributions of stochastic output variables.

A typical example of the first type is numerical integration by using random numbers (see Fig. 3.2). In this approach, a set of random points is generated in the area of interest. By computing the relation of the number of points below the function curve (red) to the total number of points (blue and red), the area below the curve (i.e., the value of the definite integral of $f(x)$ in the interval $[a, b]$) can be estimated. It is worth noting that the accuracy of the area estimate improves with a larger number of random points.

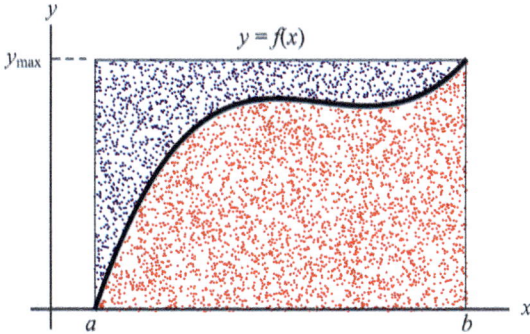

Figure 3.2 Example of using the MCM to solve a deterministic problem (definite integration).

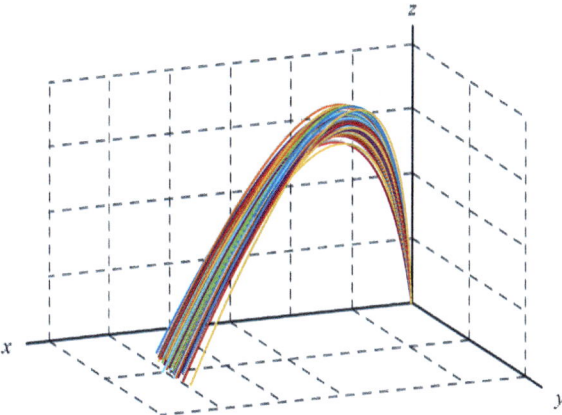

Figure 3.3 Example of using the MCM to solve a stochastic problem (impact point of a howitzer projectile).

A second example (see Fig. 3.3) illustrates the use of the MCM to solve a stochastic problem: computing the impact point of a howitzer projectile. The trajectory is determined by numerically integrating the governing differential equation. The method considers the variability of specific input parameters, such as meteorological or climatic behavior (wind speed or air density), or other aspects, such as the gunpowder charge. Random numbers from the corresponding probability distributions are generated, and the consequent trajectories are computed. Finally, the probability distribution of the impact point is estimated from the resulting trajectories. Once again, it is important to emphasize that the accuracy of the probability distribution of the impact point improves with an increased number of computed trajectories.

Two key considerations in Monte Carlo simulations should be noted. First, the quality of the outputs is heavily dependent on the actual randomness of the numbers used. Consequently, in computer simulations where pseudo-random number generators are employed, the performance of these generators becomes critical (Gentle, 2003). Secondly, adhering to the law of large numbers, the reliability of the MCM outcomes relies on conducting a large number of runs (Graham & Talay, 2013).

3.2.2 Algorithms

As was previously pointed out, there is no unique algorithm for the MCM; however, there are some general steps that are usually included whenever this approach is used. These steps can be summarized as follows (Chanan et al., 2019):

1. Define a domain and a probability distribution for each stochastic input considered in the problem.
2. Generate random inputs from the corresponding probability distributions over the preestablished domains.
3. Perform the deterministic computation of the outputs for every set of random inputs.
4. Aggregate the outputs and perform the proper analysis depending on the nature of the problematic in question (to determine either a deterministic value or obtain a probability distribution).

To show how the method works, let us examine the problem of the trajectory of charged particles in a constant magnetic field. Consider a charged particle generator, which launches the particles at a speed $v_{min} \leq v \leq v_{max}$, and an angle in the interval $\pm \theta$ with respect to the horizontal leftward direction. Both parameters can be assumed to be random with a uniform probability distribution in the given intervals. By taking into account this information, a number of initial velocity vectors can be randomly generated (see Fig. 3.4A).

The output to be computed is the point where the particle impacts the detection screen. The trajectory (with the corresponding impact point) for each velocity vector can be deterministically computed by using the numerical integration of the Lorentz force equation, $\mathbf{F} = q\mathbf{v} \times \mathbf{B}$, where q is the particle charge, $\mathbf{v}$ is the velocity vector, and the magnetic field vector is $\mathbf{B}$ (see Fig. 3.4B).

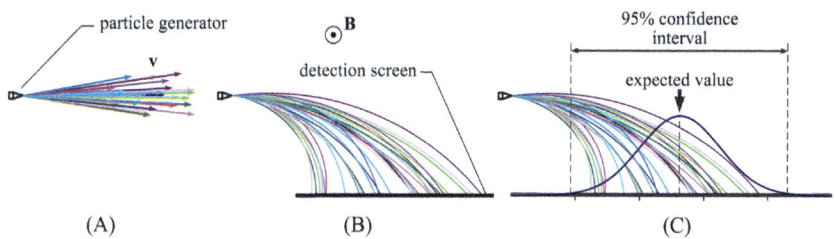

Figure 3.4 Example of the steps in the Monte Carlo method. (A) Generation of random inputs, (B) deterministic computation of outputs, (C) aggregation and analysis of outputs.

Finally, all the resulting outputs (impact point positions) are aggregated by computing the probability distribution parameters (mean and standard deviation) that can be used to determine not only the expected value but also the confidence interval (see Fig. 3.4C).

3.2.3 Advantages and shortcomings

The MCM has a number of advantages that have made it an appealing option for solving an important number of scientific and engineering problems. First and foremost, it allows to deal with problems that cannot be solved analytically. This is a key issue when faced with real-world problems that involve complex, nonlinear, and noisy systems. A second advantage is the capability of the MCM to survey the complete search space of a problem, assuming the model is correctly parameterized. Finally, the outputs of the MCM are easy to understand. By using the proper probabilistic tools, the corresponding analyses are straightforward to perform, providing valuable insights into the system's performance (Chanan et al., 2019).

In contrast, the main drawback of the MCM is its high demand for computational resources. This demand naturally increases with the complexity of the problem at hand and the number of runs required (Lofgren, 2014). It is important to highlight that, similarly to any other simulation technique, the reliability of the outcome depends entirely on the quality of the models involved and the accuracy of the input parameters (El Bakkari et al., 2012).

3.2.4 Applications to renewable energies

Due to the stochastic nature of most RES, the MCM has been widely applied to research in this field. Some of the most recent applications include solar energy (Lyudmila, 2021; Wales et al., 2022), wind energy (Díaz et al., 2022; Jang et al., 2022; Vavatsikos et al., 2022; Zhao et al., 2021), wave energy (Avila et al., 2022; Avila, Marichal, Quiza, et al., 2021; Görmüs et al., 2022), bioenergies (Benjamin et al., 2021; Khadivi & Sowlati, 2022), and hybrid or combined systems (Saeedi & Hassan Hosseini, 2022; Uwineza et al., 2021).

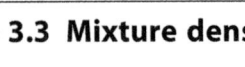

3.3 Mixture density networks

3.3.1 Foundations of machine learning

Machine learning refers to a broad group of techniques aimed at extracting useful information, such as models of patterns from existing data, in order to make computers modify or adapt their actions so that these actions get more accurate (Marsland, 2015).

Although most machine learning techniques fall under the umbrella of artificial intelligence, the overlap between the two is not absolute. For example, there are machine learning techniques (such as linear regression) that are not included in artificial intelligence. By contrast, important branches of artificial intelligence (for example, bio-inspired optimization metaheuristics) are not usually considered to be part of machine learning (Joshi, 2022).

Unlike other techniques, such as data mining, machine learning focuses on extracting information that can be generalized beyond the cases present in the data sets used in training. Consequently, validation is a mandatory component of model fitting in machine learning (Carter et al., 2023).

Some of the most popular machine learning techniques include artificial neural networks (Portillo Juan & Negro Valdecantos, 2022), fuzzy inference systems (Kar et al., 2014), support vector machines (Zendehboudi et al., 2018), decision trees (Audemard et al., 2022), and deep learning (Michelucci, 2019), all of which are widely applied for solving scientific and engineering problems.

From the point of view of the characteristics of the training process, machine learning techniques can be grouped as follows (Paluszek & Thomas, 2019):

- *Supervised learning*: Where the training set includes not only the inputs but also the desired outputs. It is especially suitable for modeling tasks.

- *Unsupervised learning*: Where the training set does not include the outputs. It is applied to clustering or classification tasks.
- *Semi-supervised learning*: Where the training set includes both labeled (inputs have the corresponding outputs) and unlabeled data. It is used when labeling relies on human actions and requires excessive time and effort.
- *Online learning*: Where the training set is continuously updated. It is widely used in dynamic systems.

3.3.2 Gaussian distribution and Gaussian mixture

A probability density function (pdf), $f_X(x\,|\,\mathbf{b})$, where $\mathbf{b}$ is a vector of parameters, is a function describing the expected probability of a continuous random variable, X, that can be seen as a limit case of a histogram (see Fig. 3.5A). For a given probability density function, $f_X(x\,|\,\mathbf{b})$, the probability of variable X being in the interval $[a, b]$ is defined as (see Fig. 3.5B):

$$P_X|_{a \leq X \leq b} = \int_a^b f_X(x|\mathbf{b})dx. \tag{3.1}$$

A related concept is the so-called "cumulative density function" (cdf), $F_X(x\,|\,\mathbf{b})$, which is defined as the probability of X having a value lower than or equal to x (i.e., being in the interval $[-\infty, x]$) (see Fig. 3.5C).

The Gaussian (also called "normal" or "bell-shaped") distribution is given by the expression:

$$f_X(x|\mu, \sigma) = \frac{1}{\sqrt{2\pi}\sigma} e^{-\frac{1}{2}\left(\frac{x-\mu}{\sigma}\right)^2}; \tag{3.2}$$

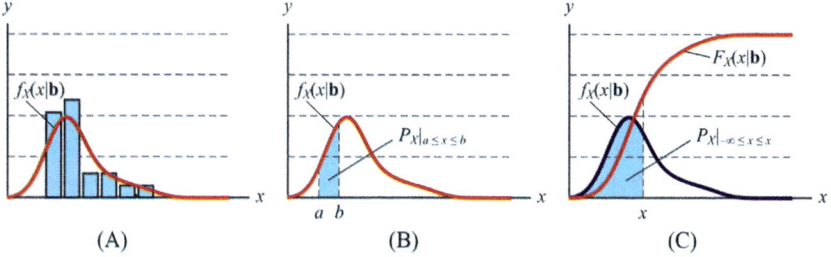

Figure 3.5 Probability function definitions. (A) probability density function, (B) probability and probability density function, (C) cumulative density function.

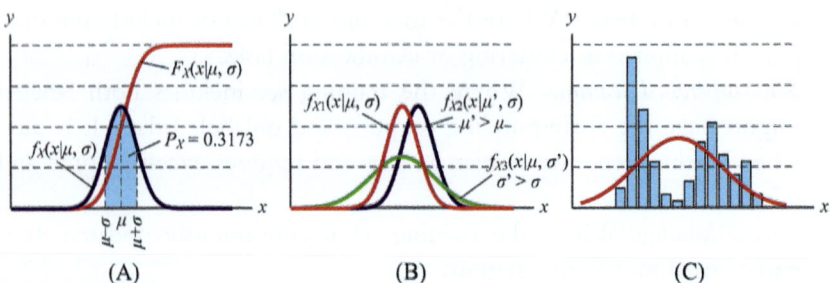

Figure 3.6 Gaussian distribution. (A) Gaussian distributions pdf and cdf, (B) effect of the parameters in the Gaussian distribution, (C) poorly fitted Gaussian distribution.

where the function parameters μ and σ are called mean value and standard deviation, respectively (see Fig. 3.6A and B). The corresponding cumulative density is:

$$F_X(x) = \frac{1}{2}\left[1 + \mathrm{erf}\left(\frac{x-\mu}{\sqrt{2}\sigma}\right)\right]; \qquad (3.3)$$

where erf(•) is the so-called "error function," which is defined by the nonelementary integral:

$$\mathrm{erf}(x) = \frac{2}{\sqrt{\pi}}\int_0^x e^{-t^2}\mathrm{d}t. \qquad (3.4)$$

The Gaussian distribution is widely used in statistical analysis due to an interesting property known as the central limit theorem, which states that when independent random variables are summed, their properly normalized sum trend toward a Gaussian distribution even if the original random variables themselves are not normally distributed (Sun et al., 2022).

The Gaussian distribution, however, cannot represent the distribution of any random variable (see Fig. 3.6C), and several other distributions have been proposed. The mixture distributions are convex combinations (i.e., the weighted sum of other distributions, with nonnegative weights that add to one) of other distributions, which are used to describe random variables, especially those where several subpopulations appear.

Gaussian mixture distributions are composed of several normal distributions, which are properly weighted:

$$f_X(x) = \sum_{i=1}^{N} \frac{w_i}{\sqrt{2\pi}\sigma_i} e^{-\frac{1}{2}\left(\frac{x-\mu_i}{\sigma_i}\right)^2}; \qquad (3.5)$$

where N is the number of components, w_i is the weight of the ith component, and μ_i and σ_i are the corresponding distribution parameters (mean and standard deviation) (Jin et al., 2021) (see Fig. 3.7).

3.3.3 MDN architecture

MDN is a type of artificial neural network whose output is a mixture of Gaussian distributions rather than a single value. An MDN consists of two sections (see Fig. 3.8). In the first place, there is a feed-forward network containing one input layer that just takes the inputs and transfers them to the next step. The so-called "hidden layer" does a weighted sum

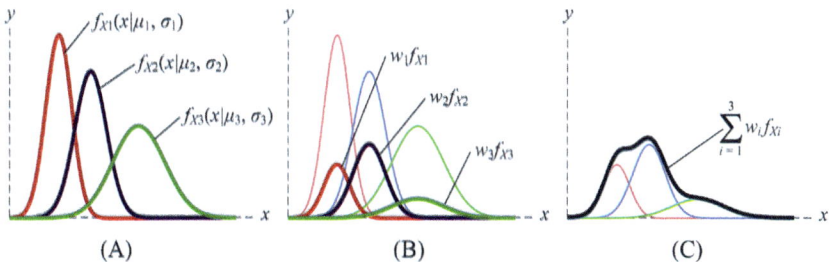

Figure 3.7 Mixture Gaussian distribution. (A) Gaussian distributions, (B) weighted Gaussian distributions, (C) mixture Gaussian distribution.

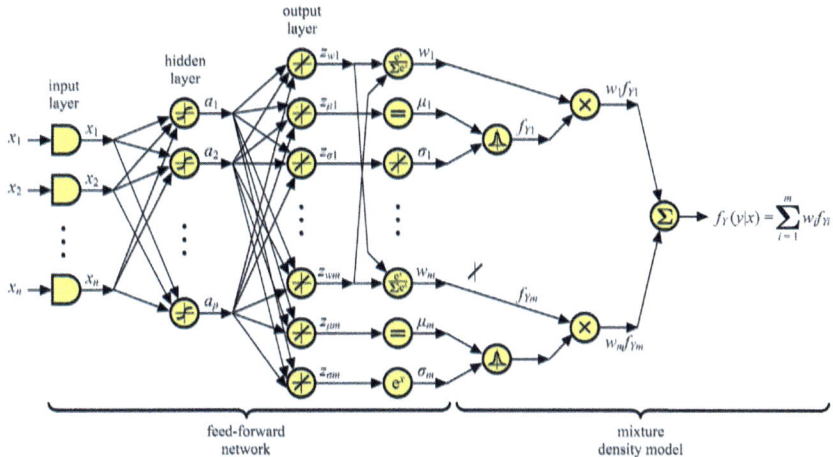

Figure 3.8 Representation of a mixture density network.

of the inputs, $\{x_1, x_2, \ldots, x_n\}$ and then applies a sigmoid activation function (Bishop, 1994):

$$a_j = \frac{1}{1 + e^{-\left(b_j + \sum_{i=1}^{n} u_{i,j} x_i\right)}}; \qquad (3.6)$$

where a_j and b_j are the output and the bias, respectively, of the j-th hidden neuron, and $u_{i,j}$ is the synapse connecting the ith input with the j-th hidden neuron.

The output layer is composed of $3m$ neurons, where m is the number of distributions used in the mixed models in the next section. Their respective output values, z_{wj}, $z_{\mu j}$, and $z_{\sigma j}$, can be the result of the linear combination of the outcomes of the hidden layers:

$$z_{wj} = c_j + \sum_{i=1}^{p} v_{i,j} a_i; \qquad (3.7a)$$

$$\mu_j = c'_j + \sum_{i=1}^{p} v'_{i,j} a_i; \qquad (3.7b)$$

$$\sigma_j = c''_j + \sum_{i=1}^{p} v''_{i,j} a_i; \qquad (3.7c)$$

where c_j, c'_j and c''_j are the biases of the outputs corresponding to the jth distribution parameters, and $v_{i,j}$, $v'_{i,j}$ and $v''_{i,j}$, are the weights connecting the ith hidden neuron to the corresponding elements of the jth distribution parameters (Herzallah & Lowe, 2004).

The distribution parameters w_j, μ_j, and σ_j are computed from the outputs of the networks by applying the following transformations. The weights are computed by a normalization through the so-called "softmax" function:

$$w_j = \frac{e^{z_{wj}}}{\sum_{i=1}^{m} e^{z_{wi}}}; \qquad (3.8a)$$

the mean values are just equal to the corresponding outputs:

$$\mu_j = z_{\mu j}; \qquad (3.8b)$$

and the standard deviations, which are scale parameters, are the exponents of the corresponding outputs

$$\sigma_j = z_{\sigma j}; \qquad (3.8a)$$

Finally, the Gaussian mixture model, with constitutes the outcome of the MDN, is composed, as explained in the previous subsection, of the weights and distribution parameters forecast by the previous section:

$$p(y|x) = \sum_{i=1}^{m} \frac{w_i(\mathbf{x})}{(2\pi)^{c/2}\sigma_i(\mathbf{x})^c} e^{-\frac{\|t-\mu_i(\mathbf{x})\|^2}{2\sigma_i(\mathbf{x})^2}} ; \tag{3.9}$$

where the vector $\mu_i(\mathbf{x})$ represents the center of the ith kernel, $\sigma_i(\mathbf{x})$ its standard deviation, and $w_i(\mathbf{x})$ the corresponding weight.

3.3.4 MDN training

The training of an MDN is similar to that used in a regular multilayer perceptron (MLP), and relies on error-backpropagation. It involves two steps. The first (forward) step computes the outputs of each layer from the first to the last. The second (backward) step updates the weights and biases based on the error of the estimations. The weights and biases in the hidden layers are updated using the gradient descent method. The two steps are repeated for the different elements in the training dataset until some termination condition (usually, a prescribed error value or some number of epochs) is reached.

The main difference between MLP and MDN training is the way the error is computed. While MLP uses the conventional least squares method, the error in MDN is computed by considering the forecast probability distributions using the expressions (Bishop, 1994):

$$\varepsilon = -\iint \ln\left\{ \sum_{i=1}^{m} \frac{w_i(\mathbf{x})}{(2\pi)^{c/2}\sigma_i(\mathbf{x})^c} e^{-\frac{\|t-\mu_i(\mathbf{x})\|^2}{2\sigma_i(\mathbf{x})^2}} \right\} p(\mathbf{x}, \mathbf{t})d\mathbf{x} \ d\mathbf{t}. \tag{3.10}$$

3.3.5 Applications to the renewable energies

Due to its capability to forecast probability distributions, MDN has been used not only to model renewable energy systems, such as wind (Abdul Majid, 2022; Christoforou et al., 2021) or wave (Avila et al., 2022) power systems, but also to predict energy needs (Al-Gabalawy et al., 2021; Brusaferri et al., 2022).

3.4 Case study

3.4.1 Formulation

To illustrate the applications of the techniques explained above, the energy conversion potential of two points near the Canary Islands has been modeled, and the corresponding generated power was forecast for different WECs. The first point under study corresponds to the Las Palmas Este (1414) buoy, which is located at 28.05° N and 15.39° W. This buoy is located less than 2 km from the shore, with a mooring depth of 30 m; consequently, it can be regarded as a nearshore point. In contrast, the second point corresponds to the offshore Gran Canaria (2442) buoy, which is located at 28.20° N and 15.78° W, around 8 km away from shore, and with a mooring depth of 780 m (see Fig. 3.9).

The study uses data recorded from 1992 to 2019 (27 years) for the Las Palmas Este buoy, and from 1997 to 2019 (22 years) for the Gran Canaria (2442) buoy. The sampling period was 3 h, and 226,679 valid records were taken for the Las Palmas Este (1414) buoy and 151,824 for the Gran Canaria (2442) buoy. Although more parameters were measured, this study only considered the measured date and time, the significant spectral height, H, and mean peak period, T (see Fig. 3.10). The respective measurement accuracy was ± 0.05 m for significant spectral height and ± 0.05 s for mean peak period. The measured data was not preprocessed. All the data

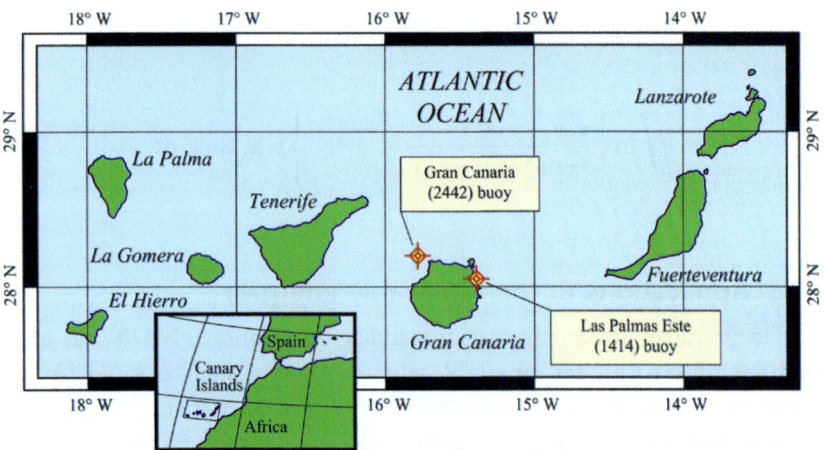

Figure 3.9 Geographic location of the buoys studied (Avila et al., 2022).

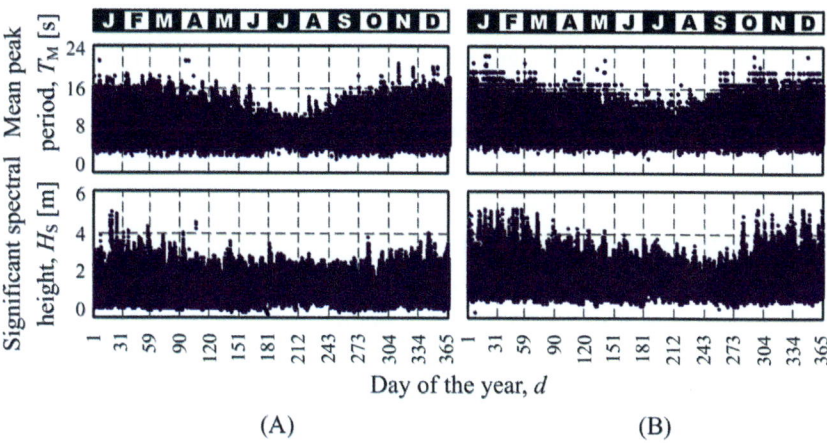

Figure 3.10 Measured wave data. (A) Las Palmas Este (nearshore buoy), (B) Gran Canaria (offshore buoy).

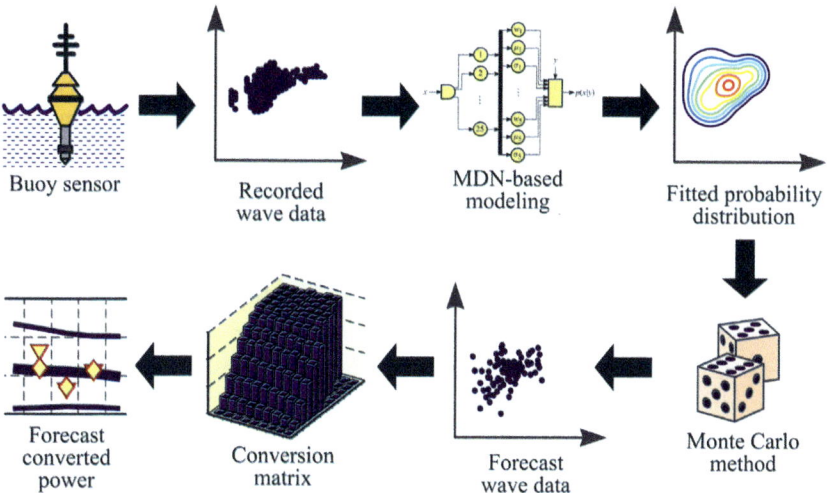

Figure 3.11 Graphical representation of the proposed approach.

was accessed through the Spanish State Ports dataset (Harbors of State of Spain, 2017, 2018; PivotBuoy, 2019).

The converted power forecasting process was organized into three steps (see Fig. 3.11). In the first place, MDN was fitted to model the bivariate probability distribution of significant spectral height and mean peak period versus the day of the year. In the second step, the MCM was used to forecast values of significant spectral height and mean peak period for each day of the year from the previous MDN-based fitted probability distributions. Finally, the generated power was forecast for different WECs, taking into account the corresponding conversion power matrices. Each step and the subsequent validation are explained in the following subsections.

3.4.2 MDN-based modeling

The MDN models were fitted using MATLAB R2020a (version 9.8.0.1323502) and the NETLAB 3.3 toolbox (Bishop & Nabney, 1997). For both models (each one corresponding to one buoy's dataset), an MDN was fitted. The input (day of the year) and the outputs (significant spectral height and mean peak period) were normalized into the interval $[-1, 1]$ using linear interpolation.

The hidden layer was integrated for 50 units (nodes) using a sigmoid activation function. The output layer, on the other hand, used linear activation functions to yield the mean and variance of the five Gaussian distributions that form the resulting mixture model.

The training process was carried out by combining the general-purpose gradient-based optimizers, as well as by sampling the subsequent parameter distribution using general- purpose Markov chain Monte Carlo sampling algorithms. Two thousand training cycles were used, with an inverse variance for weight initialization equal to 10, and 25 iterations for k-means in the initialization of the Gaussian mixture models.

Fig. 3.12 shows a graphical representation of the sampled data and the corresponding probability distributions obtained for a representative day of each month. Note that the approach used demonstrates the ability to forecast the probability distribution, even when its shape varies significantly, as observed during seasonal changes. This capability can be considered an important advantage of MDN over the use of preestablished probability distributions.

3.4.3 Monte Carlo simulation

Using the MCM, 100 values of wave parameters (i.e., significant spectral height and mean peak period) were randomly generated for each day of the year. In the first place, the bivariate mixture Gaussian probability distribution is obtained by using the MDN model described above for the corresponding day of the year. Then, the random values are generated by using the rejection method in a rectangular area (Kesemen & Tiryaki, 2018).

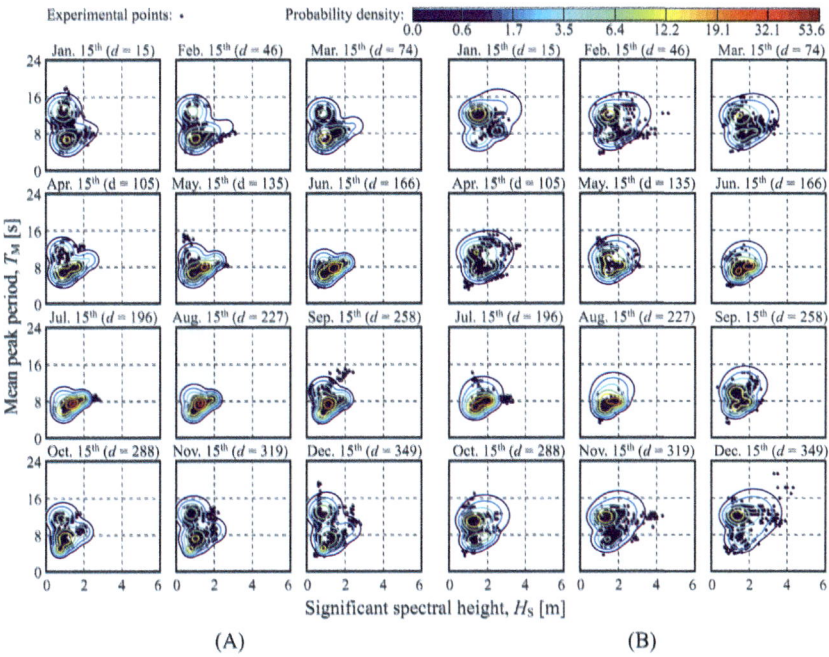

Figure 3.12 Bivariate probability distribution of the wave parameters. (A) Las Palmas Este (nearshore buoy), (B) Gran Canaria (offshore buoy).

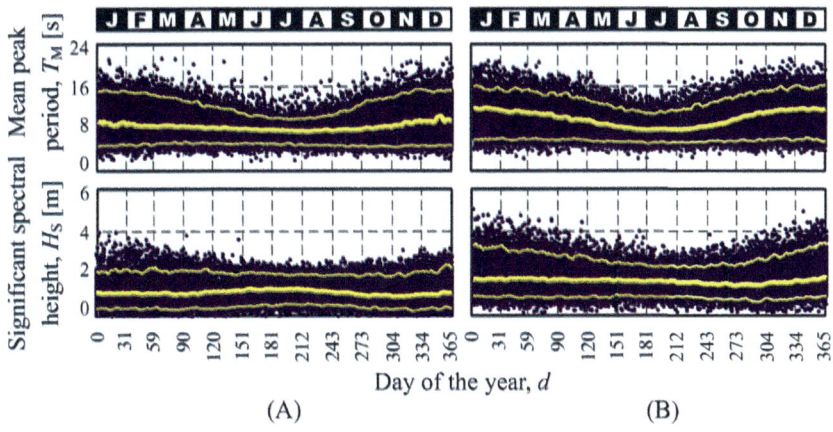

Figure 3.13 Forecast daily wave parameters. (A) Las Palmas Este (nearshore buoy), (B) Gran Canaria (offshore buoy).

Fig. 3.13 shows the forecast values of significant spectral height and mean peak period for both buoys.

With these randomly generated values of wave parameters, the daily converted power is forecast by using the conversion matrices of five commercial WECs (see Fig. 3.14). The daily power, P_D, is calculated using the expression:

$$P_D = \frac{1}{N} \sum_{i=1}^{N} M(T_{M,i}, H_{S,i});$$ (3.11)

where $N = 100$ is the number of random points, and $M(\cdot, \cdot)$ is the value of the conversion matrix for the ith wave parameter pair, $T_{M,i}$ and $H_{S,i}$. The forecast daily power is shown in Fig. 3.15.

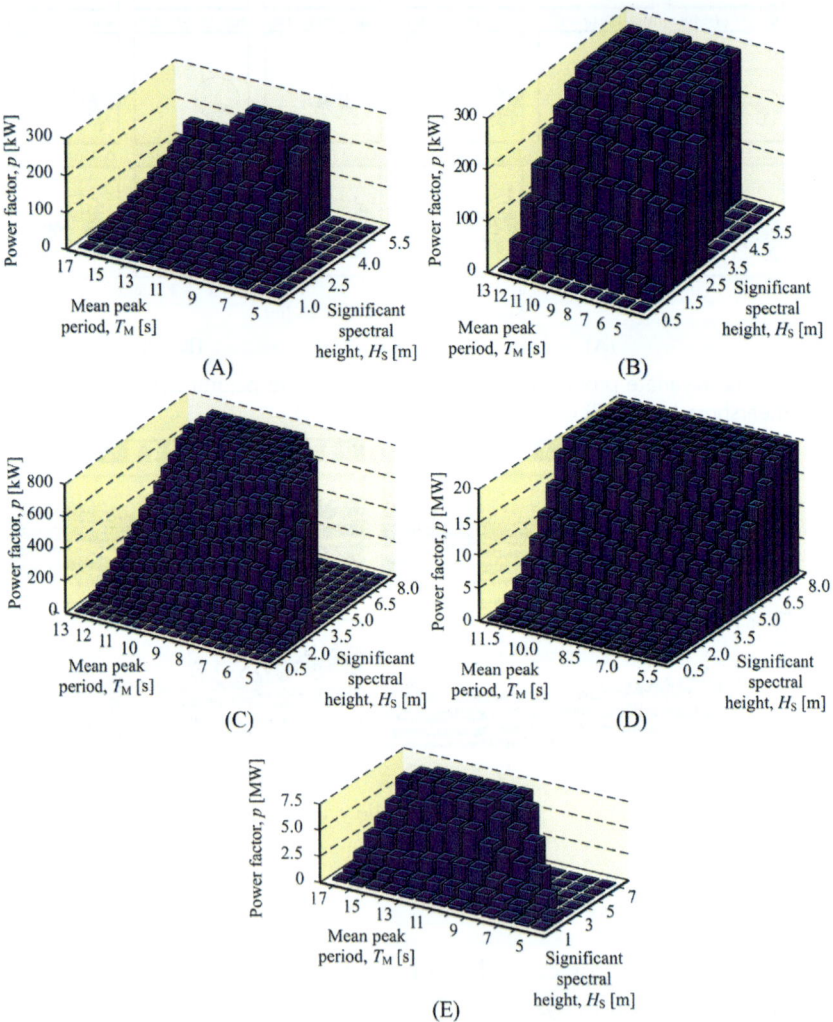

Figure 3.14 Power matrices for the WEC analyzed. (A) Aquabuoy, (B) Oyster, (C) Pelamis, (D) SSG, (E) Wave Dragon (Avila et al., 2022).

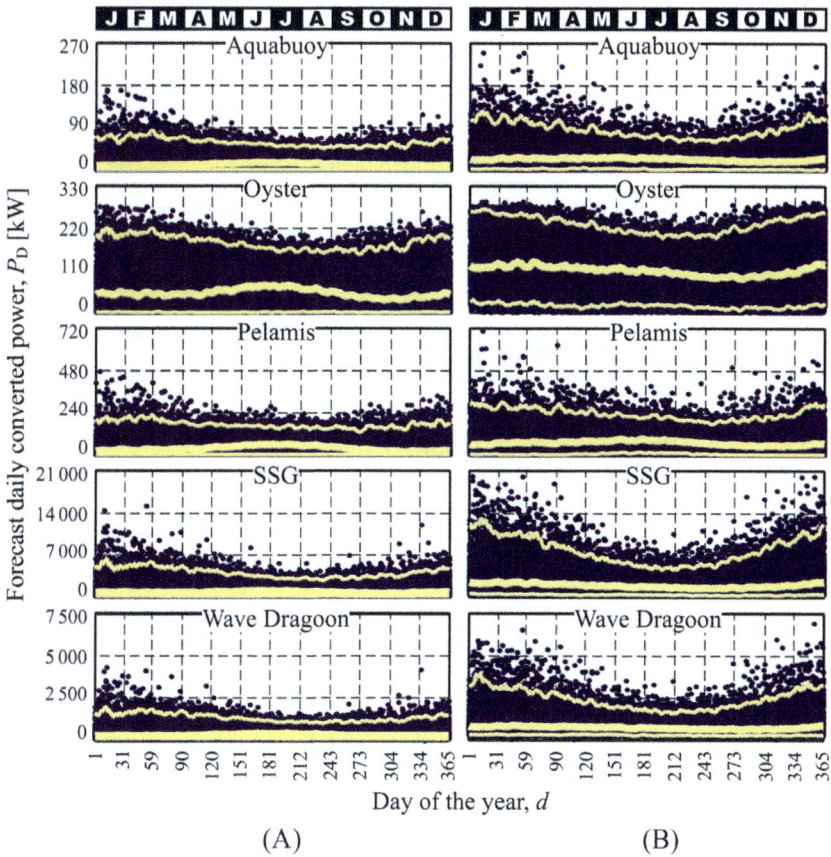

(A)   (B)

Figure 3.15 Forecast daily power for the WEC analyzed. (A) Las Palmas Este (nearshore buoy), (B) Gran Canaria (offshore buoy).

In the last step, the monthly converted energy is forecast by averaging the corresponding daily values (see Fig. 3.16). To validate the proposed approach, the monthly converted power values for 2018 and 2019, which were not used to fit the MDN-based models, are also represented.

3.4.4 Analysis of the results

In the first place, it should be noted that the Gaussian mixture probability distributions obtained from the MDN-based models (see Fig. 3.12) are more flexible than other prescribed bivariate distributions, such as bivariate Weibull. As can be seen, not only the magnitude but also the shape of the distributions changed through the year. This capability is an important

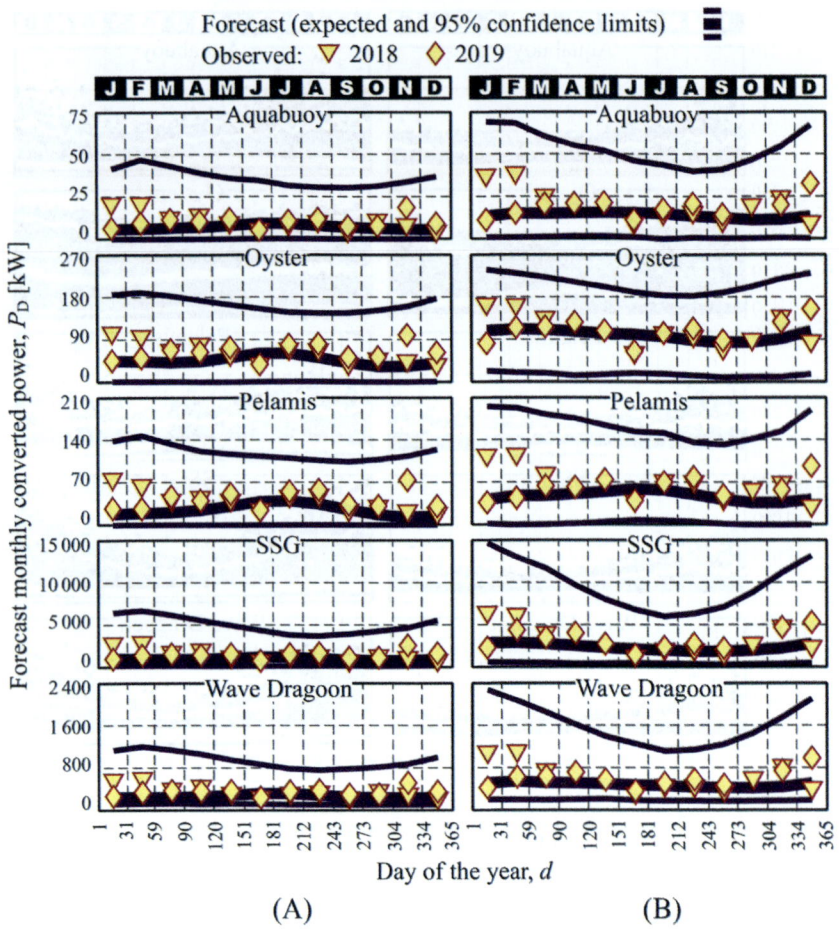

Figure 3.16 Forecast monthly power conversion. (A) Las Palmas Este (nearshore buoy), (B) Gran Canaria (offshore buoy).

asset of the proposed approach, allowing for a more fitted representation of the random behavior of the modeled data.

With regard to forecasting the wave parameters (see Fig. 3.13), a noteworthy observation is the remarkable coincidence with the observed values (see Fig. 3.10). In both cases, a noticeable seasonal behavior is evident in the mean peak periods, with higher values in winter months (from November to February) and lower values in summer (from June to August). Whereas the mean peak period performance is similar in both buoys, the significant spectral height has higher values in the offshore buoy (Gran Canaria) than in the nearshore buoy (Las Palmas Este) all year round.

Other interesting issues arise when the daily converted power is forecast (see Fig. 3.15). First, there is a significant difference between the nearshore (Las Palmas Este) and offshore (Gran Canaria) buoys, with higher values forecast for the latter, as might be expected from the higher values of significant spectral height predicted and observed for this buoy. Second, a dissimilar seasonal performance is forecast for the expected value and the upper 95% confidence limit. While in most of the cases, the former rises in the summer, the latter achieves its higher values in winter. It is important to highlight that the nominal (maximum) value of converted power is occasionally reached only at the offshore buoy (Gran Canaria). In Las Palmas Este, the nominal power is reached only for the Oyster converter, which seems to be especially suited to nearshore points.

Finally, by considering the monthly converted power (see Fig. 3.16), a seasonal behavior is also observed, but this is more evident for the upper limits of the 95% confidence intervals of the forecast.

Regarding the validation, all the observed values in 2018 and 2019 fall into the corresponding forecast intervals, many of them closer to the expected values line. This ability to accurately forecast the converted wave power at two points can be considered the most valuable result of this study.

3.5 Concluding remarks

Due to the complex nature of the phenomena involved in renewable energy systems, which include variables with a strong random component in their behavior, the use of stochastic modeling seems to be an essential tool.

In this context, the MCM can contribute its ability to use random variables to simulate stochastic processes. This simulation approach can deal with processes in which the probability distribution of the output variables cannot be determined analytically.

As for MDN, they can be used to forecast probably density functions rather than just expected values. The use of a mixture of Gaussian distributions guarantees the proper representation of random variables, even when they differ significantly from a normal probability distribution.

The presented case study shows that the combination of both tools can provide a powerful approach to solving renewable-energy-related problems involving stochastic processes. To expand on this work in the future, the proposed approach could be applied to other renewable energy technologies, such as wind or solar power systems. Also, the self-parametrization of the tools involved, including optimization techniques, is an issue that should be addressed in future research.

Acknowledgments

This research has been cofunded by FEDER funds, INTERREGMAC 2014—2020 Programme of the European Union, as part of the E5DES project (MAC2/1.1a/309). This chapter is also a partial result of the project "Architecture based on digital twins for the monitoring of industrial processes," associated with the Sectorial Program on the Cuban Software Industry and the Informatization of the Society of the Ministry of Communications of Cuba, with code PS161LH001-006.

References

Abdul Majid, A. J. (2022). Wind energy forecasting by fitting predicted probability density functions of wind speed measurements. *International Journal of Energy and Environmental Engineering*, *13*(2), 573—585. Available from https://doi.org/10.1007/s40095-022-00475-8.

ADIRA. (2008). [Autonomous Desalination system concepts for seawater and brackish water In Rural Areas with renewable energies]. *ADIRA handbook: A guide to autonomous system concepts.*

Ahmed, A., & Khalid, M. (2019). A review on the selected applications of forecasting models in renewable power systems. *Renewable and Sustainable Energy Reviews*, *100*, 9—21. Available from https://doi.org/10.1016/j.rser.2018.09.046.

Ahn, S., Haas, K. A., & Neary, V. S. (2020). Wave energy resource characterization and assessment for coastal waters of the United States. *Applied Energy*, *267*. Available from https://doi.org/10.1016/j.apenergy.2020.114922, Article 114922.

Ahn, S., Neary, V. S., Allahdadi, M. N., & He, R. (2021). Nearshore wave energy resource characterization along the East Coast of the United States. *Renewable Energy*, *172*, 1212—1224. Available from https://doi.org/10.1016/j.renene.2021.03.037.

Al-Gabalawy, M., Hosny, N. S., & Adly, A. R. (2021). Probabilistic forecasting for energy time series considering uncertainties based on deep learning algorithms. *Electric Power Systems Research*, *196*, 107216. Available from https://doi.org/10.1016/j.epsr.2021.107216.

Audemard, G., Bellart, S., Bounia, L., Koriche, F., Lagniez, J.-M., & Marquis, P. (2022). On the explanatory power of Boolean decision trees. *Data & Knowledge Engineering*, *142*, 102088. Available from https://doi.org/10.1016/j.datak.2022.102088.

Avila, D., Marichal, G. N., Hernández, A., & San Luis, F. (2021). Hybrid renewable energy systems for energy supply to autonomous desalination systems on isolated islands. In A. T. Azar, & N. A. Kamal (Eds.), *Design, analysis, and applications of renewable energy systems* (pp. 23—51). Academic Press.

Avila, D., Marichal, G. N., Quiza, R., & San Luis, F. (2021). Prediction of wave energy transformation capability in isolated islands by using the Monte Carlo method. *Journal of Marine Science Engineering*, *9*. Available from https://doi.org/10.3390/jmse9090980, Article 980.

Avila, D., Quiza, R., & Marichal, G. N. (2022). An approach for evaluating the stochastic behaviour of wave energy converters. *Applied Ocean Research*, *129*, 103372. Available from https://doi.org/10.1016/j.apor.2022.103372.

Azharul, M., Perrie, W., & Solomon, S. M. (2020). Application of SWAN model for storm generated wave simulation in the Canadian beaufort sea. *Journal of Ocean Engineering and Science*, *5*, 19−34. Available from https://doi.org/10.1016/j.joes.2019.07.003.

Benjamin, M. F. D., Andiappan, V., & Tan, R. R. (2021). Assessing the reliability of integrated bioenergy systems to capacity disruptions via Monte Carlo simulation. *Process Integration and Optimization for Sustainability*, *5*(4), 695−705. Available from https://doi.org/10.1007/s41660-021-00172-9.

Bishop, C. M. (1994). *Mixture density networks*. Aston University, [Technical report].

Bishop, C.M., & Nabney, I.T. (1997). *NETLAB Online Reference Documentation*. Retrieved 2022.12.12 from https://ccrma.stanford.edu/~unjung/nethelp/.

Brusaferri, A., Matteucci, M., Spinelli, S., & Vitali, A. (2022). Probabilistic electric load forecasting through Bayesian Mixture Density Networks. *Applied Energy*, *309*, 118341. Available from https://doi.org/10.1016/j.apenergy.2021.118341.

Carter, A., Imtiaz, S., & Naterer, G. F. (2023). Review of interpretable machine learning for process industries. *Process Safety and Environmental Protection*, *170*, 647−659. Available from https://doi.org/10.1016/j.psep.2022.12.018.

Chanan, S., Panida, J., & Joydeep, M. (2019). *Monte Carlo simulation. Electric power grid reliability evaluation: Models and methods* (pp. 165−183). IEEE. Available from https://doi.org/10.1002/9781119536772.ch6.

Chideana, M. I., Caamañoa, A. J., Ramiro-Bargueñoa, J., Casanova-Mateob, C., & Salcedo-Sanz, S. (2018). Spatio-temporal analysis of wind resource in the Iberian Peninsula with data-coupled clustering. *Renewable and Sustainable Energy Reviews*, *81*, 2684−2694. Available from https://doi.org/10.1016/j.rser.2017.06.075.

Christoforou, E., Emiris, I. Z., Florakis, A., Rizou, D., & Zaharia, S. (2021). Spatio-temporal deep learning for day-ahead wind speed forecasting relying on WRF predictions. *Energy Systems*, *14*, 473−493. Available from https://doi.org/10.1007/s12667-021-00480-6.

Corizzo, R., Ceci, M., H., F.-T., & Gama, J. (2021). Multi-aspect renewable energy forecasting. *Information Sciences*, *546*, 701−722. Available from https://doi.org/10.1016/j.ins.2020.08.003.

Díaz, H., Teixeira, A. P., & Guedes Soares, C. (2022). Application of Monte Carlo and Fuzzy Analytic Hierarchy Processes for ranking floating wind farm locations. *Ocean Engineering*, *245*, 110453. Available from https://doi.org/10.1016/j.oceaneng.2021.110453.

Eghbali, N., Hakimi, S. M., Hasankhani, A., Derakhshan, G., & Abdi, B. (2022). Stochastic energy management for a renewable energy based microgrid considering battery, hydrogen storage, and demand response. *Sustainable Energy, Grids and Networks*, *30*. Available from https://doi.org/10.1016/j.segan.2022.100652, Article 100652.

El Bakkari, B., El Bardouni, T., Nacir, B., El Younoussi, C., Boulaich, Y., Meroun, O., ... Chakir, E. (2012). Accuracy assessment of a new Monte Carlo based burnup computer code. *Annals of Nuclear Energy*, *45*, 29−36. Available from https://doi.org/10.1016/j.anucene.2012.02.011.

Gentle, J. E. (2003). *Random number generation and Monte Carlo methods*. Springer. Available from https://doi.org/10.1007/b97336.

Graham, C., & Talay, D. (2013). *Stochastic simulation and Monte Carlo methods: Mathematical foundations of stochastic simulation*. Springer-Verlag. Available from https://doi.org/10.1007/978-3-642-39363-1.

Görmüs, T., Ayat, B., & Aydogan, B. (2022). Statistical models for extreme waves: Comparison of distributions and Monte Carlo simulation of uncertainty. *Ocean Engineering*, *248*, 110820. Available from https://doi.org/10.1016/j.oceaneng.2022.110820.

Harbors of State of Spain. (2017). *Waves average. Buoy of Gran Canaria 2442 [Clima boya de Gran Canaria, 2442].* http://www.puertos.es/es-es/oceanografia/Paginas/portus_OLD.aspx.

Harbors of State of Spain. (2018). *Waves Average. Buoy of Las Palmas Este 1414 [Clima boya de Las Palmas Este, 1414].* http://www.puertos.es/es-es/oceanografia/Paginas/portus_OLD.aspx.

Herzallah, R., & Lowe, D. (2004). A mixture density network approach to modelling and exploiting uncertainty in nonlinear control problems. *Engineering Applications of Artificial Intelligence*, *17*(2), 145−158. Available from https://doi.org/10.1016/j.engappai.2004.02.001.

Jang, D., Kim, K., Kim, K.-H., & Kang, S. (2022). Techno-economic analysis and Monte Carlo simulation for green hydrogen production using offshore wind power plant. *Energy Conversion and Management*, *263*, 115695. Available from https://doi.org/10.1016/j.enconman.2022.115695.

Jin, H., Shi, L., Chen, X., Qian, B., Yang, B., & Jin, H. (2021). Probabilistic wind power forecasting using selective ensemble of finite mixture Gaussian process regression models. *Renewable Energy*, *174*, 1−18. Available from https://doi.org/10.1016/j.renene.2021.04.028.

Joshi, A. V. (2022). *Machine learning and artificial intelligence.* Springer. Available from https://doi.org/10.1007/978-3-031-12282-8.

Kar, S., Das, S., & Ghosh, P. K. (2014). Applications of neuro fuzzy systems: A brief review and future outline. *Applied Soft Computing*, *15*, 243−259. Available from https://doi.org/10.1016/j.asoc.2013.10.014.

Kesemen, O., & Tiryaki, B. K. (2018). Non-uniform random number generation from arbitrary bivariate distribution in polygonal area. *Journal of Natural and Applied Sciences*, *22*(2), 443−457. Available from https://doi.org/10.19113/sdufbed.70290.

Khadivi, M., & Sowlati, T. (2022). Biomass gasification investment: A multi-criteria decision considering uncertain conditions. *Biomass Conversion and Biorefinery*. Available from https://doi.org/10.1007/s13399-022-02700-0.

Liu, L., & Wu, L. (2021). Forecasting the renewable energy consumption of the European countries by an adjacent non-homogeneous grey model. *Applied Mathematical Modelling*, *89*, 1932−1948. Available from https://doi.org/10.1016/j.apm.2020.08.080.

Lofgren, P. (2014). On the complexity of the Monte Carlo method for incremental PageRank. *Information Processing Letters*, *114*(3), 104−106. Available from https://doi.org/10.1016/j.ipl.2013.11.006.

Lyudmila, K. (2021). Modeling of energy characteristics of parabolic concentrators based on monte carlo ray tracing method. *Applied Solar Energy*, *57*(5), 413−419. Available from https://doi.org/10.3103/S0003701X2105008X.

Marsland, S. (2015). In C. R. C. Press (Ed.), Machine learning: An algorithmic perspective *(2nd.*

Michelucci, U. (2019). *Advanced applied deep learning: Convolutional neural networks and object detection.* Apress. Available from https://doi.org/10.1007/978-1-4842-4976-5.

Nam, K., Hwangbo, S., & Yoo, C. (2020). A deep learning-based forecasting model for renewable energy scenarios to guide sustainable energy policy: A case study of Korea. *Renewable and Sustainable Energy Reviews*, *122*. Available from https://doi.org/10.1016/j.rser.2020.109725, Article 109725.

Nottona, G., Niveta, M.-L., Voyanta, C., Paolib, C., Darras, C., Mottea, F., & Fouilloy, A. (2018). Intermittent and stochastic character of renewable energy sources:

Consequences, cost of intermittence and benefit of forecasting. *Renewable and Sustainable Energy Reviews*, *87*, 96−105. Available from https://doi.org/10.1016/j.rser.2018.02.007.

Padrón, I., García, M. D., Marichal, G. N., & Avila, D. (2022). Wave energy potential of the Coast of El Hierro Island for the exploitation of a wave energy converter (WEC. *Sustainability*, *14*. Available from https://doi.org/10.3390/su141912139, Article 12139.

Paluszek, M., & Thomas, S. (2019). *MATLAB machine learning recipes: A problem-solution approach* (2nd ed.). Apress. Available from https://doi.org/10.1007/978-1-4842-3916-2.

PivotBuoy. (2019). *An advanced system for cost-effective and reliable mooring, connection, installation and operation of floating wind*. http://pivotbuoy.eu/wp-content/uploads/2019/06/D4.1_Test_site_environmental_conditions_v1.0_SENT.pdf.

Portillo Juan, N., & Negro Valdecantos, V. (2022). Review of the application of Artificial Neural Networks in ocean engineering. *Ocean Engineering*, *259*, 111947. Available from https://doi.org/10.1016/j.oceaneng.2022.111947.

Saeedi, S., & Hassan Hosseini, S. M. (2022). Stochastic coordination of the wind and solar energy using energy storage system based on real-time pricing. *Soft Computing*, *26*(18), 9607−9620. Available from https://doi.org/10.1007/s00500-022-06789-3.

Sakki, G. K., Tsoukalas, I., Kossieris, P., Makropoulos, C., & Efstratiadis, A. (2022). Stochastic simulation-optimization framework for the design and assessment of renewable energy systems under uncertainty. *Energy Reviews Renewable and Sustainable*, *168*, Article 112886. Available from https://doi.org/10.1016/j.rser.2022.112886.

Sandvik, E., Lonnum, O. J. J., & Asbjornslett, B. E. (2019). Stochastic bivariate time series models of waves in the North Sea and their application in simulation-based design. *Applied Ocean Research*, *82*. Available from https://doi.org/10.1016/j.apor.2018.11.010.

Sheng, W. (2019). Wave energy conversion and hydrodynamics modelling technologies: A review. *Renewable and Sustainable Energy Reviews*, *109*, 482−498.

Silva, A. R., Pousinho, H. M. I., & Estanqueiro, A. (2022). A multistage stochastic approach for the optimal bidding of variable renewable energy in the day-ahead, intraday and balancing markets. *Energy*, *258*. Available from https://doi.org/10.1016/j.energy.2022.124856, Article 124856.

Stevens, A. (2023). *Monte-Carlo simulation: An Introduction for engineers and scientists*. CRC Press. Available from https://doi.org/10.1201/9781003295235.

Su, C. L. (2005). Probabilistic load-flow computation using point estimate method. *IEEE Transactions on Power Systems*, *20*, 1843−1851. Available from https://doi.org/10.1109/TPWRS.2005.857921.

Sun, S., Tong, Y., Zhang, B., Yang, B., He, P., Song, W., . . . Liu, G. (2022). An adaptive optimization method for estimating the number of components in a Gaussian mixture model. *Journal of Computational Science*, *64*, 101874. Available from https://doi.org/10.1016/j.jocs.2022.101874.

Sánchez, C. (2009). *Energy technology*. Department of Energy Engineering. UNED.

Talaria, S., Shafie-khaha, M., Osórioa, G. J., J., A., & J.P.S., C. (2018). Stochastic modelling of renewable energy sources from operators' point-ofview: A survey. *Renewable and Sustainable Energy Reviews*, *81*, 1953−1965. Available from https://doi.org/10.1016/j.rser.2017.06.006.

Twidell, J., & Weir, T. (2006). Renewable energy resources (2nd ed.). Taylor & Francis.

Uwineza, L., Kim, H.-G., & Kim, C. K. (2021). Feasibility study of integrating the renewable energy system in Popova Island using the Monte Carlo model and HOMER. *Energy Strategy Reviews*, *33*, 100607. Available from https://doi.org/10.1016/j.esr.2020.100607.

Vavatsikos, A. P., Tsesmetzis, E., Koulinas, G., & Koulouriotis, D. (2022). A robust group decision making framework using fuzzy TOPSIS and Monte Carlo simulation for wind energy projects multicriteria evaluation. *Operational Research*, *22*(5), 6055—6073. Available from https://doi.org/10.1007/s12351-022-00725-x.

Wales, J. G., Zolan, A. J., Hamilton, W. T., Newman, A. M., & Wagner, M. J. (2022). Combining simulation and optimization to derive operating policies for a concentrating solar power plant. *OR Spectrum*. Available from https://doi.org/10.1007/s00291-022-00688-7.

Yadava, A. K., Sharmab, V., Malikc, H., & Chandela, S. S. (2018). Daily array yield prediction of grid-interactive photovoltaic plant using relief attribute evaluator based Radial Basis Function Neural Network. *Renewable and Sustainable Energy Reviews*, *81*, 2115—2127. Available from https://doi.org/10.1016/j.rser.2017.06.023.

Yan, R., Wang, J., S., H., Qin, Y., Zhang, J., Tang, S., . . . Zhou, L. (2023). Flexibility improvement and stochastic multi-scenario hybrid optimization for an integrated energy system with high-proportion renewable energy. *Energy*, *263*. Available from https://doi.org/10.1016/j.energy.2022.125779, Article 125779.

Zendehboudi, A., Baseer, M. A., & Saidur, R. (2018). Application of support vector machine models for forecasting solar and wind energy resources: A review. *Journal of Cleaner Production*, *199*, 272—285. Available from https://doi.org/10.1016/j.jclepro.2018.07.164.

Zhang, Y., Ch, C., H., C., Jin, X., Jia, Z., X., W., . . . Yang, T. (2022). Long-term stochastic model predictive control and efficiency assessment for hydro-wind-solar renewable energy supply system. *Applied Energy*, *316*. Available from https://doi.org/10.1016/j.apenergy.2022.119134, Article 119134.

Zhao, X., Ge, C., Ji, F., & Liu, Y. (2021). Monte Carlo method and quantile regression for uncertainty analysis of wind power forecasting based on Chaos-LS-SVM. *International Journal of Control, Automation and Systems*, *19*(11), 3731—3740. Available from https://doi.org/10.1007/s12555-020-0529-z.

CHAPTER FOUR

Profitability and performance improvement of smart photovoltaic/energy storage microgrid by integration of solar production forecasting tool

Gilles Notton, Sarah Ouédraogo, Ghjuvan Antone Faggianelli, Cyril Voyant and Jean Laurent Duchaud
Sciences for Environment Laboratory, University of Corsica Pasquale Paoli, Ajaccio, France

4.1 Introduction

As the numbers of installed intermittent energy systems continue to grow, a revolutionary transformation is underway in the global electricity distribution network. This revolution in the energy sector is mainly driven by the adoption of sustainable production methods. Consequently, there is a growing integration of intermittent and stochastic photovoltaic production into the energy mix. This shift is occurring in parallel with the rapid development of emerging technologies, such as electric vehicles. This development highlights the need for more intelligence in electrical grid management and the implementation of new associated technical solutions, contributing to the realization of the "smart grid" concept.

Initially, designed for the passive distribution of electricity from production centers to consumers distribution networks must now enable bidirectional electricity circulation throughout the entire territory. This necessitates a new model including distributed generation (DG), which require flexibility, continuous adaptability and dynamic data processing to monitor the system's state in real-time.

Mehigan et al. (2018) examined the factors that impact the operation of DG and explored how the electricity systems will evolve in the near

Intelligent Learning Approaches for Renewable and Sustainable Energy
DOI: https://doi.org/10.1016/B978-0-443-15806-3.00004-8

future. The significant and stochastic variations of the photovoltaic power further necessitate the use of energy storage in these smart networks.

The definitions of the smart grids are numerous, and may vary slightly depending on the region (Alotaibi et al., 2020). In a global context, a smart grid is characterized as an automated network that is able to store energy, exchange data, and decide which actions to realize. Table 4.1 provides a comparison of conventional electrical grids and the new smart grids (Alotaibi et al., 2020; Tuballa & Abundo, 2016), while Fig. 4.1 illustrates the present and future electrical network landscapes (Dembski, 2018).

Numerous state-of-the-art papers have been recently written, discussing the major benefits and challenges associated with the development of

Table 4.1 Short comparison between conventional and smart electrical grids (Alotaibi et al., 2020; Tuballa & Abundo, 2016).

Conventional grid	Smart grid
Uses mechanic operation	Uses digital functions
From production to consumption	In two directions
Centralized production	Distributed production
Radial connection	Scattered
Few sensors	Many sensors
Few control	Highly controlled
Mainly manual	Totally automated
Low security problems	High level of security
Slow response	Rapid response

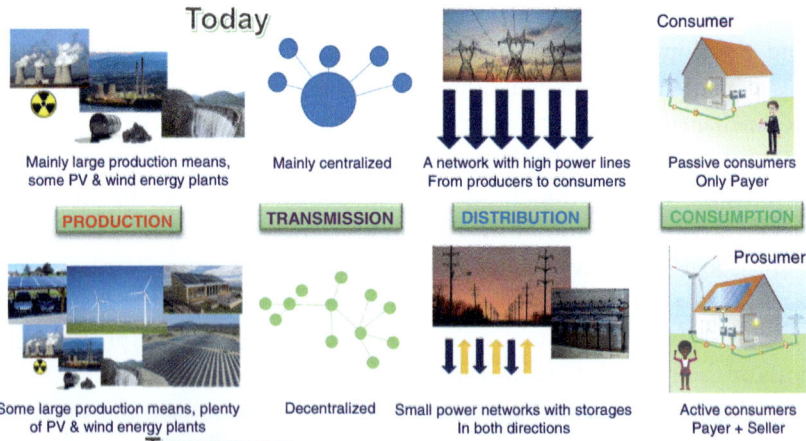

Figure 4.1 The electric grid of today and tomorrow. Adapted from (Dembski, 2018).

these innovative and highly exploratory microgrids (Alam et al., 2019; Vadari, 2020).

Such microgrids, using more and more renewable energy production, need to incorporate energy storage means to store the excess production (compared to the load) and to release it when the load supply requires it. Adding energy storage means is not sufficient; it is also crucial to know when charging and discharging this storage and how to do it with optimal energy management.

Decisions must be taken at various time horizons to control the storage according to:

• the state of the production means;
• the state of the storage means;
• the cost of the electricity produced and sold;
• the near future electrical needs and energy availability;

Having knowledge in advance of upcoming consumption and production is essential to enhance energy grid management efficiency. This foresight is fundamental for predicting decisions and activating flexibility measures.

Numerous studies were performed on demand forecasting, taking into account both consumer behavior and weather conditions (Fay & Ringwood, 2010; Hong et al., 2013; Nose-Filho et al., 2011; Shu et al., 2009). The strong link between the development of demand forecasting and smart-grid was well described by Khodayar and Wu (2015), Khan et al. (2016), and Vrablecová et al. (2018).

Due to the intermittent nature of solar resources, forecasting photovoltaic system production becomes essential to ensure higher electrical stability of the electrical grid and effective management of the energy exchanges (Boum et al., 2022). Solar forecasting is often integrated into EMS within a smart grid (Manur et al., 2020; Wan et al., 2015). Tripathy and Prusty (2021) conducted a comprehensive review of such forecasting methods for various types of renewable energy production in different smart grid contexts.

The main objective of this chapter is to show how the integration of a forecasting tool, using machine learning models applied to solar production, improves the cost profitability and energy performance of a smart photovoltaic/storage microgrid.

The available forecasting methods are initially presented according to the forecasting time horizon and temporal resolution. As machine learning methods shall be the primary focus of the following exploration, as they shall be the primary focus of the following exploration, as they are well adapted to predicting solar radiation with hourly and subhourly time steps, and for a time horizon between 15-min and 6 hours.

In the second section, nine predictive models are briefly presented, including two naive ones used as references and seven machine learning models. These models are applied to the meteorological site of the smart grid.

In section three, the smart microgrid with intermittent PV production and battery storage supplying electricity to a building and an electric vehicle located in our laboratory is presented. Following the description of the energy management algorithm based on rule-based control methods, the more efficient model of solar prediction for the site shall be introduced in the EMS algorithms of the microgrid.

In the final section, a comparison of the use of an EMS with and without a machine learning predictive tool shall be performed from a cost profitability and energy performance point of view.

4.2 Forecasting of solar radiation and PV production

In an electrical network, balance between production and consumption must be achieved at each moment. Maintaining this balance is complex even with conventional production methods; Moreover, when the electricity production is not controllable, it is clear that anticipating events in advance makes them easier to manage. Whether managed by human or machine, the entity overseeing the electrical system mustanticipate both the production and the consumption (Notton, Nivet, Voyant, et al., 2018).

The main objective of this section is to forecast the production of the PV system. It is generally more difficult to directly forecast the PV power rather than to predict the solar radiation incident on the PV plant and subsequently convert this into PV power. This difficulty arises due to the following reasons:

- the development of statistical and artificial intelligence forecasting methods requires a large dataset of past values. consequently, the PV plant must have already been installed for a very long time, which is uncommon;
- the size of the PV plant may change during its life, some technical issues may occur, or a portion of the PV plant can be disconnected. In such cases, the dataset of PV production does not consistently correspond to the same PV station.

The power produced by a PV plant depends on the received global solar irradiance, the PV cell temperature, and at a lower level on the wind speed. The PV power will be predicted in two steps: firstly, the solar irradiance will be estimated, and secondly, the forecasted solar irradiance value will be used in a PV production model. Numerous PV production models are available, even; the simplest ones, such as the well-known Evan's model (Evans, 1981) used in this study, exhibit good accuracy, which often surpasses that of the solar irradiance forecasting methods. Given that the influence of the cell temperature is less significant, and its variation over time is small, it isreasonable to assume that its value at the next time step is the same as the future one. For these reasons, accurate prediction of solar irradiance basically implies a well-forecasted PV production (Notton & Voyant, 2018).

4.2.1 Brief overview of the forecasting methods for solar radiation

The methods used to predict solar irradiance depend on the prediction time horizon and on the temporal granularity of the solar irradiance data (Fig. 4.2).

Generally, the prediction methods are classified into four categories (Diagne et al., 2013; Notton & Voyant, 2018):

- time series-based methods: These methods rely on a significant set of historically recorded solar irradiations and employ statistical or artificial intelligence (AI) methods. However, the availability of this history of measured data can often be a challenge;
- numerical weather forecast: This category uses equations of fluid dynamics to simulate the meteorological evolution within the atmosphere. However, these methods require substantial computing capacity and involve numerous parameters that are difficult to measure;

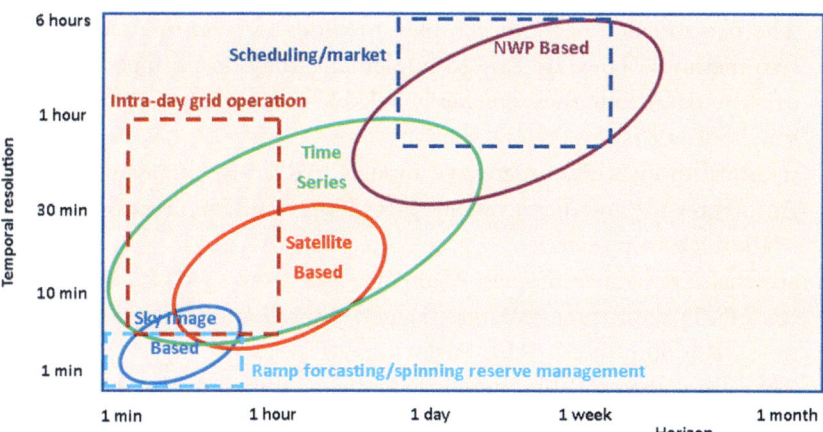

Figure 4.2 Prediction time horizon versus temporal granularity with corresponding applications (Diagne et al., 2013).

- satellite imagery-based methods: These methods leverage data obtained from satellites orbiting the Earth. Satellites capture a broad view of atmospheric conditions and solar radiation, providing a global perspective. However, limitations may include lower temporal resolution and potential issues in capturing detailed cloud cover information.
- sky image-based methods: These methods use sky eye cameras, which can estimate the future positions of clouds using associated software packages.

Each model category performs satisfactorily for its associated time horizon range. In certain instances, these models within these categories may be combined to predict future solar irradiance more accurately, leading to the development of "hybrid methods." For the specific task of predicting solar irradiance up to 6 hours, the first category of models is deemed more efficient. These methods will be briefly presented in the following parts of this chapter.

4.2.2 Time series based forecasting methods

The forecasted tilted global solar irradiation (TGSI) at time $t + h$, $\widehat{TGSI}(t + h)$ is estimated from n earlier measured TGSI, $TGSI(t - p)$, according to:

$$\widehat{TGSI}(t + h) = f\left[TGSI(t), TGSI(t - 1), TGSI(t - 2), \ldots, TGSI(t - p)\right]$$

$$(4.1)$$

Several predicting methods belonging to the first category exist:

- naive models: the simple and smart persistence are used as a reference. The performance of more complex methods are compared with these two methods. They are easy to implement (they do not need a history of solar data), but they are rarely reliable for a time-step lower than one hour at best;
- statistical models: Autoregressive models (AR), Moving average (MA), Autoregressive moving average (ARMA), Gaussian process (GP), or Markov chain models;
- artificial intelligence models: Multilayer perceptron (MLP), Regression trees (RT) with three variants (Bagged, Busted, and Pruned regression trees), Random forest (RF), Fuzzy logic (FL), ...

Nine from the available models have been applied to the dataset considered in this study, subjected to validation, and subsequently compared. The following section will provide a brief description of these nine models. However, it is essential to first preprocess (control and clean) the data.

4.2.2.1 Cleaning the data (making it stationary)

The first essential step in the elaboration of a forecasting model mainly based on time series involves verifying the quality of the solar data. The data acquisition system may encounter some synchronization issues and incorrect measurements, particularly at sunset/sunrise periods or other artifacts. The assessment of data quality has been extensively examined by Espinar et al. (2012) and El Alani et al. (2021). They have developed interesting and efficient quality control procedures for solar data including some visual aids (graphs and histograms), which helps in the interpretation of this quality check. These established procedures have been applied to assess the quality of our solar data.

After this step, a filter is applied to the data to eliminate not only the null solar data during the night but also the sunset and sunrise periods. The data corresponding to low solar altitudes during these periods are deemed unreliable for two primary reasons: potential inaccuracies in the solar measuring device (pyranometer) and the risk of obstruction, such as a mask presence. Data with a solar elevation lower than 5° are systematically removed.

AI and statistical models have good performances in forecasting stationary time series (Trapletti et al., 2000). Therefore, it is essential to address the seasonal and periodic variations in solar radiation. This periodicity must be removed, rendering the solar irradiance data stationary. The solar irradiance data are transformed into a clearness index, quantity denoted by k_t. This dimensionless quantity represents the ratio of the solar radiation TGSI measured at the ground level to solar radiation $TGSI_{CS}$ under clear sky conditions. The latter corresponds to the solar radiation received at the same moment with a clear sky, i.e., without clouds but considering the attenuation of the solar radiation through the atmosphere and depending on the climatic conditions of the site. The clearness index, k_t is defined by:

$$k_t(t) = \frac{TGSI(t)}{TGSIG_{cs}(t)} \tag{4.2}$$

Clear sky radiation models are numerous and generally use various meteorological parameters as inputs (Badescu et al., 2012) along with several geometrical dimensions (Ineichen, 2016). Among all the available clear sky models, the Solis model by to Mueller et al. (2004) and transformed by Ineichen (2008) is one of the most popular. Other noteworthy models include European solar radiation atlas model (Rigollier et al., 2000), the reference evaluation on solar transmittance 2 model (Gueymard, 2008) and McClear model (Gschwind et al., 2019; Lefèvre et al., 2013; Qu et al., 2017). The Solis model is used in this work.

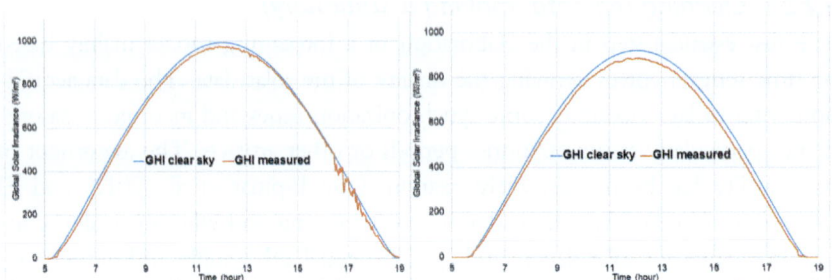

Figure 4.3 Comparison of Global solar irradiance measured and calculated by the Solis model for two days (in Ajaccio).

Fig. 4.3 compares the measured global solar irradiance with the solar irradiance estimated under clear sky conditions for two days in April (left) and September (right).

As indicated by Eq. (4.1), the prediction of the solar irradiation for a temporal horizon $t + h$ needs as inputs the $p + 1$ previous values of the solar irradiation. Hence, it is important to choose the optimal number p and several methods can be used for this purpose, such as Pearson's correlation coefficient or the Spearman coefficient between the inputs and the output. In this work, the auto mutual information method (Chow & Huang, 2005; Luo et al., 2017; Parviz et al., 2008) is used, which determines the degree of statistical dependence of the variables.

Finally, the dataset is separated randomly into two groups: a training group (80%) and a test group(20%). This step is realized k times ($k = 10$) (k-fold sampling method (Wiens et al., 2008)), so as to make the results independent from the chosen training data and to avoid seasonal effects.

4.2.2.2 Persistence and smart (or scaled) persistence
These two models are considered as naive models due to their simplicity and since these do not require a historical dataset. The performance of other methods is assessed in comparison with these two models. For the first naive model, the predicted solar irradiation at $t + h$, $\hat{TGSI}(t + h)$ is equal to the measured solar irradiation at t, TGSI(t). This hypothesis is obviously rarely accurate and hence, the "smart" persistence which considers the sun's position is preferred as a reference. The simple persistence model is improved by integrating the clear sky ratio, k_t, resulting in:

$$\hat{k_t}(t + h) = k_t(t) \rightarrow \widehat{TGSI}(t + h) = TGSI(t) \times \frac{TGSI_{CS}(t + h)}{TGSI_{CS}(t)} \qquad (4.3)$$

4.2.2.3 ARMA model
The autoregressive mobile average (ARMA) model is often used to forecast solar irradiation (Shadab et al., 2020). This model, denoted as ARMA (p,q), is composed of two components: autoregressive (AR) and moving average (MA) where p and q are the AR and MA orders, respectively.

$$\hat{k}_t(t + h) = \varepsilon(t) + \sum_{i=0}^{p} \varphi_i.k_t(t - i) + \sum_{i=0}^{q} \theta_i\varepsilon(t - i) \qquad (4.4)$$

φ and θ are computed and $\varepsilon(t)$ is the error allied to a normal distribution.

4.2.2.4 Multilayer perceptron (MLP)
MLP is an ANN model and serves as a powerful machine learning tool that is widely recognized. It proves to be invaluable in exploring relationships between variables and is often used as a predictor in numerous domains, such as economics, environmental sciences, and more. In solar forecasting, the MLP with feed-forward backpropagation was successfully applied to various sites worldwide (Shah et al., 2021; Yadav & Chandel, 2014). Fig. 4.4 provides an illustration of such an MLP ANN model.

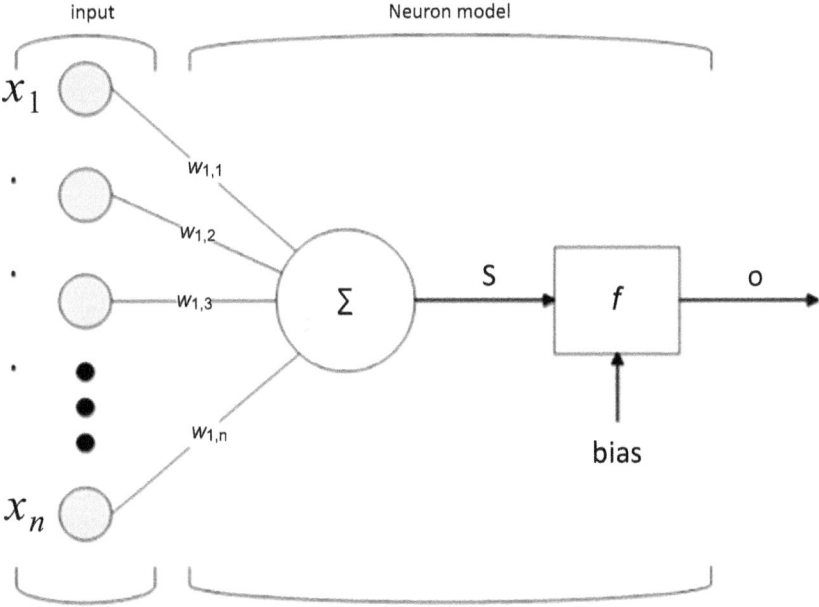

Figure 4.4 ANN model representation.

An MPL ANN model can be described by:

$$\hat{k}_t(t+h) = \sum_{j=1}^{m} \omega_j.g\left(\sum_{i=0}^{n-1} \omega_{i,j}.k_t(t-j) + b_j\right) \tag{4.5}$$

where b_j are the biases of the hidden neuron j, $\omega_{i,j}$ are the weights between the input i and the hidden node j, g is the transfer function, ω_j the weight between the output and the hidden neuron j.

4.2.2.5 Gaussian process (GP)

Each future k_t is calculated as the addition of $f(k_t(\tau))$ and an independent Gaussian noise $\mathcal{N}\left(O, \sigma_n^2\right)$ with a variance σ_n^2:

$$\hat{k}_t(t+h) = f(k_t(\tau)) + \mathcal{N}\left(O, \sigma_n^2\right) \tag{4.6}$$

4.2.2.6 Regression trees (RT)

Decision trees with continuous solar irradiation values as a target are named regression trees. Decision trees are an efficient tool for solar energy forecasting (Barrán et al., 2019; Jumin et al., 2021), and these can be expressed by:

$$\hat{k}_t(t+h) = \sum_{i=1}^{t-1} a_i.I(k_t(t-i)) \tag{4.7}$$

where a_i are constant factors, $I = 1$ if the input is used and $= 0$ otherwise.
In the RT family, some submodels can be used:

- Pruned RT: The objective is to reduce the size of the RT by cutting or removing some parts of the tree to avoid overfitting the data. Pruned trees are smaller and more easily understandable (Krueger et al., 2021).
- Boosted and Bagged RT: In the bagged configuration, numerous trees grow simultaneously and are interconnected (Voyant et al., 2018). In a boosted configuration, new trees are created to improve the precision of the previous ones (Jumin et al., 2021).
- Random Forest (RF): the output of several decision trees are gathered, to provide a unified result (Soufiane et al., 2022).

4.3 Application of the predictive methods to a Mediterranean site

In this section, the accuracy of the nine predicting methods described above is estimated on the site where the microgrid is implemented (the simple persistence will not be applied because its accuracy is not good enough, and the smart persistence is more adapted to be used as a reference model).

The validation of the work was realized on the basis of 3 years of hourly solar data (horizontal global) measured by a CM11 pyranometer in Ajaccio, Corsica, France, where our laboratory is situated (41°55′ N; 8°55′ E) (Fig. 4.5). At this site there is an insular Mediterranean climate. The measured irradiance data were used to determine the coefficients of each previous model for a prediction from h + 1 to h + 6 and to determine their accuracy.

The optimal number of input data was calculated to be $p = 8$; i.e., the eight previous solar irradiation values were used as inputs (see Eq. (4.1)). The dataset was divided into two sets: a training set used for elaborating the model and which represents 80% of the data and a testing set used to estimate the accuracy with 20% of the data.

The accuracy of the model was determined using the normalized root-mean-square error nRMSE defined by:

$$nRMSE = \sqrt{\frac{1}{N} \times \sum_{i=1}^{N} \left(\widehat{TGSI(i)} - TGSI(i) \right)^2} / \overline{TGSI} \qquad (4.8)$$

where $\overline{TGSI}$ is the mean value of TGSI.

Figure 4.5 Site of the meteorological station and of the studied microgrid.

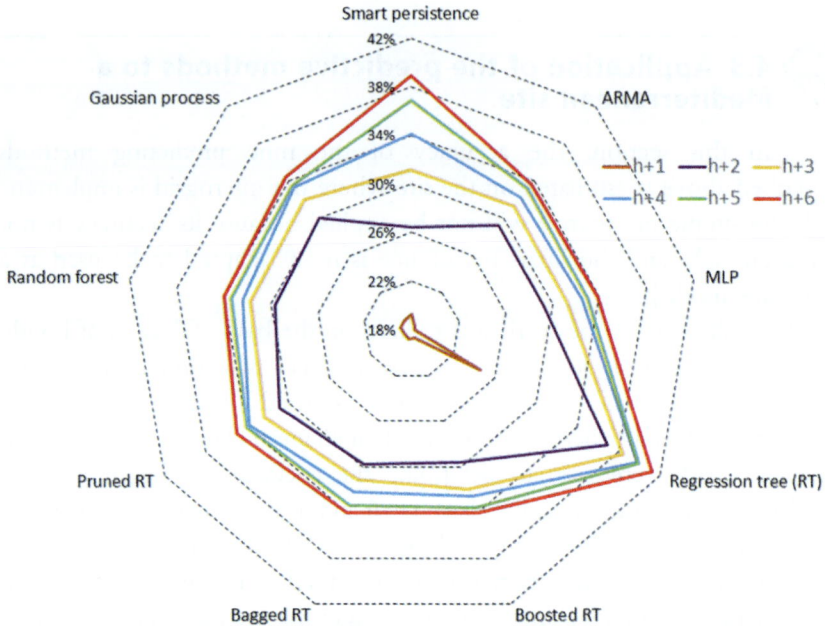

Figure 4.6 nRMSE value for each tested model and for six temporal horizons.

The nRMSE was plotted in Fig. 4.6. The simple regression tree appears to be unsuitable for the task while the smart persistence model gives correct results for h + 1 to h + 3. However, for longer term horizon, it becomes less efficient. The other models seem to have comparable performances. In Table 4.2, the best model is highlighted in red and the second model is outlined in red for each predictive horizon. One can note that the ARMA model always achieved first or second place (or third but close the second model). Hence, it was selected for the integration into the EMS of the microgrid studied in the next section.

4.4 Energy Management System in a photovoltaic microgrid with battery storage

4.4.1 A short state of art on EMS

The International Electrotechnical Commission (IEC), in the standard IEC 61970 on Energy Management System Application Program

Table 4.2 nRMSE versus forecast horizon (best model highlighted, second best model framed in red).

Horizon	h+1	h+2	h+3	h+4	h+5	h+6
Smart Persistance	19.26	26.46	31.18	34.15	36.92	38.93
ARMA	18.35	29.27	31.38	32.25	33.18	33.69
MLP	18.26	29.26	31.31	32.47	32.98	33.84
Regression Trees (RT)	24.64	36.88	38.47	39.74	39.95	41.24
Boosted RT	18.75	29.55	31.89	32.51	33.55	33.98
Bagged RT	18.76	29.80	31.10	32.17	33.35	34.02
Pruned RT	18.72	30.88	32.27	33.76	34.01	35.00
Random Forest	18.97	29.63	31.62	32.38	33.37	33.91
Gaussian Process	18.97	30.08	31.96	33.29	33.55	34.44

Interface defines an Energy Management System (EMS) as "an information system, including a software platform providing basic support services and a set of applications allowing an efficient exploitation of electricity production and transmission installations in view to provide adequate security of the energy supply at a minimum cost" (Santodomingo et al., 2016; Yaqin et al., 2005).

EMS takes the measured data in the various subsystems, analyzes these data, and with various decision-supporting tools integrating various modules, such as production and consumption forecast, electricity, market prices, etc., it takes the adequate decision in view to optimize the microgrid operation, ensuring the production/consumption balance and various technical constraints. This optimization can be relative to different objectives such as:

- optimization of the equipment sizes considering the production level to reach, the battery cycle number, etc.,
- minimization of the investment costs;
- minimization of the operation costs taking into account the variation of the purchase and sale costs of electricity;
- reduction of the environmental impact;

Various techniques and mathematic tools are available to optimize the operation of such a microgrid and are briefly reviewed in the next sections. One must keep in mind that the developed methods must be easily implemented in the microgrid management algorithm, and the decisions must also be taken quickly.

 4.5 Linear programming and Mixed-integer linear programming (MILP)

Linear Programming solves optimization problems for which the objective functions and the constraints are all linear. Sukumar et al. (2017) used this method to reduce the operational costs of a microgrid. Pascual et al. (2015) applied LP to a residential microgrid with the objective of minimizing the peaks and fluctuations of the power exchanged between the microgrid and the main network. The LP algorithm was directly integrated into the EMS. forecasting the production and the load. Tavakoli et al. (2018) maximized the microgrid capacity to absorb the power variations and prevent blackouts. While the operational costs increased by 0.19%, the system resilience saw a substantial increase of 41% during the same period. Jalilpoor et al. (2022) developed a strategy to improve the resilience of the system to extreme events such as tornados. Notably, the operational cost was reduced by 75% and 48% for perturbations with short and long durations, respectively. Atia and Yamada (2016) aimed to optimize the size of the subsystems by considering the battery degradation and the dynamic electricity tariff.

The MILP method is used when the optimization problem has some variables that are constrained to be integers. Amrollahi and Bathaee (2017) aimed to reduce the system size by reducing the gap between production and consumption through load management. In Moya et al. (2020), the objective was to minimize the electrical power provided by the main network resulting in a cost reduction of about 48% when considering the electricity tariff. The work realized by Murty and Kumar (2020) achieved reductions in the operational costs, pollutant gas emission, and power loss by, 2.25%, 2.1%, and 3.56%, respectively.

4.6 Nonlinear programming

Programming approaches such as quadratic programming are more complex and are less frequently used (Paul et al., 2018; Wibowo et al., 2016).

4.7 Dynamic programming

The main principle of dynamic programming consists in solving a complex problem by breaking it down into several subproblems, which are then solved individually. Duchaud et al. (2018) used dynamic programming method to limit the ramp rate of the grid power, optimize the energy exchanges, and control the battery charge according to various constraints. Other works considered the same strategies to reduce the energy consumption in microgrids (Bahlawan et al., 2019; Park et al., 2019).

4.8 Model predictive control (MPC)

MPC is a method that is used in several studies to minimize the cost of power exchanges between the micro-network and the main network. It is also used to enhance the utilization of the battery during peak consumption (Prodan & Zio, 2014). Numerous other studies have implemented the same method to reduce electricity costs (Bruni et al., 2016; Petrollese et al., 2016).

4.9 Rules-based control (RBC)

These strategies are widely used for optimization purposes due to their robustness and ease of implementation. These models are often implemented with two main tasks: maximizing self-consumption (Lorestani et al., 2019; Zou et al., 2022) and reducing the operational costs (Bandyopadhyay et al., 2020; Cherukuri et al., 2020; Pan et al., 2021; Zou et al., 2022).

Other optimization methods are sometimes employed for microgrid optimization. These methods can include as robust and stochastic programming (Bazmohammadi et al., 2019; Shi et al., 2019), Meta-heuristic approaches (Askarzadeh, 2018; Chalise et al., 2016) or artificial intelligence methods (Arcos-Aviles et al., 2017; Wang et al., 2019).

4.9.1 The microgrid PAGLIA ORBA

The PAGLIA ORBA platform is a microgrid using various energy storage systems and photovoltaic arrays. The goal of this platform is to optimally manage the electricity distribution between storage, loads, and the main grid. PAGLIA ORBA is sized to correspond with a small neighborhood, and it provides electricity to buildings and supplies also electric vehicles.

Two modes of operation are possible:

- the microgrid uses at a maximum level the electricity produced by the PV arrays, and the surplus of production is sent to the main grid;
- the microgrid is totally disconnected from the main grid.

A general view of the three-phase microgrid platform is shown in Fig. 4.5. For this work, only a part of this microgrid is used. The microgrid is composed by:

- three solar arrays: 3×17 kW (AC) of PV;
- one Energy storage: lead—acid batteries;
- two loads: a building and an electric vehicle.

The PV modules are integrated into a solar shade-house (Fig. 4.7) with 168 SUNPOWER E20 m-Si PV modules (327 Wp each). Each array of 56 modules is connected to an inverter (SMA SUNNY TRIPOWER 17000TL-10) (maximum efficiency = 98.2% — Peak AC power = 17 kW) (Fig. 4.7).

The energy storage is composed of 48 V lead—acid batteries (OPzV) and operated by 6 Xtender inverters/chargers (Fig. 4.7). The total battery capacity is 70 kWh.

The electrical load (Fig. 4.8), situated near the electricity production system, is in a building receiving researchers for accommodation with

Figure 4.7 PV solar shade structure, batteries, and inverters/chargers.

Figure 4.8 Loads: accommodation and electric vehicle.

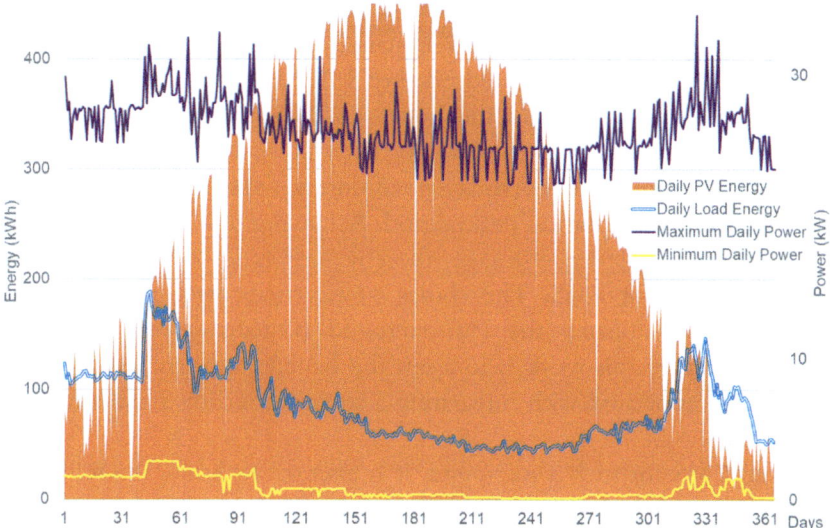

Figure 4.9 Production and Load profiles with minimum and maximum consumption powers.

seven bedrooms, a shared kitchen and a common room. The number of researchers and/or students varies according to the season. An electric vehicle with a capacity of 22 kWh is also connected to the load. This vehicle is used daily from 9 am to 4 pm. The battery of the vehicle is fully discharged at 4 pm, while charging begins at 4 pm until 6 pm. The load profile is shown in Fig. 4.9.

4.9.2 Rules based control algorithm

4.9.2.1 Assumptions

The microgrid has the capability to export electricity to the main electrical grid but with a power limitation of 24 kW. It is worth noting that one of the advantages of microgrids is to reduce the amount of electricity exported to the main grid and consequently decrease the electricity transport cost. Additionally, in this scenario, it is assumed that charging the battery from the main grid is not possible.

Two electricity tariffs currently used in France for domestic applications were employed: the first tariff considers a constant price over the day, while the second tariff considers two distinct periods. The first period corresponds to peak consumption, with a tariff of 163.1 €/MWh and the second off-peak period, with a tariff of 122.4 €/MWh, reflecting a 33% reduction. Moreover, for domestics PV power plants, the selling price (feed-in tariff) depends on the date of signature of the contract with EDF (Electricity of France), and remains constant for the subsequent 20 years. In this context, a fixed price of 137.7 €/MWh was used and is independent of the day and the hour. To increase the PV production share, it is anticipated that the PV electricity cost will vary over time by increasing the produced electricity price during consumption peak hours and decreasing it during off peak period. Achieving this objective requires the integration of energy storage, thereby increasing the importance of having an efficient EMS. The choice of electricity tariff influences the profitability of utilizing the PV microgrid (Ouédraogo et al., 2022). The objective of this work is to assess the benefits, particularity in terms of the cost, associated with integrating a solar production prediction tool in the EMS.

In an EMS for such a microgrid, it is essential to know the following information before the EMS algorithm takes a decision about the redistribution of the energy flows:

- the cost of the electricity imported from the main grid;
- the power produced by the PV plant;
- the state of charge of the battery;
- the power consumed by the building and the electric car;

These data are generally available in a smart grid, as one of the main characteristics of such an electrical system is that it can transfer not only energy but also numerical data with energy information. A forecasting tool must be developed to anticipate both production and consumption energies. In this chapter, the forecasting tool will specifically focus on predicting the photovoltaic production.

An EMS strategy based on a rule-based control (RBC) will, be developed. Initially only the real time information will be considered, and subsequently, real time and future data will be considered. The latter will be facilitated by the introduction into the algorithm of a PV production forecasting tool based successively on an ARMA model (considered as the most efficient for our meteorological site) and a "perfect" forecast (the real future values of PV production are introduced into the EMS) (Table 4.3).

4.9.2.2 RBC used in the EMS

The electrical data of the microgrid components are shown in Table 4.1. Two EMS strategies are tested and compared in terms of cost and energy performance. The second strategy includes a PV production prediction

Table 4.3 Components size.

Parameter	Value	Description
Ppv^{nom} [kW]	53	DC PV plant peak power
$Pinv^{max}$ [kW]	51	AC Inverter nominal power
Pg_{out}^{max} [kW]	24	Max AC Grid power output
Pg_{in}^{max} [kW]	24	Max AC Grid power input
C_{ESS} [kWh]	70	DC Storage capacity
Pb_{in}^{max} [kW]	14	Max DC charge power in storage at 0.2 C
Pb_{out}^{max} [kW]	35	Max DC discharge power in storage at 0.5 C
η_{in} [-]	0.9	Storage charge efficiency
η_{out} [-]	0.9	Storage discharge efficiency

tool to forecast data on PV production. In both strategies, the PV production always supplies the consumption needs first, the decision is taken according to the value of a parameter called Δ, which is the difference between the photovoltaic power (P_{PV}) and the consumed power (P_L):

$$\Delta = P_{PV} - P_L \tag{4.9}$$

4.10 RBC without PV production forecasting

The parameters taken into consideration for decision-making include electricity costs, P_{PV}, and battery state of charge (SoC). This strategy uses the electricity imported from the main grid during off-peak periods (low electricity price) and from the storage otherwise. The different rules of this strategy are as follows:

$0 \leq \Delta < Pg_{out}^{max}$: The photovoltaic plant production is consumed by the load. If there is an excess of PV production, then it charges the storage if it is not full. If not, the excess PV production is exported to the main grid (and sold).

$\Delta \geq Pg_{out}^{max}$: The PV power is consumed by the load, Pg_{out}^{max} is exported to the grid, and the excess production charges the storage. If the SOC is 1 then this energy is lost.

$\Delta < 0$ and $t \in$ [22:00−4:00], t being the time: The load receives the electricity from the grid at a low cost (off-peak hours).

$\Delta < 0$ and $t \in$ [4:00−5:00] and ($SoC \geq \frac{1}{2}$): The storage provides electricity to the load with a power Pb_{out}^{max} until SoC reaches 0.5 and the excess power is exported to the grid.

$\Delta < 0$ and $t \in [4{:}00{-}5{:}00]$ and $(SoC < \frac{1}{2})$: The missing power is supplied by the storage, and it can be completed by the main grid.

$\Delta < 0$ and $t \in [5{:}00{-}22{:}00]$: same as previous condition.

4.11 RBC with PV production forecasting

A new parameter based on the prediction of photovoltaic production is used. As mentioned previously, the ARMA model proved to be the most efficient solar prediction method for the considered site. This parameter is defined as the average photovoltaic production calculated over the next 6 hours, $mean_{6h}(Ppv)$:

$0 \leq \Delta < Pg_{out}^{max}$ and $mean_{6h}(Ppv) \geq Pg_{out}^{max}$: The photovoltaic plant provides separate electricity to the load and the excess produced electricity is exported to the main grid.

$0 \leq \Delta < Pg_{out}^{max}$ and $mean_{6h}(Ppv) < Pg_{out}^{max}$: The photovoltaic plant provides separate electricity to the load and the excess produced electricity is used to charge the storage and is lost if the SoC is equal to 1.

$\Delta \geq Pg_{out}^{max}$: At first, Pg_{out}^{max} provides electricity to the load. Secondly, it is exported to the grid. If power is available again, it is used to charge the storage.

With this new PV forecasting indicator, on sunny days, the microgrid has the capability to sell energy to the main grid before charging the battery. This approach aims to leverage the storage and recover production during the peak of PV production, thereby preventing excessive power loss when the battery is full.

Using the same strategy, the ARMA forecasting was replaced by the "perfect one" (real measured data were introduced). This substitution enables to study the impact of errors introduced by an imperfect prediction tool on the performance of the EMS.

4.11.1 Linear programming (LP) optimization

Linear programming (LP) optimization is used to determine the optimal solution to minimize the cost of the used electricity.

The considered system has the flexibility to both buy and sell electrical power. A "perfect" energy management strategy aims to minimize the cost paid, essentially maximizing the gain. If $C_{sell}(t)$ is the selling price of electricity produced and $C_{buy}(t)$ is the buying price of electricity (here varying according to an electricity "peak/off-peak" tariff), then the balance between the cost of the electricity bought and sold is:

$$C_{tot} = \sum_{t=1}^{N} \left[Pg, L(t)C_{buy}(t) - (Ppv, g(t) + Pb, g(t))C_{sell}(t) \right] dt \qquad (4.10)$$

where $Ppv, g(t)$ is the PV power to the grid, $Pb, g(t)$ is the storage power to the grid, $Pg, L(t)$ is the grid power to the load.

The microgrid CAPEX and OPEX are not considered in the analysis. The objective is to compare the efficiencies of the EMS algorithms without changing the characteristics of the electricity system.

This method yields an optimal solution provided that all the powers occurring in the system during the operation period are known in advance. It is obvious that such a strategy cannot be applied directly in a real-world scenario, because, unfortunately the electrical powers are not known in advance. However, it serves as a benchmark to assess the efficiency of the three other EMS algorithms (RBC without forecasting, RBC with ARMA and RBC with perfect forecasting) with respect to this "best strategy" that, unfortunately cannot be implemented.

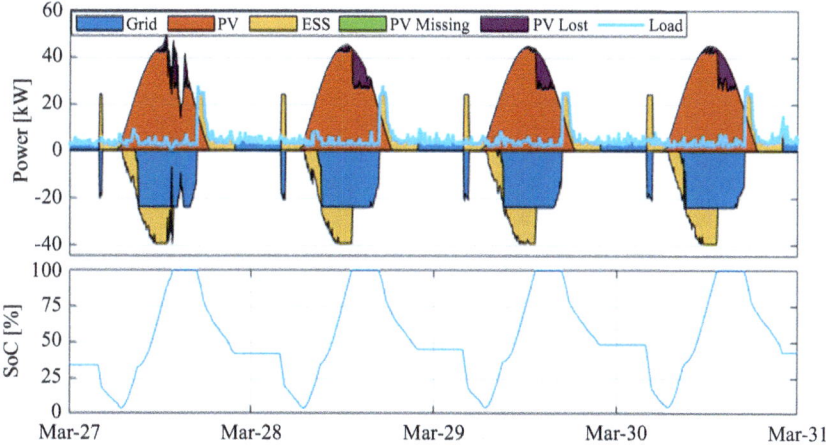

Figure 4.10 RBC without forecasting.

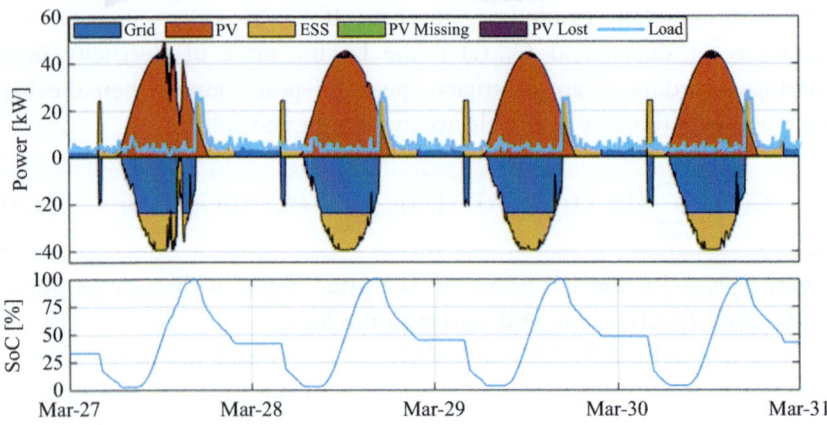

Figure 4.11 RBC with forecasting.

4.12 Results

Figs. 4.10 and 4.11 show, the results obtained using the RBC strategy without forecast and with forecast on a period of 4 days in March, a period during which the occupation level of the building is relatively low. Various electrical powers were plotted in the form of full curves:

- power provided by the grid to the system (blue positive) or sold to the grid by the PV plant (blue, negative);
- power produced by the photovoltaic array (brown);
- power charging the storage by the PV array (orange negative) or discharging the battery (orange, positive) to supply the load;
- power of the load not satisfied;
- PV power lost.

The load power is plotted in the same figure in light blue. The state of charge of the storage is presented in the second graphic below. The lost PV power can be due to three factors:

- a limitation of the grid power (this limitation was added in view to decrease the power passing over the main electrical grid, reducing the electricity transport cost);
- a limitation of the storage charge power due to the battery technology;
- a full state of charge of the storage.
  Some remarks can be made on Figs. 4.10 and 4.11:
- an effect of the two price periods for the electricity bought from the grid (peak and off-peak periods) is seen. A plateau in the SoC curve appears before 4 am on both figures.

- at the beginning of each days, a morning discharge is seen in both figures. The algorithm decides to discharge the storage. The battery will have a SoC allowing it to store the PV production in excess relative to the load during the day.
- without PV forecasting, the microgrid cannot know the PV quantity that would be produced. As soon as the PV starts to produce, the batteries start charging immediately, and when they are full, they cannot recover any more PV.
- with the forecast, one can know approximately how much PV will be produced during the day. At this moment, the batteries stop recharging in view to recover the surplus of production to decrease the lost PV energy.
- thanks to the PV prediction, a decrease in the PV lost in Fig. 4.11 can be observed. Over the 4 days, the lost PV goes from 92.06 to 11.91 kWh.

The behavior of the microgrid was simulated over one year, and the results in term of cost, PV energy lost, and battery cycles number are shown in Fig. 4.12.

The strategy without PV production forecasting conduces to an economic gain of 5472.2 €. If a PV production forecast using an ARMA model is incorporated into the energy management strategy, the gain reaches 5943.69 €

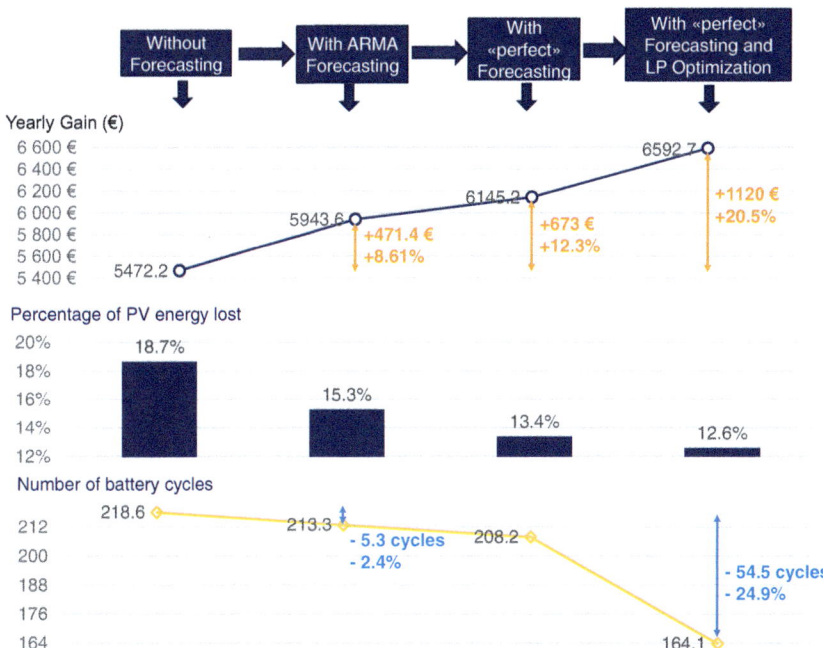

Figure 4.12 Results calculated on one year of simulation.

(+8.6%). If now a "perfect" forecast replaces the ARMA method, the gain would be 6145.2 €. Compared with the ARMA model, the improvement is about 3.4% showing that the forecasting model used is relevant. Part of the lost solar energy and the number of battery cycles decreased, improving the global performances of the microgrid. The amount of PV lost is calculated in relation to the total PV production, which is 91,126 kWh over the year. The number of cycles decreases because the battery is used a little more intelligently and avoids some unnecessary cycles. This decrease of the number of cycles also improves the battery lifetime, and consequently, the benefit of the system.

4.13 Conclusion

The main objective of this chapter consisted in highlighting the importance of solar irradiation forecasting for a microgrid EMS.

Nine prediction models based on statistical and AI methods were implemented for the microgrid site located in Ajaccio. Based on a comparison with hourly global solar irradiations (from 1 hour later to 6 hours later), it was shown that the ARMA model is the most suitable for the intended application.

Another objective was to assess the value of integrating a PV production predictor into an EMS algorithm. Four energy management strategies were developed and compared: one linear programming optimization strategy and three rule-based control strategies (RBC). The linear programming optimization strategy is considered to be the optimal solution, as the algorithm has perfect knowledge of future load and production power flows. The three RBC management strategies consider the variation in energy cost in the energy power flows management. The first RBC model does not consider energy forecasting, while the two others use, ARMA and perfect (for comparison purposes only) forecasting, respectively.

The simulations are applied to a microgrid using an hybrid photovoltaic/battery system to provide electricity to a building and an electric vehicle for one year. For the four strategies, the annual gain, photovoltaic energy lost and battery number of cycles were determined.

The conclusions of this work are:
- production forecasting with an ARMA model improves the microgrid gain by +8.6%.

- use of the perfect forecasting would improve the microgrid gain by +12.3%.
- the amount of PV energy lost was reduced by 3.4%, and the number of battery cycles performed by 2.4% with the help of a forecasting tool based on an ARMA model.
- the perfect knowledge of all power flow with the use of linear programming optimization would increase the gain by +10.92% compared to the RBC management strategy with the ARMA model.

References

Alam, M. N., Chakrabarti, S., & Ghosh, A. (2019). Networked microgrids: State-of-the-art and future perspectives. *IEEE Transactions on Industrial Informatics*, *15*−*3*, 1238−1250. Available from https://doi.org/10.1109/TII.2018.2881540.

Alotaibi, I., Abido, M. A., Khalid, M., & Savkin, A. V. (2020). A comprehensive review of recent advances in smart grids: A sustainable future with renewable energy resources. *Energies*, *13*, 6269. Available from https://doi.org/10.3390/en13236269.

Amrollahi, M. H., & Bathaee, S. M. T. (2017). Techno-economic optimization of hybrid photovoltaic/wind generation together with energy storage system in a stand-alone micro-grid subjected to demand response. *Applied Energy*, *202*, 66−77. Available from https://doi.org/10.1016/j.apenergy.2017.05.116.

Arcos-Aviles, D., Pascual, J., Guinjoan, F., Marroyo, L., Sanchis, P., & Marietta, M. P. (2017). Low complexity energy management strategy for grid profile smoothing of a residential grid-connected microgrid using generation and demand forecasting. *Applied Energy*, *205*, 69−84. Available from https://doi.org/10.1016/j.apenergy.2017.07.123.

Askarzadeh, A. (2018). A memory-based genetic algorithm for optimization of power generation in a microgrid. *IEEE Transactions on Sustainable Energy*, *9*−*3*, 1081−1089. Available from https://doi.org/10.1109/TSTE.2017.2765483.

Atia, R., & Yamada, N. (2016). Distributed renewable generation and storage system sizing based on smart dispatch of microgrids. *Energies*, *9*−*3*, 3. Available from https://doi.org/10.3390/en9030176.

Badescu, V., Gueymard, C. A., Cheval, S., Oprea, C., Baciu, M., Dumetrescu, A., Iacobescu, F., Milos, I., & Rada, C. (2012). Computing global and diffuse solar hourly irradiation on clear sky. Review and testing of 54 models. *Renewable and Sustainable Energy Reviews*, *16*−*3*, 1636−1656. Available from https://doi.org/10.1016/j.rser.2011.12.010.

Bahlawan, H., Morini, M., Pinelli, M., & Spina, P. R. (2019). Dynamic programming based methodology for the optimization of the sizing and operation of hybrid energy plants. *Applied Thermal Engineering*, *160*113967. Available from https://doi.org/10.1016/j.applthermaleng.2019.113967.

Bandyopadhyay, S., Mouli, G. R. C., Qin, Z., Elizondo, L. R., & Bauer, P. (2020). Techno-economical model based optimal sizing of PV-battery systems for microgrids. *IEEE Transactions on Sustainable Energy*, *11*−*3*, 1657−1668. Available from https://doi.org/10.1109/TSTE.2019.2936129.

Barrán, A. T., Alonso, A., & Dorronsoro, J. R. (2019). Regression tree ensembles for wind energy and solar radiation prediction. *Neurocomputing*, *326*−*327*, 151−160. Available from https://doi.org/10.1016/j.neucom.2017.05.104.

Bazmohammadi, N., Tahsiri, A., Anvari-Moghaddam, A., & Guerrero, J. M. (2019). A hierarchical energy management strategy for interconnected microgrids considering uncertainty. *International Journal of Electrical Power & Energy Systems*, *109*, 597−608. Available from https://doi.org/10.1016/j.ijepes.2019.02.033.

Boum, A. T., Foba Kakeu, V. J., Mbey, C. F., & Yem Souhe, F. G. (2022). Photovoltaic power generation forecasting using a novel hybrid intelligent model in smart grid. *Computational Intelligence and Neuroscience*7495548. Available from https://doi.org/10.1155/2022/7495548.

Bruni, G., Cordiner, S., Mulone, V., Sinisi, V., & Spagnolo, F. (2016). Energy management in a domestic microgrid by means of model predictive controllers. *Energy*, *108*, 119−131. Available from https://doi.org/10.1016/j.energy.2015.08.004.

Chalise, S., Sternhagen, J., Hansen, T. M., & Tonkoski, R. (2016). Energy management of remote microgrids considering battery lifetime. *The Electricity Journal*, *29*−6, 1−10. Available from https://doi.org/10.1016/j.tej.2016.07.003.

Cherukuri, S. H. C., Saravanan, B., & Arunkumar, G. (2020). A rule-based approach for improvement of autonomous operation of hybrid microgrids. *Electrical Engineering*, *102*−2, 989−1004. Available from https://doi.org/10.1007/s00202-020-00928-5.

Chow, T. W. S., & Huang, D. (2005). Effective feature selection scheme using mutual information. *Neurocomputing*, *63*, 325−343. Available from https://doi.org/10.1016/j.neucom.2004.01.194.

Dembski, F. (2018). Digitalization taking a quite byte. Energy atlas, facts and figures about renewables in Europe. *Revue Litteraire Mensuelle*. Available from https://gef.eu/publication/energy-atlas-2018/.

Diagne, M., David, M., Lauret, P., Boland, J., & Schmutz, N. (2013). Review of solar irradiance forecasting methods and a proposition for small-scale insular grids. *Renewable and Sustainable Energy Reviews*, *27*, 65−76. Available from https://doi.org/10.1016/j.rser.2013.06.042.

Duchaud, J. L., Notton, G., Darras, C., & Voyant, C. (2018). Power ramp-rate control algorithm with optimal State of Charge reference via Dynamic Programming. *Energy*, *149*, 709−717. Available from https://doi.org/10.1016/j.energy.2018.02.064.

El Alani, O., Ghennioui, H., Ghennioui, G., Saint-Drenan, Y. M., Blanc, P., Hanrieder, N., & Dahr, F. E. (2021). A visual support of standard procedures for solar radiation quality control. *International Journal of Renewable Energy Development*, *10*−3, 401−414. Available from https://doi.org/10.14710/ijred.2021.34806.

Espinar, B., Blanc, P., Wald, L., Hoyer-Klick, C., Schroedter-Homscheidt, M., & Wanderer, T. On quality control procedures for solar radiation and meteorological measures, from subhourly to monthly average time periods. In *EGU General Assembly 2012*, April 2012, Vienna, Austria. [Online]. Available: http://elib.dlr.de/80168/.

Evans, D. L. (1981). Simplified method for predicting photovoltaic array output. *Solar Energy*, *27*, 555−560. Available from https://doi.org/10.1016/0038-092X(81)90051-7.

Fay, D., & Ringwood, J. V. (2010). On the influence of weather forecast errors in short-term load forecasting models. *IEEE Transactions on Power Systems*, *25*, 1751−1758. Available from https://doi.org/10.1109/TPWRS.2009.2038704.

Gschwind, B., Wald, L., Blanc, P., Lefèvre, M., Schroedter-Homscheidt, M., & Arola, A. (2019). Improving the McClear model estimating the downwelling solar radiation at ground level in cloud-free conditions—McClear-v3. *Meteorologische Zeitschrift*, *28*−2, 147−163. Available from https://doi.org/10.1127/metz/2019/0946.

Gueymard, C. A. (2008). REST2: High-performance solar radiation model for cloudless-sky irradiance, illuminance, and photosynthetically active radiation—Validation with a benchmark dataset. *Solar Energy*, *82*−3, 272−285. Available from https://doi.org/10.1016/j.solener.2007.04.008.

Hong, T., Wilson, J., & Jingrui, X. (2013). Long term probabilistic load forecasting and normalization with hourly information. *IEEE Transactions on Smart Grid*, *5*−1, 456−462. Available from https://doi.org/10.1109/TSG.2013.2274373.

Ineichen, P. (2008). A broadband simplified version of the Solis clear sky model. *Solar Energy, 82−8*, 758−762. Available from https://doi.org/10.1016/j.solener.2008.02.009.

Ineichen, P. (2016). Validation of models that estimate the clear sky global and beam solar irradiance. *Solar Energy, 132*, 332−344. Available from https://doi.org/10.1016/j. solener.2016.03.017.

Jalilpoor, K., Nikkhah, S., Sepasian, M. S., & Aliabadi, M. G. (2022). Application of precautionary and corrective energy management strategies in improving networked microgrids resilience: A two-stage linear programming. *Electric Power Systems Research, 204*107704. Available from https://doi.org/10.1016/j.epsr.2021.107704.

Jumin, E., Basaruddin, F. B., Yusoff, Y. B. M., Latif, S. D., & Ahmed, A. N. (2021). Solar radiation prediction using boosted decision tree regression model: A case study in Malaysia. *Environmental Science and Pollution Research, 28*, 26571−26583. Available from https://doi.org/10.1007/s11356-021-12435-6.

Khan, A. R., Mahmood, A., Safdar, A., Khan, Z. A., & Khan, N. A. (2016). Load forecasting, dynamic pricing and DSM in smart grid: A review. *Renewable and Sustainable Energy, 54*, 1311−1322. Available from https://doi.org/10.1016/j.rser.2015.10.117.

Khodayar, M. E., & Wu, H. (2015). Demand forecasting in the smart grid paradigm: Features and challenges. *The Electricity Journal, 28−6*, 51−62. Available from https:// doi.org/10.1016/j.tej.2015.06.001.

Krueger, E., Bongale, S., & Franklin, D. (2021). Build better decision trees with pruning—Reducing overfitting and complexity of decision trees by limiting max-depth and pruning. *Towards Data Science*. Available at: https://towardsdatascience.com/build-better-decision-trees-with-pruning-8f467e73b107.

Lefèvre, M., Oumbe, A., Blanc, P., Espinar, B., Gschwind, B., Qu, Z., Wald, L., Schroedter-Homscheidt, M., Hoyer-Klick, C., Arola, A., Benedetti, A., Kaiser, J. W., & Morcrette, J. J. (2013). McClear: A new model estimating downwelling solar radiation at ground level in clear-sky conditions. *Atmospheric Measurement Techniques, 6−9*, 2403−2418. Available from https://doi.org/10.5194/amt-6-2403-2013.

Lorestani, A., Gharehpetian, G. B., & Nazari, M. H. (2019). Optimal sizing and techno-economic analysis of energy- and cost-efficient standalone multi-carrier microgrid. *Energy, 178*, 751−764. Available from https://doi.org/10.1016/j. energy.2019.04.152.

Luo, Y., Shi, Y., Zheng, Y., Gang, Z., & Cai, N. (2017). Mutual information for evaluating renewable power penetration impacts in a distributed generation system. *Energy, 141*, 290−303. Available from https://doi.org/10.1016/j.energy.2017.09.033.

Manur, A., Marathe, M., Manur, A., Ramachandra, A., Subbarao, S., & Venkataramanan, G. (2020). Smart solar home system with solar forecasting. In *2020 IEEE International Conference on Power Electronics, Smart Grid and Renewable Energy (PESGRE2020)* (pp. 1−6). Available from https://doi.org/10.1109/PESGRE45664.2020.9070340.

Mehigan, L., Deane, J. P., Gallachoir, B. P. O., & Bertsch, V. (2018). A review of the role of distributed generation (DG) in future electricity systems. *Energy, 163*, 822−836. Available from https://doi.org/10.1016/j.energy.2018.08.022.

Moya, F. D., Torres-Moreno, J. L., & Álvarez, J. D. (2020). Optimal model for energy management strategy in smart building with energy storage systems and electric vehicles. *Energies, 13−14*, 14. Available from https://doi.org/10.3390/en13143605.

Mueller, R. W., Dagestad, K. F., Ineichen, P., Schroedler Homscheidt, M., Cros, S., Dumortier, D., Kuhlemann, R., Olseth, J., Piernavieja, G., Reise, C., Wald, L., & Heinnemann, D. (2004). Rethinking satellite-based solar irradiance modelling: The SOLIS clear-sky module. *Remote Sensing of Environment, 91−2*, 160−174. Available from https://doi.org/10.1016/j.rse.2004.02.009.

Murty, V. V. S. N., & Kumar, A. (2020). Multi-objective energy management in microgrids with hybrid energy sources and battery energy storage systems. *Protection and*

Control of Modern Power Systems, *5-2*. Available from https://doi.org/10.1186/s41601-019-0147-z.

Nose-Filho, K., Lotufo, A. D. P., & Minussi, C. R. (2011). Short-Term multinodal load forecasting using a modified general regression neural network. *IEEE Transactions on Power Delivery*, *26*, 2862—2869. Available from https://doi.org/10.1109/TPWRD.2011.2166566.

Notton, G., Nivet, M. L., Voyant, C., Paoli, C., Darras, C., Motte, F., & Fouilloy, A. (2018). Intermittent and stochastic character of renewable energy sources: consequences, cost of intermittence and benefit of forecasting. *Renewable and Sustainable Energy Reviews.*, *87*, 96—105. Available from https://doi.org/10.1016/j.rser.2018.02.007.

Notton, G., & Voyant, C. (2018). Forecasting of intermittent solar energy resource. In I. Yahyaoui (Ed.), *Advances in renewable energies and power technologies* (pp. 77—109). Elsevier Science, ISBN 978-012-8131855; 01/02/2018. Available from https://doi.org/10.1016/C2016-0-04518-7.

Ouédraogo, S., Faggianelli, G. A., Notton, G., Duchaud, J. L., & Voyant, C. (2022). Impact of energy pricing policies and energy management strategies on PV/battery microgrid performances. *Renewable Energy*, *199*, 816—825. Available from https://doi.org/10.1016/j.renene.2022.09.042.

Pan, X., Khezri, R., Mahmoudi, A., Yazdani, A., & Shafiullah, G. M. (2021). Energy management systems for grid-connected houses with solar PV and battery by considering flat and time-of-use electricity rates. *Energies*, *14—16*, 16. Available from https://doi.org/10.3390/en14165028.

Park, K., Lee, W., & Won, D. (2019). Optimal energy management of DC microgrid system using dynamic programming. *IFAC-PapersOnLine*, *52—4*, 194—199. Available from https://doi.org/10.1016/j.ifacol.2019.08.178.

Parviz, R. K., Nasser, M., & Motlagh, M. R. J. (2008). Mutual information based input variable selection algorithm and wavelet neural network for time series prediction. In V. Kurkova, R. Neruda, & J. Koutník (Eds.), *Artificial Neural Networks—ICANN 2008* (pp. 798—807). Berlin: Springer.

Pascual, J., Barricarte, J., Sanchis, P., & Marroyo, L. (2015). Energy management strategy for a renewable-based residential microgrid with generation and demand forecasting. *Applied Energy*, *158*, 12—25. Available from https://doi.org/10.1016/j.apenergy.2015.08.040.

Paul, T. G., Hossain, S. J., Ghosh, S., Mandal, P., & Kamalasadan, S. (2018). A quadratic programming based optimal power and battery dispatch for grid-connected microgrid. *IEEE Transactions on Industry Applications*, *54—2*, 1793—1805. Available from https://doi.org/10.1109/TIA.2017.2782671.

Petrollese, M., Valverde, L., Cocco, D., Cau, G., & Guerra, J. (2016). Real-time integration of optimal generation scheduling with MPC for the energy management of a renewable hydrogen-based microgrid. *Applied Energy*, *166*, 96—106. Available from https://doi.org/10.1016/j.apenergy.2016.01.014.

Prodan, I., & Zio, E. (2014). A model predictive control framework for reliable microgrid energy management. *International Journal of Electrical Power & Energy Systems*, *61*, 399—409. Available from https://doi.org/10.1016/j.ijepes.2014.03.017.

Qu, Z., Oumbe, A., Blanc, P., Espinar, B., Gesell, G., Gschwind, B., Klüser, L., Lefèvre, M., Saboret, L., Schroedter-Homscheidt, M., & Wald, L. (2017). Fast radiative transfer parameterisation for assessing the surface solar irradiance: The Heliosat-4 method. *Meteorologische Zeitschrift*, *26—1*, 33—57. Available from https://doi.org/10.1127/metz/2016/0781.

Rigollier, C., Bauer, O., & Wald, L. (2000). Radiation atlas with respect to the heliosat method. *Solar Energy*, *68—1*, 33—48. Available from https://doi.org/10.1016/S0038-092X(99)00055-9.

Santodomingo, R., Uslar, M., Specht, M., Rohjans, S., Taylor, G., Pantea, S., Bradley, M., & McMorran, A. (2016). *IEC 61970 for energy management system integration. Smart grid handbook.* John Wiley & Sons, Ltd, doi.10.1002/9781118755471.sgd094.

Shadab, A., Ahmad, S., & Said, S. (2020). Spatial forecasting of solar radiation using ARIMA model. *Remote Sensing Applications: Society and Environment, 20*100427. Available from https://doi.org/10.1016/j.rsase.2020.100427.

Shah, D., Patel, K., & Shah, M. (2021). Prediction and estimation of solar radiation using artificial neural network (ANN) and fuzzy system: A comprehensive review. *International Journal of Energy and Water Resources, 5*, 219−233. Available from https://doi.org/10.1007/s42108-021-00113-9.

Shi, Z., Liang, H., Huang, S., & Dinavahi, V. (2019). Distributionally robust chance-constrained energy management for Islanded microgrids. *IEEE Transactions on Smart Grid, 10−2*, 2234−2244. Available from https://doi.org/10.1109/TSG.2018.2792322.

Shu, F., Methaprayoon, K., & Wei-Jen, L. (2009). Multiregion load forecasting for system with large geographical area. *IEEE Transactions on Industry Applications, 45*, 1452−1459. Available from https://doi.org/10.1109/TIA.2009.2023569.

Soufiane, G., Ouafia, F., & Ahmed, A. (2022). Solar radiation time-series prediction using random forest algorithm-based feature selection approach. In S. Motahhir, & B. Bossoufi (Eds.), *Digital technologies and applications. ICDTA 2022. Lecture Notes in Networks and Systems* (p. 455). Cham: Springer, doi.10.1007/978-3-031-02447-4_68.

Sukumar, S., Mokhlis, H., Mekhilef, S., Naidu, K., & Karimi, M. (2017). Mix-mode energy management strategy and battery sizing for economic operation of grid-tied microgrid. *Energy, 118*, 1322−1333. Available from https://doi.org/10.1016/j.energy.2016.11.018.

Tavakoli, M., Shokridehaki, F., Funsho Akorede, M., Marzband, M., Vechiu, I., & Pouresmaeil, E. (2018). CVaR-based energy management scheme for optimal resilience and operational cost in commercial building microgrids. *International Journal of Electrical Power and Energy Systems, 100*, 1−9. Available from https://doi.org/10.1016/j.ijepes.2018.02.022.

Trapletti, A., Leisch, F., & Hornik, K. (2000). Stationary and integrated autoregressive neural network processes. *Neural Computation, 12−10*, 2427−2450. Available from https://doi.org/10.1162/089976600300015006.

Tripathy, D. S., & Prusty, R. (2021). Chapter 10—Forecasting of renewable generation for applications in smart grid power systems. In A. Tomar, & R. Kandari (Eds.), *Advances in smart grid power system* (pp. 265−298). Academic Press, ISBN 9780128243374. Available from https://doi.org/10.1016/B978-0-12-824337-4.00010-2.

Tuballa, M. L., & Abundo, M. L. (2016). A review of the development of smart grid technologies. *Renewable and Sustainable Energy Reviews, 59*, 710−725. Available from https://doi.org/10.1016/j.rser.2016.01.011.

Vadari, M. (2020). The future of distribution operations and planning: The electric utility environment is changing. *IEEE Power & Energy Magazine, 18−1*, 18−25. Available from https://doi.org/10.1109/MPE.2019.2945344.

Voyant, C., Motte, F., Notton, G., Fouilloy, A., Nivet, M. L., & Duchaud, J. L. (2018). Prediction intervals for global solar irradiation forecasting using regression trees methods. *Renewable Energy, 126*, 332−340. Available from https://doi.org/10.1016/j.renene.2018.03.055.

Vrablecová, P., Ezzeddine, A. B., Rozinajová, V., Šárik, S., & Sangaiah, A. K. (2018). Smart grid load forecasting using online support vector regression. *Computer and Electrical Engineering, 65*, 102−117. Available from https://doi.org/10.1016/j.compeleceng.2017.07.006.

Wan, C., Zhao, J., Song, Y., Xu, Z., Lin, J., & Hu, Z. (2015). Photovoltaic and solar power forecasting for smart grid energy management. *CSEE Journal of Power and Energy Systems*, *1–4*, 38–46. Available from https://doi.org/10.17775/CSEEJPES.2015.00046.

Wang, T., He, X., & Deng, T. (2019). Neural networks for power management optimal strategy in hybrid microgrid. *Neural Computing & Applications*, *31–7*, 2635–2647. Available from https://doi.org/10.1007/s00521-017-3219-x.

Wibowo, R.S., Firmansyah, K.R., Aryani, N.K., & Soeprijanto, A. (2016). Dynamic economic dispatch of hybrid microgrid with energy storage using quadratic programming. In *2016 IEEE Region 10 Conference (TENCON)* (pp. 667–670). doi.10.1109/TENCON.2016.7848086.

Wiens, T. S., Dale, B. C., Boyce, M. S., & Kershaw, G. P. (2008). Three-way k-fold cross-validation of resource selection functions. *Ecological Modelling*, *212*, 244–255. Available from https://doi.org/10.1016/j.ecolmodel.2007.10.005.

Yadav, K., & Chandel, S. S. (2014). Solar radiation prediction using Artificial Neural Network techniques: A review. *Renewable and Sustainable Energy Reviews.*, *33*, 772–781. Available from https://doi.org/10.1016/j.rser.2013.08.055.

Yaqin, Y., Zhongxi, W., & Chunming, H. (2005). A Novel implementation of IEC 61970 CIS based on Ice middleware. In *2005 IEEE/PES Transmission & Distribution Conference & Exposition: Asia and Pacific* (pp. 1–4). Available from https://doi.org/10.1109/TDC.2005.1546839.

Zou, B., Peng, J., Li, S., Li, Y., Yan, J., & Yang, H. (2022). Comparative study of the dynamic programming-based and rule-based operation strategies for grid-connected PV-battery systems of office buildings. *Applied Energy*, *305*117875. Available from https://doi.org/10.1016/j.apenergy.2021.117875.

Applications of intelligence learning approaches for renewable and sustainable energy

CHAPTER FIVE

Intelligent learning models for renewable energy forecasting

Esteban Jove, Álvaro Michelena, Miriam Timiraos, Víctor Caínzos López, Hector Quintian and Jose Luis Calvo-Rolle
Department of Industrial Engineering, University of A Coruña, CTC, CITIC Research, Rúa Mendizábal, Ferrol, A Coruna, Spain

5.1 Introduction

Environmental care is presently a prominent trend and an important issue for society and governments. Moreover, for obvious reasons, there is a clear inclination to prioritize environmental care. In reality, achieving a scenario with no impact is extremely challenging, if not impossible. Nonetheless, aspects such as sustainability and striving for the maximum possible reduction in environmental impact are very important (Kuwae & Hori, 2018).

In terms of energy needs, renewable energies play a key role in contributing to a reduction in environmental impact and emissions (Karunathilake et al., 2019). however, it is important to consider the impact of the powerplant implementation itself when based on renewable sources, as there is usually not a zero impact (Varun et al., 2009).

Achieving a null impact remains unattainable, even with the alternatives and use of renewable energies. Therefore, there is a legal obligation to optimize and plan installations for maximum efficiency (Wei et al., 2010). In addition, the facilities' efficiency must be measured according to the right ratios and criteria to ensure the desired minimum impact (Giacone & Mancò, 2012).

The most significant environmental impact and economic investment occur during the construction of the powerplant. Typically, depending on the technology, the powerplant requires significantly less expense (Boyle, 2004). In certain circumstances, it is more beneficial to keep the installation operational rather than shutting it down (Keeney, 2013). When this occurs, energy storage emerges as a recommended solution (Dunn et al., 2011).

Intelligent Learning Approaches for Renewable and Sustainable Energy
DOI: https://doi.org/10.1016/B978-0-443-15806-3.00005-X

The country's electric sector is complex for many reasons (Montero-Sousa, Casteleiro-Roca, et al., 2017a). The most crucial challenge lies in ensuring that energy generation matches energy demand. The multitude of technologies for energy production, each with its unique characteristics, further complicate the system (Montero-Sousa, Casteleiro-Roca, et al., 2017b). Even the growth of renewable energy-based power plants has played a role in disrupting the stability of the system.

Additionally, incorporating renewable energies or other potential energy sources in all types of grid-connected buildings, even when the purpose is not energy generation, poses challenges in network management. Hence, it is crucial to effectively develop tools to manage diverse energy generation and consumption points. The Smartgrid concept (Amin, 2013) thus emerges for all of the aforementioned reasons, emphasizing the need to measure the generation, the consumption, predict both generation and consumption, with the aim of making informed decisions, and ultimately enhance the overall system efficiency in every way (Vilar-Martinez et al., 2014). Nevertheless, it is a challenging task to match the generation with the demand, making energy storage a desirable option in such cases.

In some geographical areas, buildings have electricity energy needs but are not connected to the electric grid. These cases do not usually have an easy solution because connecting a building to the electricity network implies a high cost which may never be amortized. An alternative, possibly more feasible, is implementing energy storage systems (Montero-Sousa, Fernández-Serantes, et al., 2017).

For all the reasons mentioned above, the justification for implementing energy storage systems in various electric power scenarios is evident. There are many technologies for this purpose, although some of them are relatively old and inefficient, such as pumping water to a water dam (Yang & Jackson, 2011). However, due to this great need under the current scenario, there have been many research and development projects carried out in recent years (Dunn et al., 2011). The electric car is one of the reasons why electric energy storage systems are being developed so quickly (Westbrook, 2001). Researchers are focusing on various technologies, especially battery technologies and fuel cells (Hall & Bain, 2008). In the case of fuel cells, which is the focus of this work, there are also many proposals regarding primary fuels (Tao et al., 2008). Hydrogen emerged as one of the earliest used fuels due to various factors. This gas is readily obtainable through a straightforward hydrolysis process, although its storage may present challenges compared to other options. Nevertheless, hydrogen remains one of the most viable alternatives for such purposes (Dicks & Rand, 2018).

Considering the current trajectory of the electric sector and the goal of enhancing grid efficiency, it is imperative to rely on accurate forecasts to make informed decisions (Potter et al., 2009). This forecasting enables strategic actions, such as purchasing or selling energy at optimal prices, storing energy at convenient times, and more. Reliable forecasts are especially important for fuel cell systems that rely on hydrogen and its generation for energy storage.

Various alternatives can be considered when conducting the modeling process, each with potentially different performance outcomes. One commonly employed technique that relies on traditional regression models is multiple regression analysis (MRA) (Turrado et al., 2014). There are several applications where this method or ones with small variations are used (Jin et al., 2012; Osborn et al., 2014). The MRA technique, however, has some limitations in several cases (Cho & Awbi, 2007; Turrado et al., 2014). One of the reasons why the performance is not suitable is the non-linearity of the problem to be modeled.

Intelligent systems are used in some different applications with very satisfactory performance overall (Moreno-Fernandez-de-Leceta et al., 2018; Rincon et al., 2018; Segovia et al., 2018; Wojciechowski, 2018). Of course, the nonlinearity problem could be solved in many cases with the use of soft-computing techniques (Fontenla Romero & Calvo Rolle, 2018; Nieves et al., 2013; Suárez Sánchez et al., 2011). The problem when the system is nonlinear could be the same, even by using simple, intelligent systems. In such cases, it is possible to divide the problem with clustering techniques, such as K-Means (Ghaseminezhad & Karami, 2011; Martínez-Rego et al., 2011; Casteleiro-Roca et al., 2017; Quintián et al., 2014).

This chapter deals with different forecasting approaches applied in a wide variety of applications related to renewable energies. The next section describes seven case studies in the fields of energy generation, energy storage, and energy demand, where forecasting methods are implemented. Section 5.3 presents the clustering, intelligent regression techniques, and time series techniques needed to achieve the objective. Section 5.4 develops the approaches and results, while and the last section discusses the conclusions.

5.2 Cases of study

As already mentioned in the introduction, the use of renewable and sustainable energies is playing a key role, promoted by the governments of many countries, as well as private institutions. This fact, combined with the current development of new artificial intelligence methods and techniques, has resulted in the funding of a large number of projects and research lines related to the implementation of intelligent techniques in the field of renewable energies in recent years. Therefore, a wide variety of research studies can be found in which intelligent techniques are proposed and applied to solve problems or tasks related to a multitude of aspects related to the generation, use, or distribution of all types of renewable energies.

5.2.1 Bioclimatic house

This chapter describes different approaches and techniques applied to various renewable energy systems in a bioclimatic house dedicated to harness renewable sources. The building (Fig. 5.1) in Galicia, Spain, was established to promote energy efficiency and the use of renewable and clean energy systems. The design and materials ensure thermal isolation and natural lighting. In addition, different renewable energy systems cover the demand for electric power, domestic hot water (DWH), and heating, all of which are subjected to forecasting and are described in the next subsections.

5.2.1.1 Geothermal system

One of the presented works predicts a geothermal heat pump installation placed in a bioclimatic house. Geothermal heat pumps are cooling, heating, and DWH systems that take advantage of the energy taken from the

Figure 5.1 General view of bioclimatic house.

ground (Hepbasli & Ozgener, 2004). Although the description of the facility under study is designed for residential purposes, geothermal systems can also be presented in industrial processes.

The main principle of heat pumps is the thermodynamic Carnot cycle, which ensures energy transfer between hot and cold sources, generating power useful for conditioning buildings (Sarbu & Sebarchievici, 2014). The operation principle of the heat pump can be perfectly applied to geothermal systems where the ground serves as one of the sources. The key parts of such installations are mainly the heat exchanger placed under the ground, the heat pump, and the consumption points (DHW, conditioning) (Baruque et al., 2019).

The ground heat exchanger can be placed with a vertical or horizontal orientation. The first configuration is more efficient since deeper points ensure more energy transfer. However, the cost of installation is significantly higher than horizontal configurations.

The case study is designed to extract heat from the ground and use it to fulfill the energy needs of users. The process is structured in four parts, as shown in Fig. 5.2.

1. A horizontal heat exchanger, 500 meters in length, is placed at a depth of 2 meters. It is divided into four independent circuits to reduce and isolate the impact of water leakage. The four branches are connected to a collector that unites the four flows and drives them to a heat pump.

2. The heat pump is the component that exploits the heat absorbed by the exchanger. This is done following the process shown in Fig. 5.2, which is based on the thermal cycle of a gas. It presents four stages: the ground heat evaporates the refrigerant, and then it is compressed to the condensation stage. The energy released produced in the condenser can be used to heat water to the inertial accumulator. Finally, the refrigerant is cooled back at the expansion valve. This element presents a great nominal heating power of 8.4 kW and nominal electrical power consumption of 1.9 kW.

3. An inertial accumulator of 0.8 m^3 receives the heated water from the heat pump and is designed to store the energy collected from the ground.

4. The system must supply DHW and an underfloor heating system. According to Spanish law, the DHW is sized to cover 220 liters per day, and the heating system is designed to keep the house between 18°C and 22°C. This corresponds to a water temperature oscillating from 35°C to 40°C.

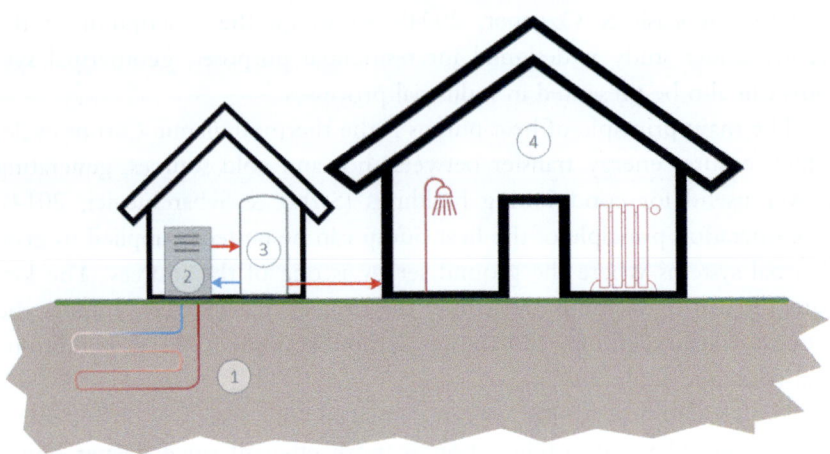

Figure 5.2 *Scheme of heat pump geothermal system.* The system includes a geothermal heat exchanger, a heat pump, and a thermal accumulator. It is used for DHW and heating.

Due to the importance of predicting energy consumption, the temperature of the heat exchanger is registered along the five loops, and the temperature at the input and output of the heat pump is monitored as well. A total amount of 29 sensors registered the state of the facility with 10 minutes of sample time during one year. This empirical dataset is used to apply intelligent techniques for energy forecasting, which are especially useful in this kind of system for aspects such as design, motorization, optimization, estimation, and management (Jha et al., 2017).

5.2.1.2 Solar thermal system
Another application related to the use of renewable energy and the acquisition of heat is solar thermal technology. Solar thermal systems are in charge of transferring solar energy received through radiation to a fluid (Tian & Zhao, 2013). This objective is achieved by a solar collector whose location, orientation, and weather conditions determine the amount of heat received. Then, accurate forecasting must take into consideration all these variables (Law et al., 2014).

Depending on the application and the temperature needed by the user, the collector may present different materials, locations, and configurations. For example, swimming pools are conditioned using low-efficiency plastic pipe collectors. In these cases, the consumption fluid is the one that flows through the collector. In other applications, such as

DHW, a carrier fluid circulates through solar panels, and the heat is then transferred to the water through a heat exchanger. Depending on its working principle, materials, ambient temperatures, or orientation, this kind of collector can present different efficiencies (Law et al., 2014).

This technology, along with the geothermal exchanger, supplies the DHW and heating to the bioclimatic building where the forecasting is carried out (Casteleiro-Roca et al., 2020). Fig. 5.3 depicts a general scheme with the installation's components. In addition to the pumps and valves that ensure the proper system behavior, the main components are described below:

1. Four sensors (RTD -PT1000) measure the heat carrier fluid's panel input and output temperatures.
2. 2 A flow meter measures the fluid the pump boosts through the solar collectors.
3. Eight solar panels (20 m^2) set in two parallel strings to receive the solar irradiance and transfer it to the fluid.
4. An accumulator with an embedded heat exchanger transfers energy from the carrier fluid to stored water up to 1000 liters.

As the radiance value is especially important to estimate the energy generated, a PYR-P sensor is placed on the roof.

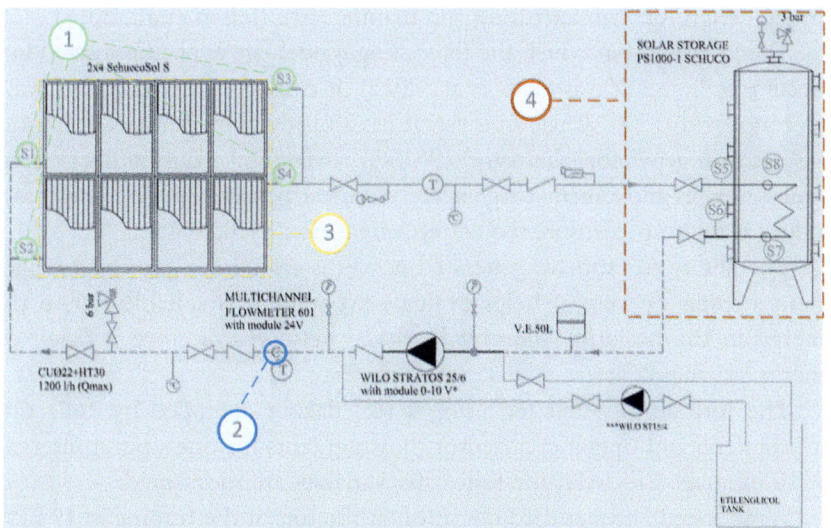

Figure 5.3 *Detailed scheme of the solar thermal system.* Eight solar collectors are in charge of transmitting the solar heat and transferring it to an accumulator.

5.2.1.3 Wind turbine

Wind energy is one of the most important elements to the energy transition and thus is experiencing a significant growth worldwide. This technology converts the rotational motion of the wind turbine's rotor into electricity using a generator, making wind the crucial factor in energy production. Studies have shown a proportional relationship between the power generated by the generator and the cube of wind speed (Tavner et al., 2006). However, the wind is a variable resource influenced by wind and factors such as surface variations and seasonal changes. Additionally, wind speed fluctuates continuously due to turbulence, which must be considered in energy production planning to prevent potential damage to the plant (Infield & Freris, 2020).

The evolution of wind energy trends towards optimizing current wind turbine designs to maximize performance and increase power generation (Pryor & Barthelmie, 2010). Depending on their design, turbines can generate power ranging from kilowatts for small consumers, to megawatts for large populations. In this context, there are two main types of turbines: horizontal-axis wind turbines (HAWT), commonly used in current applications, and vertical-axis wind turbines (VAWT), capable of generating electricity with lower wind speeds. Although VAWTs need to be placed close to the ground and may experience minor electricity generation outages, they are known for their superior performance and cost-effective, simple design for manufacturing and maintenance (Johari et al., 2018).

The installation in which the forecasting models are applied is located in a bioclimatic house (Zayas-Gato et al., 2022). It consists of a two-bladed wind generator with 1500 W rated power. It has a neodymium permanent magnet synchronous generator (alternator) (PMSG) covered by a carbon fiber casing, which uses its movement to generate electrical power. This housing has a rudder at the rear to ensure the correct orientation of the system.

The ability to estimate generated power is essential in energy management systems because it helps evaluate whether it is suitable to use the energy in the building, transmit it to the network, or store it based on energy pricing.

The forecast is generated using a real dataset compiled by collecting atmospheric and operational sensor measurements for one year at intervals of 10 minutes. Precisely, the following variables are monitored:
- Wind speed and standard deviation at the top of the turbine at 10 m.
- Wind direction and standard deviation at the top of the turbine at 10 m. Wind gusts over 10 m are recorded too.

- Temperature measurements are taken at different heights, including 1.5 m, 0.1 m, and just above the ground at -0.1 m. Additionally, rainfall temperature is measured at 1.5 m height.
- Sunshine hours and global radiation.
- Atmospheric pressure and reduced atmospheric pressure
- Relative humidity and rainfall samples at 1.5 m.
- Electrical variables such as current, voltage, power, and generated energy

The provided research demonstrates that evaluating the atmospheric conditions at a particular moment is practical to predict the power generated by the small wind system in a brief period with reasonable accuracy.

5.2.2 Energy storage systems

On the other hand, giving importance to renewable energies described in previous sections, entails the challenges of seasonality and intermittency. Energy generation through these means depends on temporal factors such as wind in wind power, sunlight in solar power, water in hydroelectric power, etc. Due to this high seasonality, there are significant differences in production between months or even between days. To meet demand, it is necessary to implement energy storage systems that can take advantage of production surpluses during production peak periods and that serve to ensure supply during those times when production does not meet demand. In this regard, two alternatives are currently being widely employed: the generation of hydrogen and the use of batteries.

5.2.2.1 Batteries

Batteries offer an alternate method for storing electric energy. These devices can store energy in an electrochemical medium and release it through electrochemical processes (Chukwuka & Folly, 2012). Lithium-ion (Li-ion) batteries typically consist of five layers: a negative collector for negative current, a negative electrode known as the "anode," a membrane-like separator, a positive electrode known as the "cathode," and a positive current collector. Li-ion batteries are available in a wide range of shapes and sizes, including cylindrical, coin, and prismatic. Cylindrical and coin batteries are frequently used in small-scale equipment, such as smart watches, calculators, and electronic toys, whereas prismatic batteries are used in high-capacity applications such as the electric mobility field (Yi et al., 2013).

The case study pertains to a LiFePO4 battery cell (Caínzos López et al., 2022). Within the charging cycle, the electrons move from the cathode to the anode, facilitated by an external power source to induce this flow. This results in an increase in the state of charge (SOC) within the battery due to the occurring reactions. Conversely, the electrons flow in the opposite direction within the electric circuit during the discharge process. As a result, the conventional current flow depicted in electric diagrams is from the positive to the negative electrode, contrary to the actual movement of electrons. In addition to these fundamental requirements, considerable attention is being directed toward endowing the separator with specific beneficial functionalities. It presents a promising approach for improving battery development, especially in the context of high-energy rechargeable next-generation batteries (Fernández–Serantes et al., 2014).

One of the main reasons for a decrease in battery capacity is temperature, among other factors. For example, battery degradation in electric vehicles can be optimized by reducing the power used to charge it. With a low-power charger, the battery does not suffer an increase in temperature (Hoke et al., 2014). On the other hand, using high-power chargers leads to a significant rise in temperature, which causes a substantial reduction in battery capacity (Neubauer & Wood, 2015).

The battery model is generally determined after a capacity confirmation test. The following stages are repeated a substantial number of cycles during the test:

1. The tester applies a continuous current to the battery, causing the voltage to increase from 3 V to 3.65 V gradually.
2. At this point, the current flow ceases, and the voltage decreases until it reaches the nominal value of 3.3 V.
3. The test requests a constant current until the voltage drops to 2 V.
4. Once the voltage reaches 2 V, the current flow is halted, resulting in the voltage rising back up to 3 V. This process can be repeated starting from step 1.

Fig. 5.4 depicts the test configuration. Since the temperature affects the operational state of the battery, two temperature sensors, as well as voltage and current sensors, monitor the process.

5.2.2.2 Hydrogen fuel cell

Another highly useful and widely employed technique for energy storage is hydrogen fuel cells. Fuel cell systems are currently the subject of

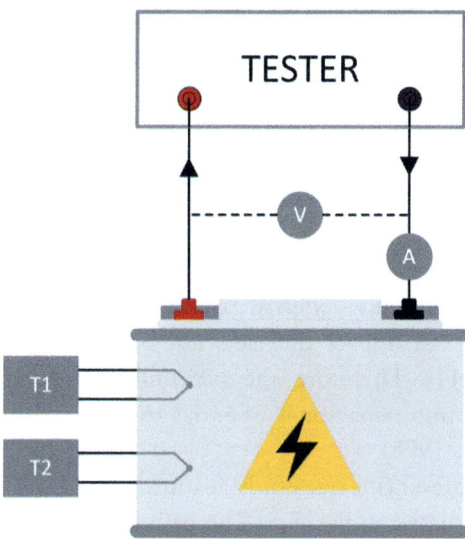

Figure 5.4 *Scheme of capacity confirmation test.* A tester is in charge of demanding and supplying current to the tester. The temperature at two points is also measured.

ongoing research, with the potential for improved performance in the coming years. These systems are considered a high-reliability option for applications that require consistent power output, such as electric vehicles and space applications. They are notable for their clean energy characteristics (Andújar & Segura, 2009). Internally, a fuel cell comprises multiple interconnected individual cells that form a stack. Within this stack, an electrochemical reaction occurs when hydrogen (H_2) is combined with oxygen (O_2) in a given environment, generating electrical power. A control system and some subsystems, such as cooling and gas conditioning, ensure the entire operation is carried out safely and efficiently. (De las Heras et al., 2018).

In some sustainable technologies, such as photovoltaic or wind generation, the production of power relies on the availability of primary energy sources like the sun or wind. In contrast, fuel cells solely require hydrogen (H_2) as an energy source, making them adaptable for installation wherever needed. Among fuel cell types, the proton exchange membrane fuel cell (PEMFC) stands out for its high efficiency, with notable features, such as high energy density, low volume, and weight, compared to other fuel cell technologies. Additionally, PEMFC operates at a low temperature (below 100°C), resulting in shorter heating times upon startup. PEMFC, furthermore, exhibits a wide power range that can be tailored to various applications (Andújar & Segura, 2009), ranging from high-power systems for connection to the electrical grid in power stations (Moreira & da Silva, 2009), to smaller mobile units (Ross, 2003). Notably, the stored energy or power output of PEMFC is solely dependent on the available amount of hydrogen (H_2).

The fuel cell output is considered unregulated power since it is produced by an electrochemical process. A system is required to regulate the inflow of hydrogen (H_2) and oxygen (O_2) and ensure the system's overall safety. Additionally, external factors, such as temperature, pressure, and other agents can impact the electrical output values of the fuel cell (Segura et al., 2011). Therefore, the availability of a predictive model for the system's dynamic behavior is essential to enhance fuel cell efficiency (Van Bussel et al., 1998).

This chapter presents a case study featuring a (PEMFC), with its schematic diagram depicted in Fig. 5.5. The cathode of the PEMFC combines ions and electrons to produce pure water as the residue of the overall reaction. Under normal conditions, a single fuel cell can generate a voltage of 1.2 V. To create high-power systems, multiple single cells can be connected in series or parallel to form a stack.

On a test bench, data for forecasting cell activity is collected. The stack's maximum output power is 3.4 kW, with average voltage and current values of 45.33 V and 75 A. The stack has an internal refrigeration system that uses air for cooling it. The hydrogen inflow pressure is about 1.36 bar. The monitor system records all of the critical stack variables, including as temperature, voltage, current, and hydrogen flow.

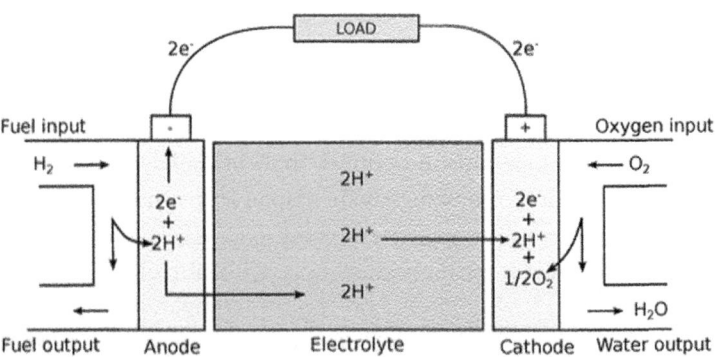

Figure 5.5 *Scheme of fuel cell.* Fuel and oxygen are inputs to the cell. As a result, water is obtained at the output.

5.2.3 Demand forecasting

In addition to predicting renewable energy generation, the accurate prediction of energy demand plays a key role in implementing optimal energy management systems. This chapter demonstrates the application of forecasting techniques to model the consumption behavior of a tourist building. Advanced energy management systems rely on generation or load forecasts, and the availability of reliable predictions is essential to enhance overall system efficiency (Serale et al., 2018).

The work presented in this chapter (Casteleiro-Roca et al., 2019) considers an intelligent prediction model for tourism buildings, a premium five-star hotel on the Atlantic Ocean south of Tenerife, Canary Islands (Spain). Tourism activities have a significant energy demand, necessitating effective energy management solutions to minimize environmental impact and operating costs. HVAC, lights, DHW, swimming pool heating, kitchens, and leisure activities are the activities that consume the most energy in the hotel business sector. According to the UNWTO, air conditioning uses most of the required energy (often more than 40%). DHW systems consume around 15% of total energy, whereas lighting entails a much more variable energy demand (15%—40%), depending on the type of hotel (Vega Lara et al., 2016).

The intricacy of hotel structures, which often encompass diverse facilities and services, necessitates the use of specialized tools for effective energy management. In addition to enhancing the establishment's profitability, these tools can also facilitate progress toward achieving environmental sustainability goals in the industry. For example, centralized control systems, commonly referred to as building management systems (BMS), enable the management of various installation elements such as heating, ventilation, air conditioning (HVAC), DHW, and more. BMS relies on computerized techniques to define the predetermined requirements of the entire building and to control diverse systems to achieve the desired operational state. To make informed decisions, BMS relies on data from multiple sensors, including predictions of future trends related to building facilities, meteorological conditions, energy production and consumption, and other important factors.

A dataset containing power demand, daily mean temperature, and occupation rate was used to create a consumption profile. The dataset contains samples for one year, with a sample rate of one hour. The hotel uses a variety of energy sources, including grid power, thermal solar, and supplemental gas systems. However, the largest source of energy use is grid-supplied electricity. Based on the average ambient temperature and electrical energy consumption statistics from 2017, each room is anticipated to require 118 kWh of power per day.

5.3 Forecasting/modeling techniques

This section describes the different techniques considered to achieve accurate forecasting models in the case studies presented in the previous section.

5.3.1 Clustering

A common approach in forecasting problems consists of grouping the data depending on the characteristics. The different clustering techniques considered are described below.

5.3.1.1 K-Means

The K-Means algorithm is an unsupervised technique that is extensively used in fields such as machine learning, image processing, and pattern recognition. This method seeks to partition the dataset into homogenous groups of data with similar characteristics, known as clusters (Kaski et al., 2005).

From a given set $X = \{x_1, x_2, \ldots, x_N\}$ where $x_i \in R^n$, the K-Means algorithm splits the data into k subsets $P_1, P_2, \ldots, P_k$, with their centroids $C = \{c_1, c_2, \ldots, c_k\}$, where $c_j \in R^n$, based on a clustering error criteria. This error is often calculated as the sum of the values of all Euclidean distances between each point $x_i \in R^n$ to its centroid c_j, according to the following equation, where $I(A) = 1$ only if A is true.

$$E(c_1, \ldots c_k) = \sum_{i=1}^{N} \sum_{j=1}^{k} I(x_i \in P_j) ||x_i - c_i|| \tag{5.1}$$

The user must manually select the number of clusters, a decision that significantly impacts the clustering outcome and relies on the shape and distribution of the data. In the training phase, the centroids are initially assigned randomly, and their positions greatly influence the clustering error. Subsequently, the centroid positions are iteratively adjusted until the clustering error is minimized (Likas et al., 2003). Finally, the training process concludes when the cluster centroids no longer change, indicating convergence (Machón-González et al., 2010).

The disadvantage of this strategy is its extreme sensitivity to the centroids' initial position. A frequent solution to this problem is to repeat the training procedure while altering the locations of the original centroids. After computing the position of each centroid, a new test data set belongs to the cluster with the shortest centroid Euclidean distance. This approach performs better when the training dataset is equally distributed in hyperspace, and the data groups are hyperspherical. Fig. 5.6 depicts how the K-Means method divides a random dataset into three groups. The centroid of each cluster is represented by a cross, and three test data points are denoted by diamonds.

5.3.2 Intelligent regression techniques

In this chapter, one of the strategies introduced involves the k-means algorithm to identify distinct groups and then develop intelligent models for each group to enhance regression performance. The contemplated intelligent techniques to obtain the models are described in this subsection.

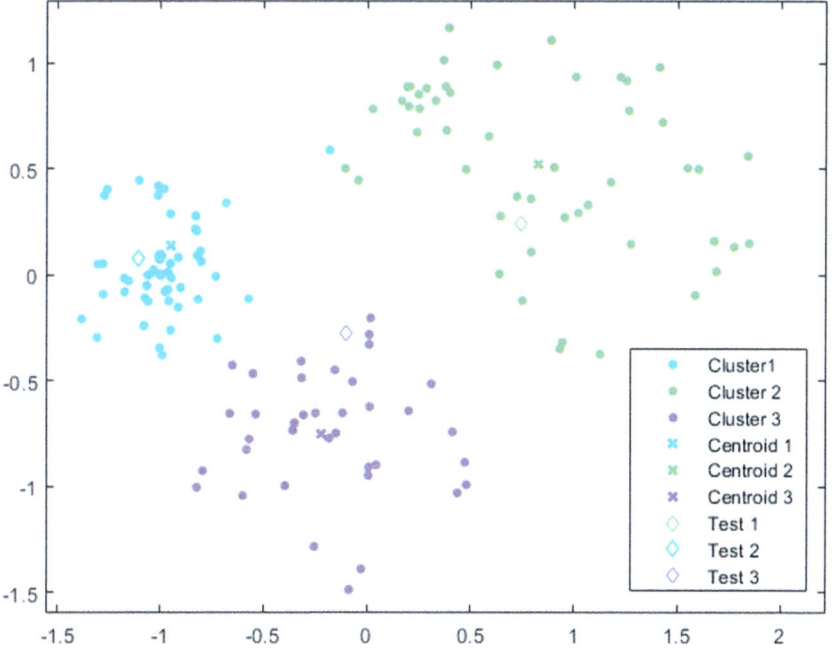

Figure 5.6 *Clusters, centroids, and test samples over a dataset in 2D.* Representation of different data points classified in different clusters.

5.3.2.1 Multilayer perceptron

Artificial neural networks (ANN) have found a wide variety of applications in diverse problem domains, including pattern recognition, prediction, and image processing (Jain et al., 1996). The first artificial neuron was proposed in 1940 (McCulloch & Pitts, 1943). Based on the biological neuron, this model aims to estimate an output function y with n inputs, assigning different weights to each input and adding the result together with an independent value known as bias, u. An activation function f_θ is then set on the output (Jain et al., 1996).

The initial approach of using artificial neural networks (ANN) has evolved into more intricate structures with diverse nonlinear activation functions. Among the commonly used ANNs today, the multilayer perceptron (MLP) is prominent, which configuration is depicted in Fig. 5.7. The MLP follows a feed-forward structure, wherein input patterns are propagated through the hidden layers without internal feedback (Jove et al., 2020).

The MLP structure consists of an input layer, an output layer, and one or more hidden layers with interconnected neurons. Each layer has the following main features:

- Input layer: the number of neurons matches the number of features in the input pattern. This layer forwards the input vector to the first hidden layer.

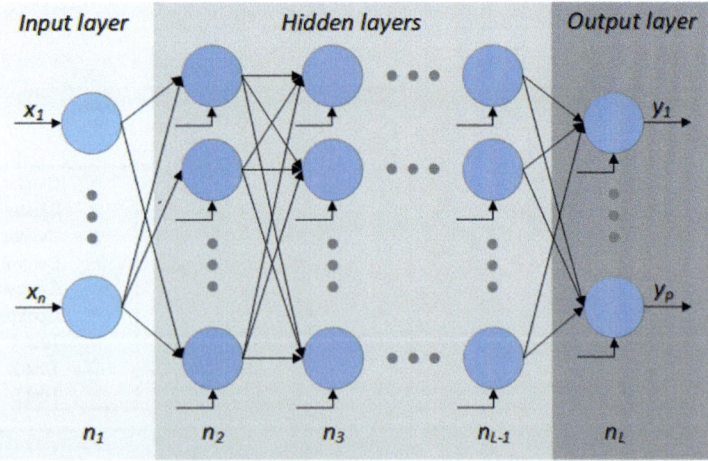

Figure 5.7 *MLP structure.* The typical structure of an MLP has one input layer, one output layer, and one or more hidden layers.

- Hidden layers: the number of neurons is variable and can be adjusted by the user.
- Output layer: the number of neurons matches the number of features in the output pattern.

Each neuron is connected to others through weighted links. Various activation functions can be configured for output neurons, such as hyperbolic tangent, sigmoid, gaussian, linear, or step functions. Hyperbolic tangent is commonly used in hidden layers for regression tasks, while linear activation function is typically applied at the output layer for regression problems.

Similar to the first artificial neuron, hidden and output layers neurons have bias inputs and receive products of previous layer neurons and their weighted connections. These inputs are aggregated and then passed through the activation function (Garcia et al., 2013; Gonzalez-Cava et al., 2018).

In the case of an MLP with one hidden layer, the output vector of the hidden layer h is computed through the next equation (Hemalatha & Rani, 2017).

$$h = f_{\theta 1}(x \cdot W_1 - u_1) \tag{5.2}$$

where $f_{\theta 1}$ defines the hidden layer activation function, x corresponds to the input vector, W_1 is the weight matrix from the input to the hidden layer, and u_1 represents the bias vector of the neurons in the hidden layer.

The output vector of the MLP y is obtained according to the following equation:

$$y = f_{\theta 2}(h \cdot W_2 - u_2) \tag{5.3}$$

where $f_{\theta 2}$ defines the activation function of the output layer, y represents the output of the hidden layer, W_2 is the weight matrix of the hidden layer to the output, and u_2 corresponds to the bias vector of the neurons of the output layer.

During the training step, the weight matrices and bias vectors are modified to minimize the difference between the output pattern and the MLP outcome (Rynkiewicz, 2012).

5.3.2.2 Support vector regression

The support vector machine (SVM) is a widely used machine learning method for classification that can also be employed for support vector regression (SVR) taks. The core concept of SVM involves mapping the data into a high-dimensional feature space F through a nonlinear transformation. Subsequently, the main goal is to identify a hyperplane that effectively separates the different classes.

The kernel is the transformation used to map data into a high-dimensional feature space. It might be polynomial, radial basis function (RBF), or Gaussian, as shown in the following equations (Wornyo & Shen, 2019).

$$k(x_i, x_j) = (ax_i^T x_j + b)^c; k(x_i, x_j) = e^{\frac{||x_i - x_j||}{\mu}}; k(x_i, x_j) = e^{\frac{||x_i - x_j||}{2\sigma^2}} \quad (5.4)$$

where $(x_i - x_j)$ define the feature vector of the input space and $a, b, c, \mu, \sigma \in R$ correspond to the kernel parameters.

When the SVM is applied to solve regression tasks, linear regression is implemented after the nonlinear transformation to the feature space.

In regression and classification problems, the least-square SVM (LSSVM) formulation successfully controls nonlinear Karush-Kun-Tucker systems. LSSVM training is easier than SVM training since it requires solving linear equations rather than quadratic programming to maximize the hyperplane parameter (Wornyo & Shen, 2019). LSSVM, like SVM, is also used for regression (LSSVR), with the regression function illustrated below.

$$y = f(X) = \omega^T \delta(x) + b$$

where x defines the training data sample $\in R^n$, y corresponds to the output $\in R$, δ is the high-dimensional mapping $R^n \to R^p$, and ω and b are the parameters of the hyperplane.

5.3.2.3 Ridge regression

Ridge regression was previously developed by Hoerl and Kennard (Hoerl & Kennard, 1970). Ridge regression mitigates specific issues associated with ordinary least squares (OLS) by applying a penalty on the magnitude of coefficients. In multicollinearity, OLS estimates are unbiased but exhibit significant variances, leading to potential deviation from the true value. By introducing a controlled level of bias to the regression estimates, ridge regression decreases the standard errors, with the aim of yielding a more reliable forecast overall.

A penalized residual sum of squares is minimized by the ridge coefficients: $min_\omega ||X_\omega - y||2^2 + \alpha||\omega||2^2$. Here, $\alpha \geq 0$ corresponds to a complexity parameter that governs a model's shrinkage degree. A higher value of α increases shrinkage, leading to more robust coefficients in collinearity. This parameter has been effectively used in various applications, including temperature and time series analysis and building energy consumption modeling.

5.3.2.4 Decision trees

The decision tree is a commonly used technique for both classification and regression tasks. This family model was first presented by Stone (Loh, 2011). Decision trees have been used for forecasting the weather, other meteorological phenomena, and energy use (Petre, 2009).

In this scenario, a variant known as classification and regression trees (CART) is employed, which generates a multiway tree by selecting the categorical feature at each node that yields the highest information gain for categorical targets. The CART model can also accommodate numerical target variables, making it suitable for regression tasks. It builds binary trees by identifying the feature and threshold that result in the most significant information gain at each node.

5.3.3 Time series

Instead of using the idea of considering the data as individual points, a very useful approach consists of considering the data as a series. Time series techniques aim to capture the model inertia by considering sample trends.

5.3.4 Autoregressive integrated moving average

Autoregressive integrated moving average (ARIMA), a classical model for time series analysis, was introduced by Box and Jenkins in 1970 (Box et al., 2015) and has gained widespread recognition. This technique has found applications in various fields, such as the economy, including stock market forecasting, supply chain management, and electricity price prediction, as well as in temperature analysis and natural phenomena forecasting (Iqelan, 2015), among other domains.

The ARIMA process comprises three key phases: an initial stage of data preparation, followed by parameter estimation and model selection, and finally, a model checking and forecasting stage. Data preparation involves applying transformations, such as square roots or logarithms, to stabilize the variance in time series data that exhibit changing variation with level. Subsequently, differencing is applied to the data to eliminate apparent patterns, such as trends or seasonality. Parameter estimation aims to determine the ARIMA model coefficients that best fit the data, while model checking involves testing the validity of the assumptions above. Once the model is selected, estimated, and checked, generating forecasts becomes relatively straightforward. In our case, the chosen model is applied to the test data, and the accuracy of the predictions is analyzed.

An ARIMA model is often denoted by $ARIMA(p, d, q)$ where:
- p corresponds to the order of the autoregressive part.
- d defines the order of the differentiation.
- q is the moving average process's order.

5.3.5 Time delay neural networks

The concept of neural networks with a temporal dimension was first introduced by (Waibel et al., 1989). These unique neural networks belong to the family of nonlinear autoregressive neural networks known as time delay neural networks (TDNN). TDNN architectures are designed to incorporate data from different time instances, making them suitable for time series analysis. They typically consist of a feed-forward schema with a single hidden layer.

In the input layer of TDNN, the number of step delays can be configured, allowing for training with a specific time step. The middle layer acts as a filter for the inputs before passing them to the output layer. How the output layer is connected back to the input layer determines the variants of this model in time series analysis.

Similar to other neural networks, TDNN receives information from the input layers and transmits it to the output layers through interconnected layers, resembling the perceptron architecture. However, the key difference is that TDNN incorporates a delay component. As a result, the input layers' older activation and connection values must be stored to analyze patterns or time-invariant structures. This is achieved by creating a copy of the previous values and their outgoing connections at each time step before updating the original values. TDNN has found wide applications in energy generation and consumption forecasting, among other areas (Sekine & Kim, 2007).

The next subsections provide an extended overview of the TDNN schema variations employed.

5.3.5.1 Nonlinear input-output

The focused time delay neural network (FTDNN) is a feed-forward network with a tapped delay line at the input. It belongs to the general class of dynamic networks known as focused networks, where the dynamics are limited to the input layer of a static multilayer feed-forward network. Although FTDNN is a simple model, it may not be one of the best-performing ones, as it lacks the inclusion of previous states of the dynamic system, which could provide valuable information for predicting future states.

Fig. 5.8 presents a schematic diagram of the FTDNN, where the first green square represents the input layer, the blue represents the hidden layers, and the green represents the output layer.

5.3.5.2 Nonlinear autoregressive
In this case, the network training simply uses the estimated series. The feed-forward network uses previous series readings to forecast future ones, and it also uses outputs created in earlier stages as inputs for subsequent states, as illustrated in Fig. 5.9. This enables the system to remember previous conditions, improving its prediction capability for future states. A delay ensures the network has sufficient dimensions to optimize its parallel computing capabilities. This involves feeding the network with numerous earlier stages of the series.

5.3.5.3 Nonlinear autoregressive with external input
The proposed model combines features from the two previous topologies but uniquely incorporates both past samples of the target time series to be predicted and previously generated outputs from the network. Additionally, it includes external attributes of the system that were sampled simultaneously with the analyzed series, Fig. 5.10. This approach enables the network to assimilate data that may not directly correlate with the modeled system but can still exert a discernible influence on it.

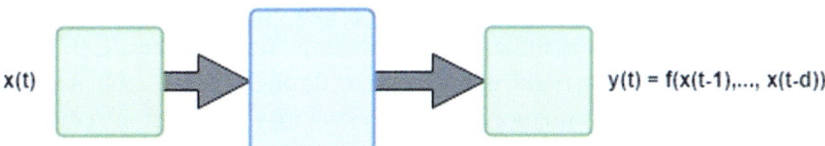

Figure 5.8 *NIO Scheme.* NIO scheme has three forwarded parts.

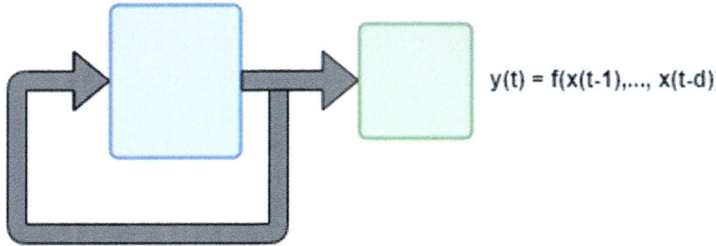

Figure 5.9 *NAR Scheme.* NAR scheme has two parts, with a feed-forward connection in the first one.

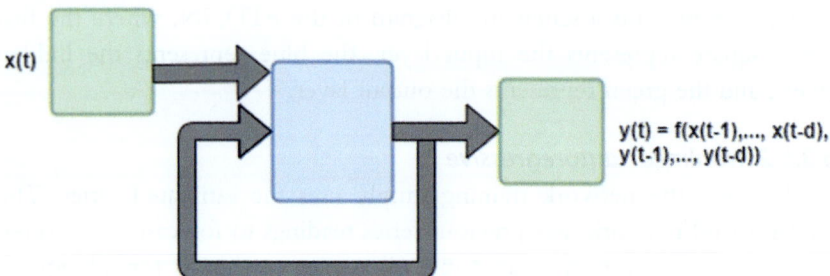

Figure 5.10 *NARX Scheme.* NAR scheme has three parts, with a feed-forward connection in the second one.

5.3.6 Long short-term memory

In short, a long short-term memory network (LSTM) is capable of holding a relevant piece of data in the input sequence and preserving it for several instants of time; therefore, it can have both short-term, like basic recurrent neural networks (RNN), and long-term memory. In this sense, they can add or remove information relevant to the processing of the sequence through the cell state.

Operationally, LSTM Networks have been developed to overcome the vanishing gradient problem in the standard RNN by improving the gradient flow within the network (Hochreiter & Schmidhuber, 1997). This was achieved by replacing the hidden layer with an LSTM unit. As shown in Fig. 5.11, an LSTM unit is composed of a set of gates and internal connections (Gers et al., 2000; Hochreiter & Schmidhuber, 1997).

The cell state $C(t)$ transfers information throughout the sequence and serves as the network's memory, to which data may be added or removed.

$$C(t) = \sigma(f(t) \odot C(t-1) + i(t)) \tag{5.5}$$

where $C(t)$ defines the cell state at the current time step t, σ indicates the sigmoid activation function, $f(t)$ corresponds the output of the forget gate, $C(t-1)$ is the cell state at previous time step $t-1$, and $i(t)$ defines the output of the input gate. The operator $\odot$ refers to Hadamard product (MILLION, 2007).

The forget gate $f(t)$, decides what is relevant to keep from previous time steps computing the sigmoid activation function of the weighted input sequence and feedback connection. Its value between 0 and 1, represents the extent to which the processed information can be dispensed with. Thus, values closer to 1 will be more likely neglected. With the

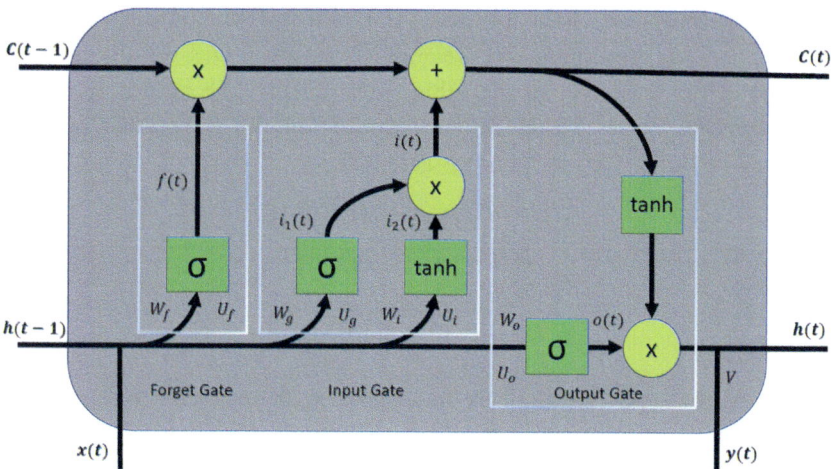

Figure 5.11 *A LSTM unit scheme.* U and W represent the weights of inputs and recurrent connections for the internal layers within each gate. The circle operators are componentwise.

output of the forget gate being $f(t)$ at the current time step t, $x(t)$ represent the inputs to the cell at the current time step t, and $h(t-1)$ the hidden state at the previous time step $t-1$ fed back into the cell. U and W are the weights of the input and recursive state, respectively.

$$f(t) = \sigma(x(t)U_f + h(t-1)W_f) \qquad (5.6)$$

The input gate $i(t)$, manages what information is relevant to add from the current time step. Likewise, it performs separately the sigmoid and hyperbolic tangent activation function of the weighted sequence of inputs and the previous hidden state $h(t-1)$. The first component, $i_1(t)$, represents the likelihood of preserving the processed data within the cell state, whereas $i_2(t)$, regulates the network, getting values between -1 and 1. In this way, both are the main components of the input gate $i(t)$ at current time step t, and along with the forget gate are responsible for updating the internal state of the cell, $C(t)$. Note that tanh refers to the hyperbolic tangent activation function.

$$i_1(t) = \sigma(x(t)U_i + h(t-1)W_i) \qquad (5.7)$$

$$i_2(t) = \tanh(x(t)U_g + h(t-1)W_g) \qquad (5.8)$$

$$i(t) = i_1(t) \odot i_2(t) \qquad (5.9)$$

The output gate, $o(t)$, computes the value of the output at current time step t, from merging the weighted inputs $x(t)$ and the hidden state $h(t-1)$ in the same way as the other gates. Finally, the output of the cell $h(t)$, concludes with the Hadamard product of the hyperbolic tangent-activated cell state $C(t)$ and output gate $o(t)$.

$$o(t) = \sigma(x(t)U_o + h(t-1)W_o) \tag{5.10}$$

$$h(t) = \tanh(C_t) \odot o(t)$$

5.3.7 GRU

GRU networks perform similarly to LSTMs, with an update and reset gate determining what information is sent to the output (Gers et al., 2000). These gates can be trained to selectively retain valuable information or discard irrelevant data from previous time steps in the prediction process. Fig. 5.12 summarizes the computation within a GRU unit (Chung et al., 2014).

The reset gate $r(t)$, decides how much of the information from the previous time steps can be forgotten by means of computing the sigmoid activation function. The concept is comparable to the forget gate $f(t)$ of an LSTM, where $x(t)$ represented the inputs to the cell at current time step t, and $h(t-1)$ the hidden state at the previous time step $t-1$ fed back into the cell. With U and W representing the weights of the input and recursive state, respectively.

$$r(t) = \sigma(x(t)U_r + h(t-1)W_r) \tag{5.11}$$

The update gate $z(t)$, selects how much of the information from the previous time steps must be saved, following an equation similar to the reset gate $r(t)$ equation, but with its own U and W weights. In this case, the level of activation determines the likelihood of retaining the data processed. Note that in Fig. 5.12, at this point, the connections follow the same path.

$$z(t) = \sigma(x(t)U_z + h(t-1)W_z) \tag{5.12}$$

The memory is responsible for keeping information along the entire sequence and represents the current $\left(\hat{h}\right)$, and the final memory (h) of the network. Both, dependent of the aforementioned gates and internal connections.

$$\hat{h}(t) = \tanh(x(t)U_h + (r(t) \odot h(t-1))W_h) \tag{5.13}$$

$$h(t) = (1 - z(t)) \odot h(t-1) + z(t) \odot \hat{h}(t)) \tag{5.14}$$

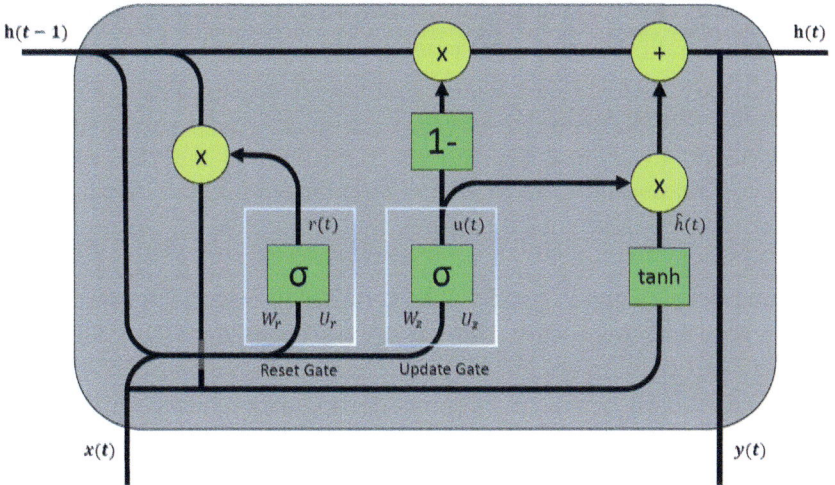

Figure 5.12 *A GRU unit scheme* . *U* and *W* represent the weights of inputs and recurrent connections for the internal layers within each gate. The circle operators are componentwise.

5.4 Approaches and results over cases studies

This section details the different approaches to achieve energy forecasting over the cases of study in Section 5.2. Each uses one or more techniques presented in Section 5.3, leading to accurate prediction.

5.4.1 Geothermal system

The main objective of the work shown in (Baruque et al., 2019) is to achieve an accurate prediction system for the output of a heat exchanger using previous readings from some of the sensors located along it. It is located on the power system described in Section 5.2. Different regression techniques have been used to obtain an accurate prediction model for these measures. These are:

- ARIMA: a well-known technique for time series analysis, which is detailed in Section 5.3. In the tests presented, ARIMA(0,0,1) and ARIMA(0,0,0) were used, with similar results.
- Ridge regression: is a technique that addresses ordinary least squares problems by imposing a penalty on the size of the coefficients, it is detailed in Section 5.3. In the case presented in the text, a Python implementation was used.

- Decision Trees: is a popular method for classification and regression (Section 5.3). The implementation used in this case is a variant called "CART," which creates multidirectional trees and supports numerical variables for regression tasks.
- Multilayer perceptron (MLP): supervised machine learning technique used in classification and regression tasks (Section 5.3). In this particular case, the Python implementation found in a specific reference has been used.
- Time delay neural networks (TDNN): TDNNs are particularly useful for modeling time series data, as they are capable of handling time delays between inputs and outputs. Three of the variants of this type of neural network are discussed: nonlinear input-output (NIO) (Fig. 5.8), nonlinear autoregressive (NAR) (Fig. 5.9), and nonlinear autoregressive with external input (NARX) (Fig. 5.10). These techniques have been detailed in Section 5.3.

The study carried out involved testing temperature readings from sensors located in a circuit and ground temperature sensor over a period of one year (2012), with readings taken every 10 minutes, resulting in a total of 52,645 data samples of 29 features. After discarding samples with errors, the final dataset consisted of 52,639 samples. The exploratory analysis showed that some sensors were highly correlated with each other, and the remaining efforts were focused on predicting the readings of a particular sensor to anticipate the system's future state. Linear Discriminant Analysis was performed, and a clear cluster structure was obtained from this analysis, classifying 95.1% of the cases correctly (Fig. 5.13).

Two different batches of data were used for the experiments, with the first batch using the complete set of readings to train different regression models and the second batch using only readings when the pump was not turned off in a steady state. The aim was to assess whether using all data was necessary to build an accurate prediction system or if fewer samples would be enough. The objective was to predict the temperature registered on sensor S28, given different subsets of data selected from the dataset. Several types of models were used, including statistical models such as Decision Tree or Ridge Regression, more accurate models such as the multilayer perceptron (MLP), and more advanced connectionist models on the family of the time-dependent neural networks (TDNN). Parameters were adjusted by a random search method.

In the experiments, different regression models were compared to model the future behavior of the system, taking into account the previous

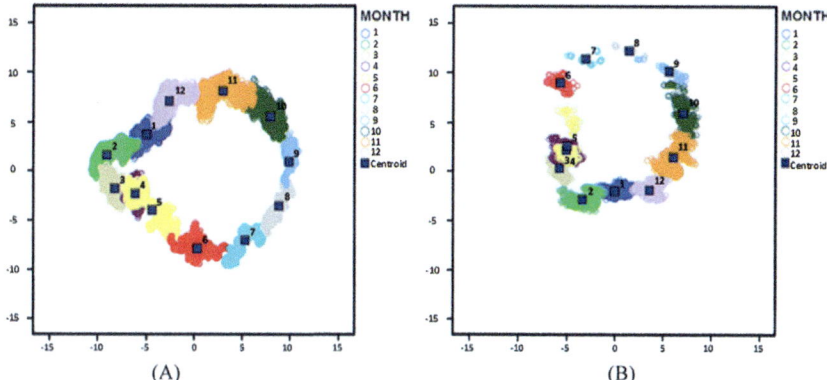

Figure 5.13 *Cluster distribution obtained from LDA (A) with the complete dataset (separated by months) and (B) with only the samples when the pump is working (separated by months).* A clear distribution by months is represented.

states of the same system. To do this, in all experiments, the dataset was divided into 12 folds of contiguous samples in time. Fig. 5.14 summarizes the configuration of each model.

Two sets of experiments were performed to assess the impact of varying data amounts used for training a regression model. In one set of experiments, the model was trained solely on a specific fold and tested on the subsequent one in chronological order. In the other set, the model was trained on all available data up to a given time point and tested on the subsequent fold in chronological order.

In the first of the data sets, the results are calculated based on the available data. The results obtained are shown in Table 5.1 and Fig. 5.15, where bold rows indicate the models with lowest MAE.

When using the data summary to train the models, the error tends to decrease for the TDNN models, while it stays about the same for the other models. For this reason, data from different years are included in the training sets. Figs. 5.16 and 5.17 show the values obtained from each model together with the real value to be predicted.

The network's ability to generalize improves when analyzing data from several months earlier through training. Conversely, the Ridge model displays different outcomes: using data from only one month may result in less overfitting, despite less accurate regression in both comparisons. This suggests that less sophisticated models benefit from using less recent data.

After deleting the data when the pump is off in a steady state, the error measurements are recalculated. In this recalculation of errors, the results

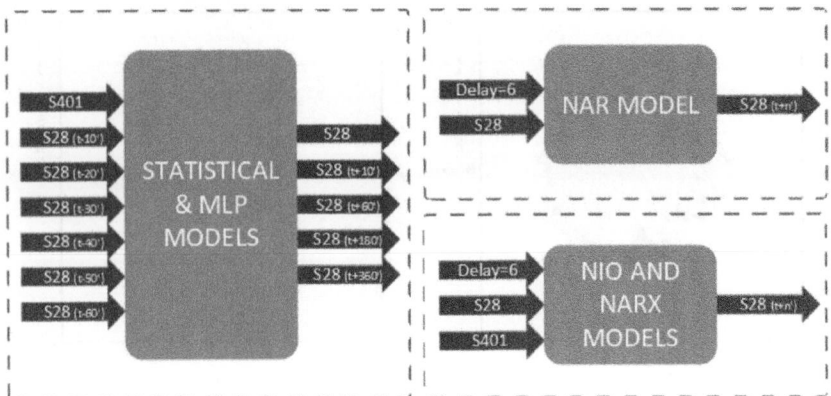

Figure 5.14 *Model configuration depending on the technique applied.* The number of inputs varies depending on the technique.

Table 5.1 Error measures obtained for the studied regression models. Both data samples gathered when the exchanger is functioning and not functioning are used for the error calculation.

Time horiz.	Model	Trained by month				Trained with the accumulated			
		MAE	MSE	MDAE	R^2	MAE	MSE	MDAE	R^2
1 h	Ridge	0.61	1.26	0.23	0.96	0.70	1.49	0.38	0.94
	ARIMA	1.00	3.83	0.37	0.86	1.36	6.75	0.56	0.76
	DectTree	0.95	2.40	0.36	0.91	0.97	2.74	0.32	0.90
	MLP	0.70	1.57	0.25	0.94	0.97	1.98	0.41	0.93
	NAR	**0.18**	**0.25**	**0.06**	**0.99**	**0.23**	**0.30**	**0.11**	**0.99**
	NARX	1.60	6.34	0.7.8	0.77	0.36	0.36	0.26	0.99
3 h	Ridge	1.03	3.01	0.49	0.89	0.96	2.57	0.52	0.91
	ARIMA	1.86	8.60	0.83	0.66	2.25	8.66	1.60	0.66
	DectTree	1.91	10.62	0.81	0.62	1.73	9.44	0.59	0.66
	MLP	1.18	3.23	0.61	0.88	1.25	3.16	0.77	0.89
	NAR	**0.20**	**0.27**	**0.06**	**0.99**	**0.22**	**0.26**	**0.10**	**0.99**
	NARX	1.02	3.22	0.43	0.88	0.23	0.30	0.15	0.99
6 h	Ridge	1	5.86	0.68	0.79	1.52	5.86	0.87	0.79
	ARIMA	17.67	14.34	18.12	0.49	3.20	15.02	2.60	0.47
	DectTree	2.75	17.91	1.31	0.36	2.32	14.49	0.95	0.48
	MLP	2.11	10.40	1.81	0.63	1.91	9.13	1.01	0.67
	NAR	**0.18**	**0.29**	**0.06**	**0.99**	**0.28**	**0.53**	**0.09**	**0.98**
	NARX	0.74	1.81	0.30	0.93	0.52	0.63	0.30	0.98

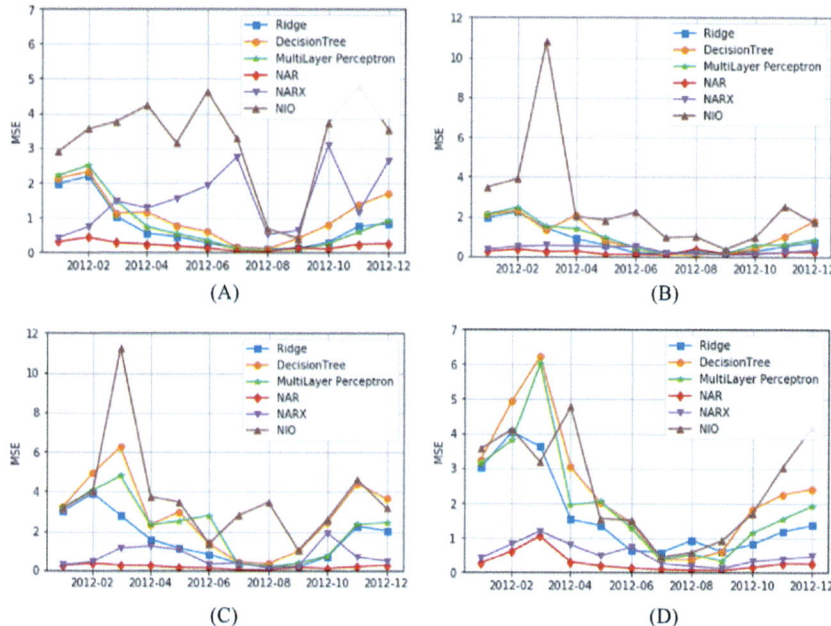

Figure 5.15 *MSE obtained for each of the compared models in each month of the year. (A) MSE for 1-hour time horizon, when models are trained only with data from the previous month. (B) MSE for 1-hour time horizon, when models are trained with all data available up to each month. (C) MSE for 6-hour time horizon, when models are trained only with data from the previous month. (D) MSE for 6-hour time horizon, when models are trained with all data available up to each month.* The figure represents the forecasted value for different scenarios.

obtained are the same as those shown in Table 5.1. Therefore, the subset of the predictions made when the pump is off in a steady state is also filtered. The error metrics are calculated again and the results obtained are shown in Table 5.2.

The final experiment involved retraining the models using the same methodology as the initial experiments but included only those samples in each fold where the pump remained in a steady state without being turned off. The results obtained when the models with this subset are shown in Table 5.3.

All results show a significantly larger error compared to when the models are trained on the entire dataset, even when considering only data samples when the pump is not stopped. The ARIMA model could not be computed with this set of data settings, as it needs data evenly spaced in time to be trained.

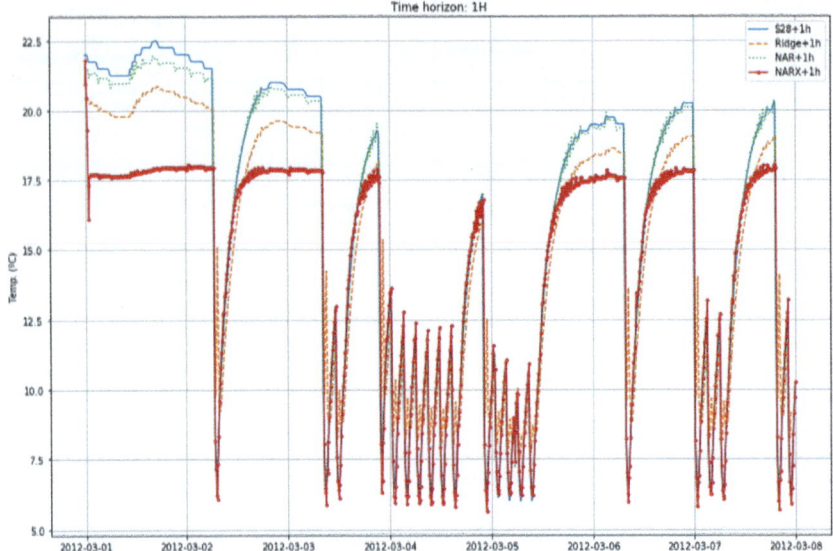

Figure 5.16 *Example of the regression values obtained for each of the models. Results when models are trained with data obtained from all previous time instants.* The figure represents the forecasted values on a test series.

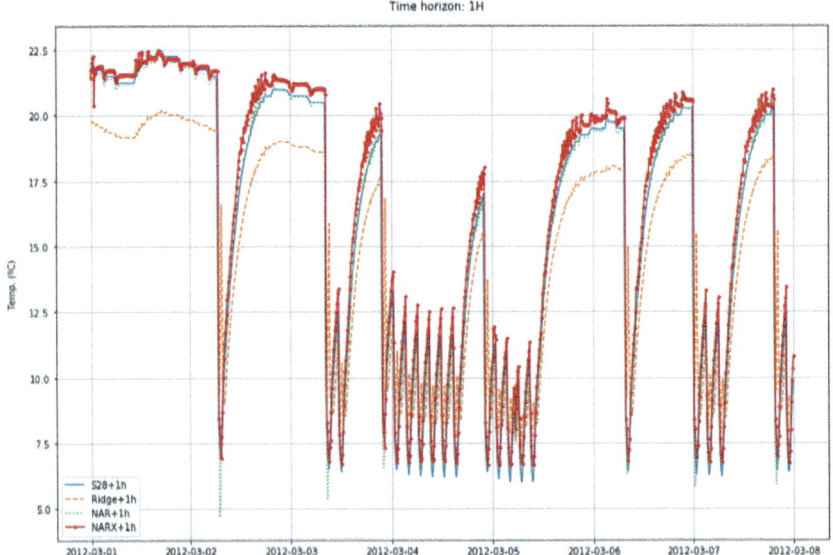

Figure 5.17 *Example of the regression values obtained for each of the models. Results when models are trained with data obtained only from the previous month.* The figure represents the forecasted values on a test series.

Table 5.2 Error measures obtained for the studied regression models. Only samples of the dataset where the heat pump is working are used for the error calculation.

Time horiz.	Model	Trained by month				Trained with the accumulated			
		MAE	MSE	MDAE	R^2	MAE	MSE	MDAE	R^2
1 h	Ridge	1.41	3.62	1.09	0.67	1.69	4.72	1.31	0.57
	ARIMA	3.97	24.21	3.22	0.17	3.84	22.17	3.21	0.17
	DectTree	2.00	6.30	1.70	0.43	1.93	5.90	1.48	0.46
	MLP	1.43	3.82	1.06	0.65	1.92	5.61	1.66	0.49
	NAR	**0.59**	**1.63**	**0.18**	**0.85**	**0.60**	**1.66**	**0.21**	**0.85**
	NARX	2.13	11.17	0.94	−0.02	0.73	1.68	0.40	0.85
3 h	Ridge	2.12	6.45	2.02	0.413	1.56	3.73	1.48	0.66
	ARIMA	4.52	30.03	3.78	0.18	4.13	25.66	3.46	0.06
	DectTree	3.26	15.26	3.09	−0.39	2.60	9.69	2.41	0.12
	MLP	2.12	7.13	1.76	0.35	1.77	4.67	1.70	0.57
	NAR	**0.59**	**1.55**	**0.23**	**0.86**	**0.62**	**1.62**	**0.25**	**0.85**
	NARX	1.20	3.34	0.79	0.69	0.69	1.59	0.37	0.85
6 h	Ridge	2.72	10.05	2.77	0.08	2.44	8.17	2.45	0.25
	ARIMA	3.98	22.62	3.56	−1.06	3.40	15.77	3.35	−0.43
	DectTree	14.77	35.50	14.74	0.14	4.29	26.54	3.86	0.13
	MLP	3.18	14.85	2.90	−0.35	2.94	11.78	2.85	0.07
	NAR	**0.63**	**1.78**	**0.23**	**0.84**	**0.56**	**1.51**	**0.19**	**0.86**
	NARX	1.06	2.85	0.63	0.74	0.79	1.78	0.42	0.84

Table 5.3 Error measures obtained for the studied regression models. Only on samples of the dataset where the heat pump is working, both for training and for testing.

Time horiz.	Model	Trained by month				Trained with the accumulated			
		MAE	MSE	MDAE	R^2	MAE	MSE	MDAE	R^2
1 h	Ridge	**1.72**	**4.45**	**1.52**	**0.58**	**1.79**	**4.92**	**1.61**	**0.54**
	ARIMA	−	−	−	−	−	−	−	−
	DectTree	1.81	4.77	1.70	0.55	1.86	5.50	1.68	0.48
	MLP	1.96	5.98	1.71	0.44	1.94	6.07	1.68	0.43
	NAR	2.29	10.47	1.67	0.02	2.16	9.09	1.53	0.14
	NARX	2.89	13.42	2.32	−0.26	2.51	14.25	1.86	−0.33

One of the initial findings suggests that system modeling can be achieved with a minimal number of sensor readings, with only two being sufficient, as long as the measurements are taken at a reasonable frequency. In addition, multiple modeling experiments were conducted, resulting in an average prediction error of less than 1°C over a full year. Another finding is that the system performs well regardless of whether the pump is running or not, and the user is unaware of this as the system adjusts its behavior without requiring such information. Time-dependent neural networks (TDNN) have proven to be the most effective models as they are specifically designed to predict future states by analyzing differences between past states of the data based on variations over time.

5.4.2 Solar thermal system

The work shown in (Casteleiro-Roca et al., 2020) aims to forecast solar thermal energy to optimize the energy system described in Section 5.2. The predictive model presents the topology shown in Fig. 5.18. The input variables are the heat carrier fluid flow rate (etilenglicol), the input temperature at the two strings shown in Fig. 5.3, and the radiation measured at the rooftop.

The approach of this work is based on the use of regression techniques. For improving prediction accuracy, these techniques are combined with a clustering algorithm in charge of dividing the dataset into different subsets with similar behavior. Then, the regression technique is applied to each subset, implementing a hybrid model topology, as shown in Fig. 5.19.

The process employed to train the hybrid intelligent model is depicted in Fig. 5.20. Initially, clusters were created, followed by a regression phase for each subset. Next, the regression models were trained using K-fold cross-validation, where each data cluster was divided into k groups for training and testing. After testing all the k models, the entire cluster data was used for testing, and the error of the specific regression technique was calculated with all the data. In the third step, the best regression algorithm was selected for each cluster, taking into account the error obtained during the training phase.

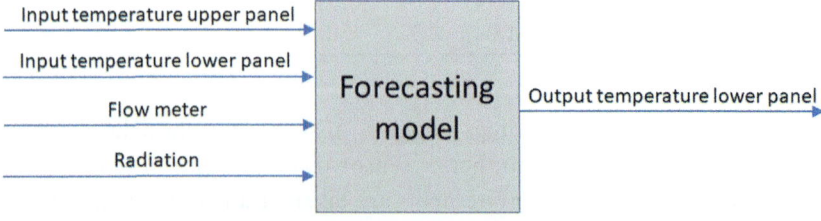

Figure 5.18 *General topology of forecasting model.* The model has four inputs and one output.

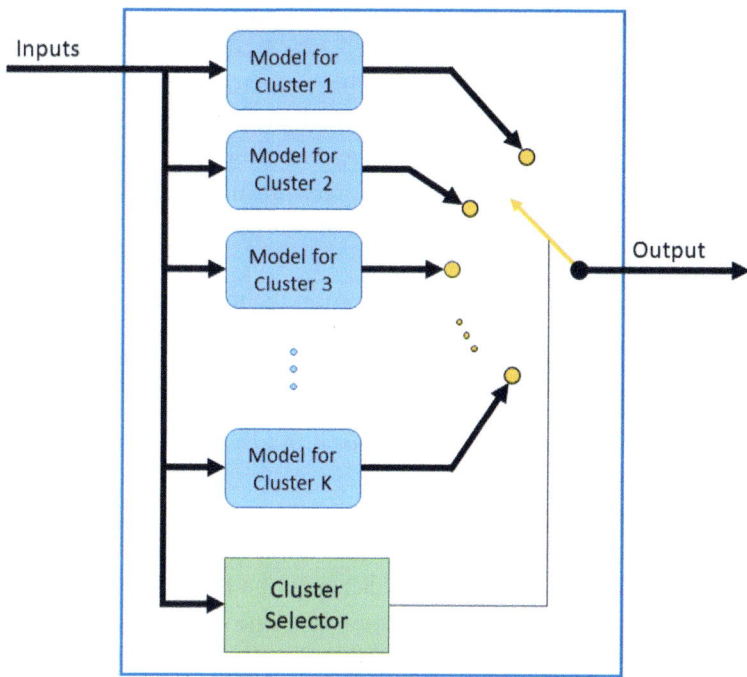

Figure 5.19 *Example of a hybrid model.* The dataset is divided into different clusters. The model belonging to each cluster is selected depending on the input sample.

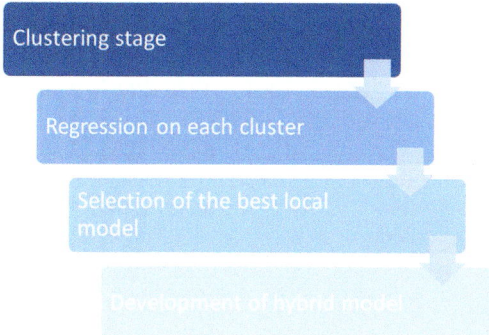

Figure 5.20 *Process followed to achieve a hybrid model.* The hybrid model procedure is divided into four stages.

Furthermore, since different regression algorithms were used, choosing the best one among them was necessary. These regression techniques had multiple tuning parameters, resulting in different algorithm configurations.

The last step in creating the hybrid model involved selecting the best hybrid setup. In accomplishing this, a separate dataset was isolated from the initial regression phase to test all the hybrid structures. The best configuration was chosen based on the error obtained with this validation dataset.

During the experiments, the dataset, comprised of 24983 samples, is divided into a variable number of clusters: from one to 10. Hence, 55 different models can be implemented, and the number of samples of each group is presented in Table 5.4

On each subset, three different regression techniques are applied: MLP, SVR, and polynomial regression. Each technique was, further, tested with different hyperparameters, such as the number of neurons. After computing the errors achieved, the topology with the lowest MSE is the one that divides the model internally into nine local models, with the configuration shown in Fig. 5.21.

Finally, the model is validated with different randomly selected subsets that were not included during the validation stage. Fig. 5.22 represents two examples where the output temperature predicted by the model is represented with green dashed lines and the real temperature in a blue continuous line.

Table 5.4 Number of samples for each cluster.

Number of clusters →	1	2	3	4	5	6	7	8	9	10
Samples in Cl 1	24983	6875	5701	653	652	619	616	615	613	611
Samples in Cl 2		18109	6988	5063	4287	2405	2316	1817	1463	1314
Samples in Cl 3			12295	7001	4928	2687	2349	2306	1899	1446
Samples in Cl 4				12267	6496	4251	2646	2333	2195	2088
Samples in Cl 5					8617	6499	2668	2633	2271	2244
Samples in Cl 6						8522	6361	4379	2643	2246
Samples in Cl 7							8029	4757	2626	2635
Samples in Cl 8								6144	4117	2708
Samples in Cl 9									6146	4171
Samples in Cl 10										5520

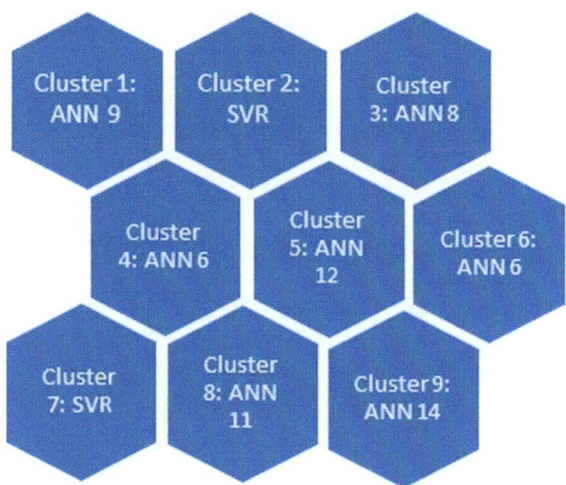

Figure 5.21 *Final hybrid model configuration.* The final model has two SVR local models and seven ANN models.

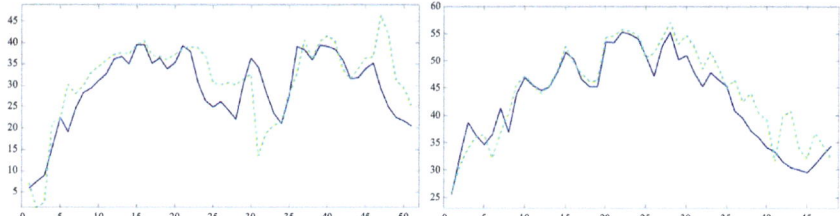

Figure 5.22 *Representation two validation sets.* During two time series tests, the forecasted value is near the real value.

5.4.3 Wind turbine

The work shown in (Zayas-Gato et al., 2022) aims to predict the active power generated by a low-power wind turbine in the bioclimatic house described in Section 5.2. The prediction model has the topology shown in Fig. 5.23. The input variables are the wind speed at two different levels: ground level and 10 meters, the wind gust speed at 10 meters and the active power.

The approach of this work is based on the use of regression techniques. For improving prediction accuracy, these techniques are, however, combined with a clustering algorithm in charge of dividing the dataset into different subsets with similar behavior. Then, the regression

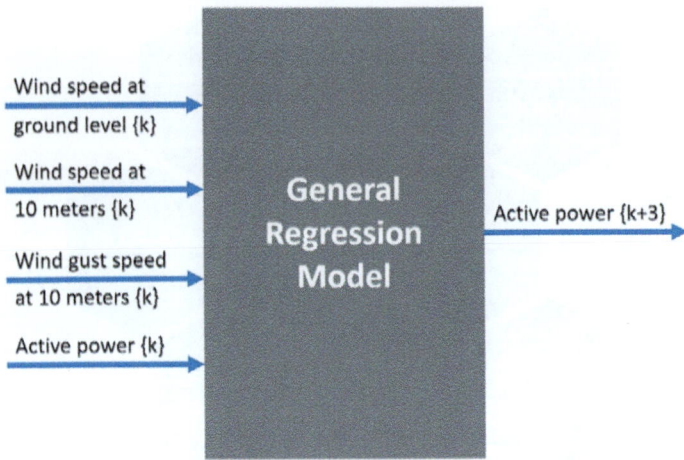

Figure 5.23 *Model approach.* The model has four inputs and one output.

technique is applied to each subset, implementing a hybrid model topology, as shown in Fig. 5.19.

Fig. 5.24 shows a diagram of the K-fold cross-validation procedure. The data within each cluster is divided k times, where k-1 subsets are added to create the training data, and the remaining subset is used for the test data. The model is trained using the training data and tested using the test data, with both actual and predicted output variables being stored. This process is repeated until all subsets have been used for testing, and then the error value is calculated using all samples in the cluster. The next step involves selecting the best local model for each cluster, which is determined by using the error value calculated with K-fold cross-validation for all regression algorithms and internal configurations. As the local models do not use all the cluster data, it is necessary to retrain all the models once the regression algorithms have been selected. Finally, the last step in creating the hybrid model is selecting the hybrid topology.

During the experiments, the dataset, comprised of 30,272 samples, is divided into a variable number of clusters: from one to 10. Hence, 55 different models can be implemented, and the number of samples of each group is presented in Table 5.5.

On each subset, three different regression techniques are applied: MLP, SVR, and polynomial regression. Each technique was, further, tested with different hyperparameters, such as the number of neurons. After computing the errors achieved, the best results have been obtained

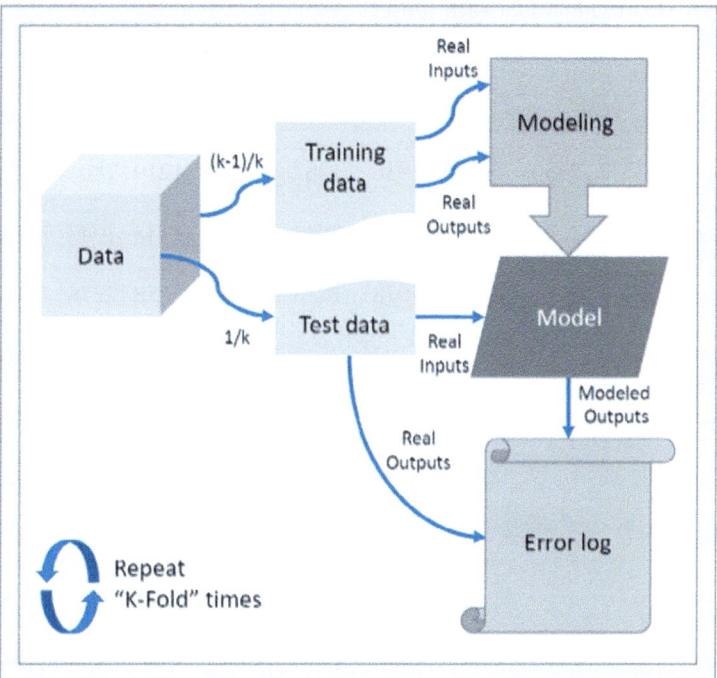

Figure 5.24 *K-Fold training and test data selection.* The K-Fold procedure divides the data into K groups, using 1/K for test and the remaining for training. It is repeated K times.

with a hybrid model with four clusters, which achieve the results in Table 5.6. This hybrid model includes four local models with ANN with one, two, three, and four neurons for the specific clusters.

Finally, the model is validated with a different randomly selected subset that was not included during the validation stage. Fig. 5.25 represents the predicted power generated by the model, which is represented with green dashed lines and the real power in a blue continuous line.

5.4.4 Batteries

The work shown in (Caínzos López et al., 2022) is to develop an intelligent model that can predict the state of charge (SOC) of power stacks. The model was generated using a dataset that includes all the operating points observed during a capacity validation test conducted in an actual system.

Table 5.5 No of samples for each subset.

Number of clusters →	1	2	3	4	5	6	7	8	9	10
Samples in Cl 1	30272	7850	3694	1916	1282	1219	1219	780	546	485
Samples in Cl 2		22422	11286	5749	3644	1621	1148	1047	727	703
Samples in Cl 3			15293	10787	7218	2391	3258	2276	1769	1719
Samples in Cl 4				11821	7541	6382	4124	3670	2140	2004
Samples in Cl 5					10587	7436	4423	4367	3433	2008
Samples in Cl 6						10324	6589	4403	3677	3196
Samples in Cl 7							9603	5280	4387	3705
Samples in Cl 8								8550	5238	4446
Samples in Cl 9									8356	4525
Samples in Cl 10										7482

Table 5.6 Performance values for the best hybrid configuration.

No. of configuration models	MSE	MAE
4 configurations	$1.261e + 4$	73.822
(with normalized values)	0.0033	0.0376

The suggested models will provide forecasts for battery capacity as single output predictions or multiple time step predictions, which are categorized into two different approaches—single output (SS) and autoregressive (AR)—as illustrated in Fig. 5.26.

The process for developing reliable prediction models for the incoming data is summarized in Fig. 5.27. The model configurations presented in Fig. 5.26 were assessed using the cross-validation (CV) resampling technique to determine long-term generalization, with the mean square error (MSE) metric being used as the objective function to minimize.

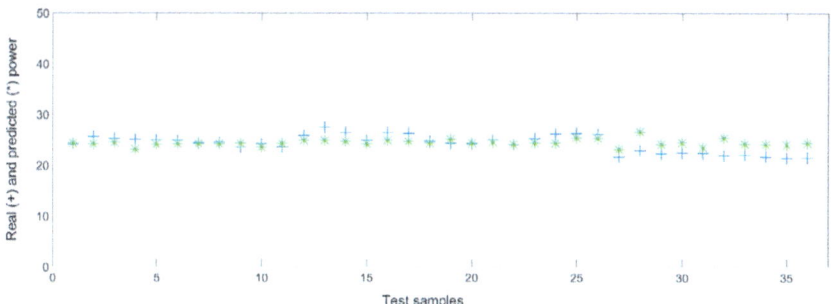

Figure 5.25 *Test results.* The forecasted values of power present very low error.

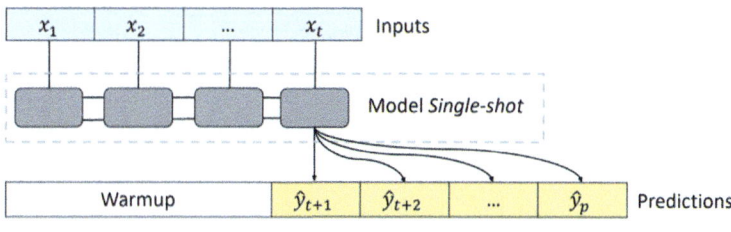

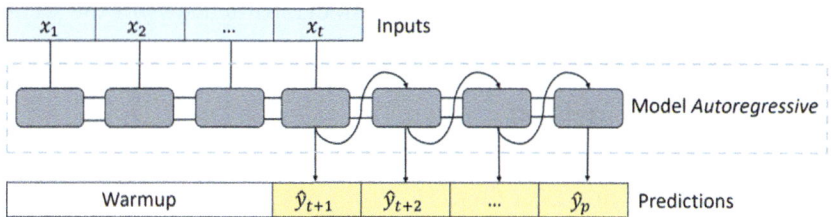

Figure 5.26 *Model configuration.* Model configurations are considered to forecast multiple time steps: single-shot (SS) and autoregressive (AR).

The optimal hyperparameters were used to create the ultimate models, which were then trained on the complete set of training data. Multiple simulations were conducted using the previously mentioned hyperparameters, which include 15, 30, 60, and 120 samples. In real time, this corresponds to retrieving historical data and predicting future trends at intervals of 15 seconds, 30 seconds, 1 minute, and 2 minutes, based on the 1 Hertz sampling frequency.

Looking at Fig. 5.28, the SOC value to confirm the battery capacity can be predicted in real time.

As depicted in Fig. 5.29, it is apparent that the SS method outperforms the AR approach, particularly for larger window sizes, with an evident

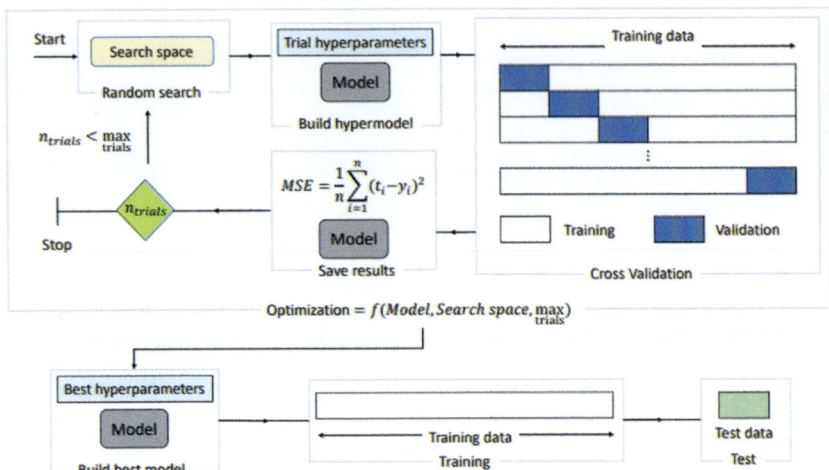

Figure 5.27 *Process to develop prediction* optimization, training, and performance stage procedure for the models evaluated for each configuration and case of study.

inflection point around the 60-second mark. The assessment of performance metrics such as MAE, MSE, MAPE, and R^2 demonstrates that the SS model performs superiorly to the AR model, particularly for the wider window ranges.

5.4.5 Hydrogen fuel cell

The work shown in (Alaiz-Moretón et al., 2019) aims to predict the hydrogen consumption of a fuel cell in an energy storage system as the one described in Section 5.2. The prediction model has the topology shown in Fig. 5.30. The input variables are the power generated, the hydrogen inlet flow, and the power desired.

The approach of this work is based on the use of regression techniques. For improving prediction accuracy, these techniques are combined with a clustering algorithm in charge of dividing the dataset into different subsets with similar behavior. Then, the regression technique is applied to each subset, implementing a hybrid model topology, as shown in Fig. 5.19.

The diagram followed to create the hybrid model is shown in Fig. 5.20. To perform the third step, the selection of the best local model, K-fold cross-validation is used to split the data subsets (cluster data) for training and testing.

During the experiments, the dataset, comprised of 187 samples, is divided into a variable number of clusters: from two to 9. Table 5.7 displays the

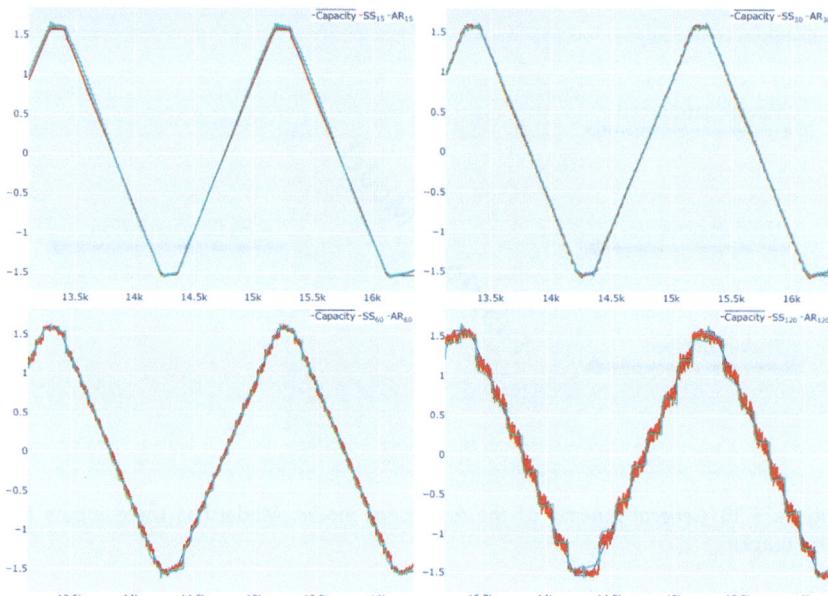

Figure 5.28 *Results of the prediction* predicted scaled values for the capacity of the battery over the hold-out test set for the SS and AR model configurations in each window scenario.

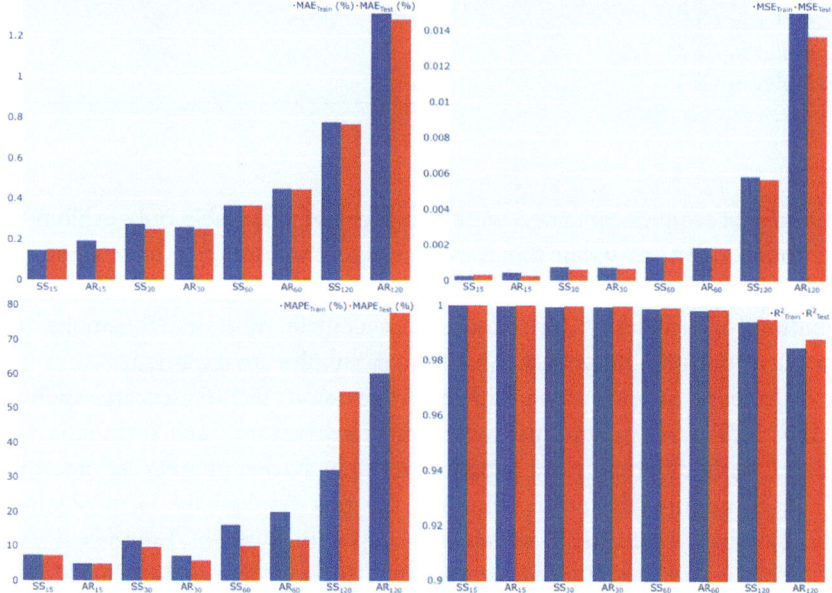

Figure 5.29 *Bar chart representing metric results for model and window case of study.* The SS method outperforms the AR approach, particularly for larger window sizes.

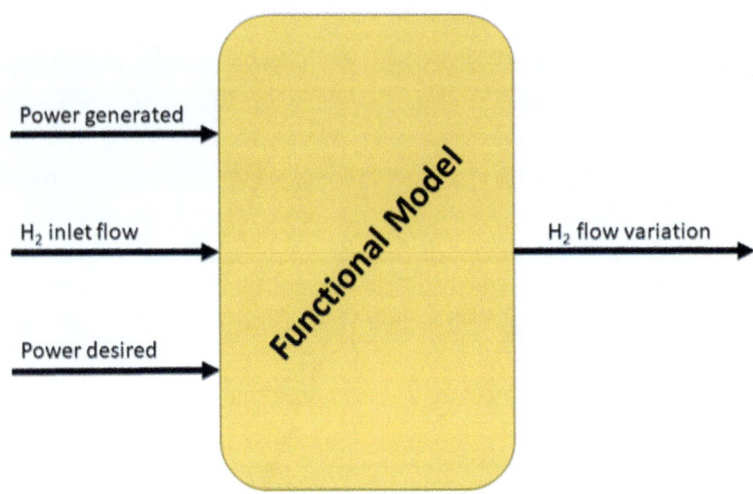

Figure 5.30 *General schema of the functional model.* Model has three inputs and one output.

Table 5.7 Number of samples in each created cluster.

Number of clusters →	Global	2	3	4
Samples in Cl 1	187	92	52	42
Samples in Cl 2		95	67	45
Samples in Cl 3			68	47
Samples in Cl 4				53

number of samples contained within each cluster. The table only exhibits up to four clusters, as when the K-Means algorithm attempts to separate the data into additional groups, the resulting clusters end up with an insufficient amount of samples. Only clusters that contain at least 15 samples are retained, whereas the ones that have fewer samples are discarded.

On each subset, three different regression techniques are applied: MLP, SVR, and polynomial regression; furthermore, each technique was tested with different hyperparameters, such as the number of neurons. After computing the errors achieved, the best results have been obtained with a best hybrid model, which is created with three local models.

The hybrid model is validated with a different randomly selected subset that was not included during the validation stage. Table 5.8 shows the numerical results that validate the model.

5.4.6 Demand forecasting

The work shown (Casteleiro-Roca et al., 2019) is a new methodology to predict the energy load in a hotel based on intelligent techniques. The hybrid model includes as inputs the own information on energy demand, occupancy rate, and temperature. The hybrid model has the topology shown in Fig. 5.31. Validation was performed using real hotel data and compared with time series models.

The approach of this work is based on the use of regression techniques. For improving prediction accuracy, these techniques are, however, combined with a clustering algorithm in charge of dividing the dataset into different subsets with similar behavior. Then, the regression technique is applied to each subset, implementing a hybrid model topology, as shown in Fig. 5.19.

The training of the regression algorithm was made using K-fold cross-validation. Fig. 5.24 shows the training phase generally. This step is repeated 16 times for each cluster; 15 different configurations of the MLP (with different neurons number in the hidden layer) and LSSVR.

During the experiments, the dataset, comprised of 354 samples, is divided into a variable number of clusters: from one to 10. Table 5.9 displays the number of samples contained within each cluster. The outcome of the algorithm produced seven distinct clusters. There were two options evaluated: using a universal model that incorporates all the data or implementing six specific models that partition the data into two to seven groups.

Table 5.8 Mean squared error for each model. Validate results.

Metric	Global	Hybrid model (local models)		
		2	3	4
MSE	24.0758	33.8591	23.8121	398.9072

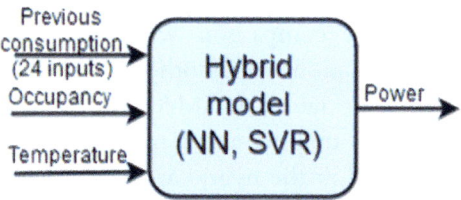

Figure 5.31 *Hybrid model to predict the next day power.* Model has three inputs and one output.

Table 5.9 Number of samples per cluster in the model to predict the power at 15:00.

Number of clusters →	Global	2	3	4	5	6	7
Samples in Cl 1	354	147	91	52	40	36	21
Samples in Cl 2		207	91	65	43	43	39
Samples in Cl 3			172	97	65	59	47
Samples in Cl 4				140	87	60	57
Samples in Cl 5					119	76	60
Samples in Cl 6						80	61
Samples in Cl 7							69

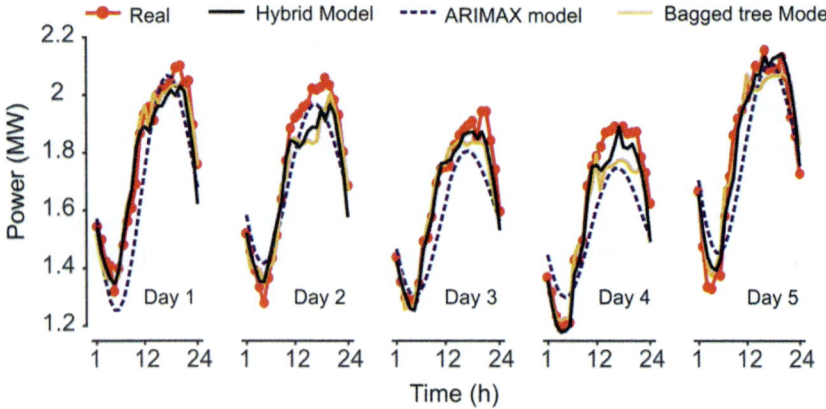

Figure 5.32 *Predicted power demand by the model.* Validation of the hybrid intelligent model, the ARIMAX model, and the bagged decision tree model.

On each subset, different regression techniques are applied. Each technique was further tested with different hyperparameters, such as the number of neurons. After computing the errors achieved, the best results have been obtained with the hybrid model.

For validation of the results, a 24-h forecast of the 5 days of validation with the hybrid model is performed. Fig. 5.32 (solid continuous line).

The study includes a comparison with two forecasting methods: ARIMAX and bagged decision trees. Both methods show acceptable performance with low error rates. ARIMAX offers higher performance, while the bagged decision tree presents a much better mean square error; however, when compared to the hybrid approach in this paper, as can be seen, the hybrid clearly outperforms the ARIMAX approach and performs better than the bagged decision tree method.

5.5 Conclusion

The use of renewable and sustainable energies is playing a key role and is being promoted by governments of many countries, as well as private institutions. This fact, together with the current development of new methods and techniques of artificial intelligence, has caused a great wave of research studies in which intelligent techniques are proposed and applied to solve problems or tasks related to many aspects related to the generation, use, or distribution of all types of renewable energies.

Two interesting concepts have been fed back to this world of renewable energies. By conducting various studies on the different energies available in the bioclimatic house under study, several prediction models for the different renewable energies corresponding to each one have been obtained, which can be merged in order to give rise to and relate to the second concept of feedback, the energy storage systems.

These prediction models for the three available energies: solar, geothermal, and wind, have prompted the need to store this produced energy, in this case, in batteries and a hydrogen fuel cell. For these two storage systems, prediction models have been obtained, which, together with the previous ones, can lead to a brilliant combination.

As a final conclusion, what has been sought is to intimately relate, through individual study and the creation of prediction models, the renewable energies present in the bioclimatic house with energy storage systems.

Acknowledgments

Álvaro Michelena's research was supported by the Spanish Ministry of Universities (https://www.universidades.gob.es/) under the "Formación de Profesorado Universitario" grant with reference FPU21/00932.

Míriam Timiraos's research was supported by the *Xunta de Galicia* (Regional Government of Galicia) through grants to industrial Ph.D. (http://gain.xunta.gal) under the *Doutoramento Industrial 2022* grant with reference: 04_IN606D_2022_2692965.

CITIC, as a Research Center of the University System of Galicia, is funded by Consellería de Educación, Universidade e Formación Profesional of the Xunta de Galicia through the European Regional Development Fund (ERDF) and the Secretaría Xeral de Universidades (Ref. ED431G 2019/01).

References

Alaiz-Moretón, H., Jove, E., Casteleiro-Roca, J. L., Quintián, H., García, H. L., Benítez-Andrades, J. A., Novais, P., & Calvo-Rolle, J. L. (2019). Bioinspired hybrid model to predict the hydrogen inlet fuel cell flow change of an energy storage system. *Processes*, 7(11). Available from https://doi.org/10.3390/pr7110825.

Amin, M. (2013). The smart-grid solution. *Nature*, *499*(7457), 145−147. Available from https://doi.org/10.1038/499145a.

Andújar, J. M., & Segura, F. (2009). Fuel cells: History and updating. A walk along two centuries. *Renewable and Sustainable Energy Reviews*, *13*(9), 2309−2322. Available from https://doi.org/10.1016/j.rser.2009.03.015.

Baruque, B., Porras, S., Jove, E., & Calvo-Rolle, J. L. (2019). Geothermal heat exchanger energy prediction based on time series and monitoring sensors optimization. *Energy*, *171*, 49−60. Available from https://doi.org/10.1016/j.energy.2018.12.207.

Box, G. E., Jenkins, G. M., Reinsel, G. C., & Ljung, G. M. (2015). *Time series analysis: Forecasting and control*. John Wiley & Sons.

Boyle, G. (2004). *Renewable energy: Power for a sustainable future*. The Open University.

Casteleiro-Roca, J.-L., Calvo-Rolle, J., Méndez Pérez, J., Roqueñí Gutiérrez, N., & de Cos Juez, F. (2017). Hybrid intelligent system to perform fault detection on BIS sensor during surgeries. *Sensors*, *17*(12), 179. Available from https://doi.org/10.3390/s17010179.

Casteleiro-Roca, J. L., Chamoso, P., Jove, E., González-Briones, A., Quintián, H., Fernández-Ibáñez, M. I., Vega, R. A. V., Piñón Pazos, A. J., López Vázquez, J. A., Torres-Álvarez, S., Pinto, T., & Calvo-Rolle, J. L. (2020). Solar thermal collector output temperature prediction by hybrid intelligent model for smartgrid and smartbuildings applications and optimization. *Applied Sciences (Switzerland)*, *10*(13). Available from https://doi.org/10.3390/app10134644.

Casteleiro-Roca, J.-L., Gómez-González, J., Calvo-Rolle, J., Jove, E., Quintián, H., Gonzalez Diaz, B., & Mendez Perez, J. (2019). Short-term energy demand forecast in hotels using hybrid intelligent modeling. *Sensors*, *19*(11), 2485. Available from https://doi.org/10.3390/s19112485.

Caínzos López, V., Jove, E., Zayas Gato, F., Pinto-Santos, F., Piñómn-Pazos, A. J., Casteleiro-Roca, J. L., Quintian, H., & Calvo-Rolle, J. L. (2022). Intelligent model for power cells state of charge forecasting in EV. *Processes*, *10*(7). Available from https://doi.org/10.3390/pr10071406.

Cho, Y., & Awbi, H. B. (2007). A study of the effect of heat source location in a ventilated room using multiple regression analysis. *Building and Environment*, *42*(5), 2072−2082. Available from https://doi.org/10.1016/j.buildenv.2006.03.008.

Chukwuka, C., Folly, K.A. (2012). South Africa Batteries and super-capacitors. *12 2012/12 IEEE Power and Energy Society Conference and Exposition in Africa: Intelligent Grid Integration of Renewable Energy Resources*, PowerAfrica 2012. Available from https://doi.org/10.1109/PowerAfrica.2012.6498634.

Chung, J., Gulcehre, C., Cho, K., & Bengio, Y. (2014). Empirical evaluation of gated recurrent neural networks on sequence modeling. *arXiv preprint arXiv:1412.3555*.

De Las Heras, A., Vivas, F. J., Segura, F., & Andújar, J. M. (2018). From the cell to the stack. A chronological walk through the techniques to manufacture the PEFCs core. *Renewable and Sustainable Energy Reviews*, *96*, 29−45. Available from https://doi.org/10.1016/j.rser.2018.07.036.

Dicks, A. L., & Rand, D. A. J. (2018). *Fuel cell systems explained*. Wiley. Available from https://doi.org/10.1002/9781118706992.

Dunn, B., Kamath, H., & Tarascon, J. M. (2011). Electrical energy storage for the grid: A battery of choices. *Science (New York, N.Y.)*, *334*(6058), 928−935. Available from https://doi.org/10.1126/science.1212741.

Fernández-Serantes, L.A., Estrada Vázquez, R., Casteleiro-Roca, J.L., Calvo-Rolle, J.L., Corchado, E. (2014). Hybrid intelligent model to predict the SOC of a LFP power cell type. 1 2014/01 *Lecture Notes in Computer Science (including subseries Lecture Notes in Artificial Intelligence and Lecture Notes in Bioinformatics)*, Springer Verlag, Spain http://springerlink.com/content/0302-9743/copyright/2005/8480

Fontenla Romero, O., & Calvo Rolle, J. L. (2018). Artificial intelligence in engineering: Past, present and future. *DYNA*, *93*(1), 350−352. Available from https://doi.org/10.6036/8639.

Garcia, R. F., Rolle, J. L. C., Gomez, M. R., & Catoira, A. D. (2013). Expert condition monitoring on hydrostatic self-levitating bearings. *Expert Systems with Applications*, *40* (8), 2975−2984. Available from https://doi.org/10.1016/j.eswa.2012.12.013.

Gers, F. A., Schmidhuber, J., & Cummins, F. (2000). Learning to forget: Continual prediction with LSTM. *Neural Computation*, *12*(10), 2451−2471. Available from https://doi.org/10.1162/089976600300015015.

Ghaseminezhad, M. H., & Karami, A. (2011). A novel self-organizing map (SOM) neural network for discrete groups of data clustering. *Applied Soft Computing*, *11*(4), 3771−3778. Available from https://doi.org/10.1016/j.asoc.2011.02.009.

Giacone, E., & Mancò, S. (2012). Energy efficiency measurement in industrial processes. *Energy*, *38*(1), 331−345. Available from https://doi.org/10.1016/j.energy.2011.11.054.

Gonzalez-Cava, J. M., Reboso, J. A., Casteleiro-Roca, J. L., Calvo-Rolle, J. L., & Méndez Pérez, J. A. (2018). A Novel fuzzy algorithm to introduce new variables in the drug supply decision-making process in medicine. *Complexity* (2018). Available from https://doi.org/10.1155/2018/9012720.

Hall, P. J., & Bain, E. J. (2008). Energy-storage technologies and electricity generation. *Energy Policy*, *36*(12), 4352−4355. Available from https://doi.org/10.1016/j.enpol. 2008.09.037.

Hemalatha, K., & Rani, K. U. (2017). Advancements in multi-layer perceptron training to improve classification accuracy. *International Journal on Recent and Innovation Trends in Computing and Communication*, *5*(6), 353−357.

Hepbasli, A., & Ozgener, L. (2004). Development of geothermal energy utilization in Turkey: A review. *Renewable and Sustainable Energy Reviews*, *8*(5), 433−460. Available from https://doi.org/10.1016/j.rser.2003.12.004.

Hochreiter, S., & Schmidhuber, J. (1997). Long short-term memory. *Neural Computation*, *9*(8), 1735−1780. Available from https://doi.org/10.1162/neco.1997.9.8.1735.

Hoerl, A. E., & Kennard, R. W. (1970). Ridge regression: Applications to nonorthogonal problems. *Technometrics*, *12*(1), 69−82. Available from https://doi.org/10.1080/00401706. 1970.10488635.

Hoke, A., Brissette, A., Smith, K., Pratt, A., & Maksimovic, D. (2014). Accounting for lithium-ion battery degradation in electric vehicle charging optimization. *IEEE Journal of Emerging and Selected Topics in Power Electronics*, *2*(3), 691−700. Available from https://doi.org/10.1109/JESTPE.2014.2315961.

Infield, D., & Freris, L. (2020). *Renewable energy in power systems*. John Wiley & Sons.

Iqelan, B. M. (2015). Time series modelling of monthly temperature data of Jerusalem/Palestine. *MATEMATIKA: Malaysian Journal of Industrial and Applied Mathematics*, 159−176.

Jain, A. K., Mao, J., & Mohiuddin, K. M. (1996). Artificial neural networks: A tutorial. *Computer*, *29*(3), 31−44. Available from https://doi.org/10.1109/2.485891.

Jha, S. K., Bilalovic, J., Jha, A., Patel, N., & Zhang, H. (2017). Renewable energy: Present research and future scope of Artificial Intelligence. *Renewable and Sustainable Energy Reviews*, 77, 297−317. Available from https://doi.org/10.1016/j.rser.2017. 04.018.

Jin, R., Cho, K., Hyun, C., & Son, M. (2012). MRA-based revised CBR model for cost prediction in the early stage of construction projects. *Expert Systems with Applications*, *39*(5), 5214−5222. Available from https://doi.org/10.1016/j.eswa.2011.11.018.

Johari, M. K., Jalil, M. A. A., & Shariff, M. F. M. (2018). Comparison of horizontal axis wind turbine (HAWT) and vertical axis wind turbine (VAWT). *International Journal of Engineering and Technology(UAE)*, *7*(4), 74−80. Available from https://doi.org/10.14419/ijet.v7i4.13.21333.

Jove, E., Casteleiro-Roca, J. L., Quintián, H., Simić, D., Méndez-Pérez, J. A., & Calvo-Rolle, J. L. (2020). Anomaly detection based on one-class intelligent techniques over a control level plant. *Logic Journal of the IGPL*, *28*(4), 502−518. Available from https://doi.org/10.1093/JIGPAL/JZZ057.

Karunathilake, H., Hewage, K., Mérida, W., & Sadiq, R. (2019). Renewable energy selection for net-zero energy communities: Life cycle based decision making under uncertainty. *Renewable Energy*, *130*, 558−573. Available from https://doi.org/10.1016/j.renene.2018.06.086.

Kaski, S., Sinkkonen, J., & Klami, A. (2005). Discriminative clustering. *Neurocomputing*, *69* (1−3), 18−41. Available from https://doi.org/10.1016/j.neucom.2005.02.012.

Keeney, R. L. (2013). *Siting energy facilities*. Academic Press.

Kuwae, T., & Hori, M. (2018). *The future of blue carbon: Addressing global environmental issues* (pp. 347−373). Springer Nature. Available from https://doi.org/10.1007/978-981-13-1295-3_13.

Law, E. W., Prasad, A. A., Kay, M., & Taylor, R. A. (2014). Direct normal irradiance forecasting and its application to concentrated solar thermal output forecasting - A review. *Solar Energy*, *108*, 287−307. Available from https://doi.org/10.1016/j.solener.2014.07.008.

Likas, A., Vlassis, N., & Verbeek, J. J. (2003). The global k-means clustering algorithm. *Pattern Recognition*, *36*(2), 451−461. Available from https://doi.org/10.1016/S0031-3203(02)00060-2.

Loh, W. Y. (2011). Classification and regression trees. *Wiley Interdisciplinary Reviews: Data Mining and Knowledge Discovery*, *1*(1), 14−23. Available from https://doi.org/10.1002/widm.8.

Machón-González, I., López-García H., & Luís Calvo-Rolle J. (2010, December). *A hybrid batch SOM-NG algorithm*. Proceedings of the International Joint Conference on Neural Networks, Spain. Available from https://doi.org/10.1109/IJCNN.2010.5596812.

Martínez-Rego, D., Fontenla-Romero, O., & Alonso-Betanzos, A. (2011). Efficiency of local models ensembles for time series prediction. *Expert Systems with Applications*, *38* (6), 6884−6894. Available from https://doi.org/10.1016/j.eswa.2010.12.036.

McCulloch, W. S., & Pitts, W. (1943). A logical calculus of the ideas immanent in nervous activity. *The Bulletin of Mathematical Biophysics*, *5*(4), 115−133. Available from https://doi.org/10.1007/BF02478259.

Million, E. (2007). *The Hadamard product*. Course Notes. 3.

Montero-Sousa, J. A., Casteleiro-Roca, J. L., & Calvo-Rolle, J. L. (2017a). Evolución del sector eléctrico tras la Segunda Guerra Mundial. *Dyna (Spain)*, *92*(3), 280−284. Available from https://doi.org/10.6036/8121.

Montero-Sousa, J. A., Casteleiro-Roca, J. L., & Calvo-Rolle, J. L. (2017b). The electricity sector from inception to the World War II. *Dyna (Spain)*, *92*(1), 43−47. Available from https://doi.org/10.6036/7947.

Montero-Sousa, J. A., Fernández-Serantes, L. A., Casteleiro-Roca, J. L., Vilar-Martínez, X. M., & Calvo-Rolle, J. L. (2017). Gestión de almacenamiento energético para instalaciones de generación-distribución. *Dyna (Spain)*, *92*(2), 140−141. Available from https://doi.org/10.6036/8172.

Moreira, M. V., & da Silva, G. E. (2009). A practical model for evaluating the performance of proton exchange membrane fuel cells. *Renewable Energy, 34*(7), 1734−1741. Available from https://doi.org/10.1016/j.renene.2009.01.002.

Moreno-Fernandez-de-Leceta, A., Lopez-Guede, J. M., Ezquerro Insagurbe, L., Ruiz de Arbulo, N., & Graña, M. (2018). A novel methodology for clinical semantic annotations assessment. *Logic Journal of the IGPL*. Available from https://doi.org/10.1093/jigpal/jzy021.

Neubauer, J.S., & Wood, E. (2015, April 14). SAE International United States Will Your Battery Survive a World with Fast Chargers? *SAE Technical Papers* http://papers.sae. org/2015

Nieves, J., Santos, I., & Bringas, P. G. (2013). Utilización de algoritmos de meta-clasificación para la mejora de los Modelos Predictivos de Control. *Dyna (Spain), 88* (3), 290−298. Available from https://doi.org/10.6036/5426Spain.

Osborn, J., Guzman, D., de Cos Juez, F. J., Basden, A. G., Morris, T. J., Gendron, E., Butterley, T., Myers, R. M., Guesalaga, A., Sanchez Lasheras, F., Gomez Victoria, M., Sanchez Rodriguez, M. L., Gratadour, D., & Rousset, G. (2014). Open-loop tomography with artificial neural networks on CANARY: On-sky results. *Monthly Notices of the Royal Astronomical Society, 441*(3), 2508−2514. Available from https://doi.org/10.1093/mnras/stu758.

Petre, E. G. (2009). A decision tree for weather prediction. *Universitatea Petrol-Gaze din Ploiesti, 61.*

Potter, C.W., Archambault, A., & Westrick, K. (2009, September). Australia Building a smarter smart grid through better renewable energy information. *2009 IEEE/PES Power Systems Conference and Exposition, PSCE 2009* 10.1109/PSCE.2009.4840110.

Pryor, S. C., & Barthelmie, R. J. (2010). Climate change impacts on wind energy: A review. *Renewable and Sustainable Energy Reviews, 14*(1), 430−437. Available from https://doi.org/10.1016/j.rser.2009.07.028.

Quintián, H., Calvo-Rolle, J. L., & Corchado, E. (2014). A hybrid regression system based on local models for solar energy prediction. *Informatica, 25*(2), 265−282. Available from https://doi.org/10.15388/informatica.2014.14.

Rincon, J. A., Julian, V., Carrascosa, C., Costa, A., & Novais, P. (2018). Detecting emotions through non-invasive wearables. *Logic Journal of the IGPL*. Available from https://doi.org/10.1093/jigpal/jzy025.

Ross, D. (2003). Power struggle [power supplies for portable equipment]. *IEE Review, 49* (7), 34−38. Available from https://doi.org/10.1049/ir:20030705.

Rynkiewicz, J. (2012). General bound of overfitting for MLP regression models. *Neurocomputing, 90,* 106−110. Available from https://doi.org/10.1016/j.neucom.2011.11.028.

Sarbu, I., & Sebarchievici, C. (2014). General review of ground-source heat pump systems for heating and cooling of buildings. *Energy and Buildings, 70,* 441−454. Available from https://doi.org/10.1016/j.enbuild.2013.11.068.

Segovia, F., Górriz, J. M., Ramírez, J., Martinez-Murcia, F. J., & García-Pérez, M. (2018). Using deep neural networks along with dimensionality reduction techniques to assist the diagnosis of neurodegenerative disorders. *Logic Journal of the IGPL*. Available from https://doi.org/10.1093/jigpal/jzy026.

Segura, F., Andújar, J. M., & Durán, E. (2011). Analog current control techniques for power control in PEM fuel-cell hybrid systems: A critical review and a practical application. *IEEE Transactions on Industrial Electronics, 58*(4), 1171−1184. Available from https://doi.org/10.1109/TIE.2010.2049710.

Sekine, H., & Kim, C.H., 2007. *2007 Proceedings of the 14th International Conference on Intelligent System Applications to Power Systems 978−986 Application of Neural Network to One-Day-Ahead 24 hours Generating Power Forecasting for Photovoltaic System.*

Serale, G., Fiorentini, M., Capozzoli, A., Bernardini, D., & Bemporad, A. (2018). Model Predictive Control (MPC) for enhancing building and HVAC system energy efficiency: Problem formulation, applications and opportunities. *Energies*, *11*(3), 631. Available from https://doi.org/10.3390/en11030631.

Suárez Sánchez, A., Riesgo Fernández, P., Sánchez Lasheras, F., De Cos Juez, F. J., & García Nieto, P. J. (2011). Prediction of work-related accidents according to working conditions using support vector machines. *Applied Mathematics and Computation*, *218* (7), 3539−3552. Available from https://doi.org/10.1016/j.amc.2011.08.100.

Tao, H., Duarte, J. L., & Hendrix, M. A. M. (2008). Line-interactive UPS using a fuel cell as the primary source. *IEEE Transactions on Industrial Electronics*, *55*(8), 3012−3021. Available from https://doi.org/10.1109/TIE.2008.918472.

Tavner, P., Edwards, C., Brinkman, A., & Spinato, F. (2006). Influence of wind speed on wind turbine reliability. *Wind Engineering*, *30*(1), 55−72. Available from https://doi.org/10.1260/030952406777641441.

Tian, Y., & Zhao, C. Y. (2013). A review of solar collectors and thermal energy storage in solar thermal applications. *Applied Energy*, *104*, 538−553. Available from https://doi.org/10.1016/j.apenergy.2012.11.051.

Turrado, C. C., López, M. D. C. M., Lasheras, F. S., Gómez, B. A. R., Rollé, J. L. C., & Juez, F. J. D. C. (2014). Missing data imputation of solar radiation data under different atmospheric conditions. *Sensors (Switzerland)*, *14*(11), 20382−20399. Available from https://doi.org/10.3390/s141120382.

Van Bussel, H. P. L. H., Koene, F. G. H., & Mallant, R. K. A. M. (1998). Dynamic model of solid polymer fuel cell water management. *Journal of Power Sources*, *71*(1−2), 218−222. Available from https://doi.org/10.1016/S0378-7753(97)02744-4.

Varun., Prakash, R., & Bhat, I. K. (2009). Energy, economics and environmental impacts of renewable energy systems. *Renewable and Sustainable Energy Reviews*, *13*(9), 2716−2721. Available from https://doi.org/10.1016/j.rser.2009.05.007.

Vega Lara, B. G., Castellanos Molina, L. M., Monteagudo Yanes, J. P., & Rodríguez Borroto, M. A. (2016). Offset-free model predictive control for an energy efficient tropical island hotel. *Energy and Buildings*, *119*, 283−292. Available from https://doi.org/10.1016/j.enbuild.2016.03.040.

Vilar-Martinez, M., Montero-Sousa, A., & Calvo-Rolle, L. (2014). *Casteleiro-Roca, Expert system development to assist on the verification of\ TACAN\ system performance*. 89.

Waibel, A., Hanazawa, T., Hinton, G., Shikano, K., & Lang, K. J. (1989). Phoneme recognition using time-delay neural networks. *IEEE Transactions on Acoustics, Speech, and Signal Processing*, *37*(3), 328−339. Available from https://doi.org/10.1109/29.21701.

Wei, M., Patadia, S., & Kammen, D. M. (2010). Putting renewables and energy efficiency to work: How many jobs can the clean energy industry generate in the US? *Energy Policy*, *38*(2), 919−931. Available from https://doi.org/10.1016/j.enpol.2009.10.044.

Westbrook, M.H. (2001). *The Electric Car: Development and future of battery, hybrid and fuel-cell cars*.

Wojciechowski, S. (2018). A comparison of classification strategies in rule-based classifiers. *Logic Journal of the IGPL*, *26*(1), 29−46. Available from https://doi.org/10.1093/jigpal/jzx053, http://jigpal.oxfordjournals.org/.

Wornyo, D. K., & Shen, X.-J. (2019). Coupled least squares support vector ensemble machines. *Information*, *10*(6), 195. Available from https://doi.org/10.3390/info10060195.

Yang, C. J., & Jackson, R. B. (2011). Opportunities and barriers to pumped-hydro energy storage in the United States. *Renewable and Sustainable Energy Reviews*, *15*(1), 839−844. Available from https://doi.org/10.1016/j.rser.2010.09.020.

Yi, J., Kim, U. S., Shin, C. B., Han, T., & Park, S. (2013). Three-dimensional thermal modeling of a lithium-ion battery considering the combined effects of the electrical and thermal contact resistances between current collecting tab and leadwire. *Journal of the Electrochemical Society*, *160*(3), A437–A443. Available from https://doi.org/10.1149/2.039303jes.

Zayas-Gato, F., Jove, E., Casteleiro-Roca, J. L., Quintián, H., Pérez-Castelo, F. J., Piñón-Pazos, A., Arce, E., & Calvo-Rolle, J. L. (2022). Intelligent model for active power prediction of a small wind turbine. *Logic Journal of the IGPL*. Available from https://doi.org/10.1093/jigpal/jzac040.

Yuan, Xiao, X.S., Shin, C.H., Han, T., & Park, C. (2012). Three-dimen-sional thermal imaging for evaluating the combined effect of the elevated ... and the air chamber between thermal conductivity and between ... of the thermoplastic. Scrip. Soure. Adv. ... Available from: https://doi.org/10.1021/...

... Jové, L., Casanova-Batlle, T.T., Guix, A., ... T., Cardona, J., Ribas, ..., Arús, R., & Calvo-Solís, A.A. (2020). Intelligent model for evaluation ... prediction of a wind wind turbine. Jour. Journal of ... Available from: https://doi.org/10.1016/...

CHAPTER SIX

Intelligent learning approach for multienergy load forecasting

Zhang Ge, Zhu Songyang and Bai Xiaoqing
Key Laboratory of Power System Optimization and Energy Saving Technology, Guangxi University, Nanning, P.R. China

6.1 Introduction

With the rapid advancements in science and technology, coupled with the acceleration of industrialization, the demand for energy in human society has also been on a steady rise. However, nonrenewable energy, faces issues such as uneven distribution, low efficiency, and serious environmental pollution, along with the risk of depletion due to continuous consumption. Therefore, using nonrenewable energy not only meets human needs but also aggravates the environmental deterioration. These factors have led to an urgent need for renewable and clean energy sources. To ensure sustainable development, it is imperative to explore and establish an environmentally friendly, sustainable, and efficient energy system (Bie et al., 2017).

The National Energy Administration of China has previously proposed that the key objectives for developing the current power market. These objectives include exploring the construction of an energy Internet platform, meeting the grid connection requirements for new and distributed power generation, and enhancing the coordination capabilities of distributed power generation and distribution networks (Xing, 2006). The concept of the "smart grid" was introduced in literature to explore efficient and large-scale utilization of renewable and clean energy. The advantage of the smart grid lies in its ability to control power distribution more flexibly, reliably, efficiently, and automatically (Chengshan et al., 2015). An increasing number of scholars are directing their attention to the energy Internet to promote the complementary use of electric energy with other sources and improve the overall efficiency. This approach

Intelligent Learning Approaches for Renewable and Sustainable Energy
DOI: https://doi.org/10.1016/B978-0-443-15806-3.00006-1

157

comprehensively considers the coupling of multiple energy sources (Wang Zhang et al., 2020).

The energy Internet integrates clean energy, ultrahigh voltage power grid, and smart grid technologies (Li, 2018), achieving a high degree of integration between information flow and energy flow. The coupling between the source and load ends of the energy Internet is characterized by a close connection, with two-way uncertainty between the supply and demand sides. Furthermore, the coupling relationship is further complicated by the diverse range of energy load types, which encompass electric, cold, thermal, and other energy forms (Jianzhong, 2016). The basic composition of energy Internet can be divided into three parts: source, network, and load. In this context, "source" refers to the origin point facilitating large-scale renewable energy consumption and economical energy production. The "network" is critical to realize large-scale and long-distance transmission of various energy sources, while the term "load" refers to customer demand for various energy forms (Bie et al., 2017; Xiaping Comprehensive, 2019; Zeng et al., 2016). The Energy Internet is centered around the power system and built upon the foundation of the Smart Grid. It aims to achieve multienergy complementarity, maximize renewable energy consumption, and reduce environmental pollution.

Energy conversion, distribution, and organic coordination are essential aspects of the integrated energy system, which is recognized as one of the key physical infrastructures of the energy Internet (Yu et al., 2016). The scale of the integrated energy system can be divided into multiple levels, including the user-level integrated energy system, also known as the integrated energy microgrid. The integrated energy microgrid focuses on the collaborative optimization of different energy sources (Yu et al., 2016), emphasizing the physical aspects of the energy supply system, energy exchange links, and terminal energy units (Jia et al., 2015). It plays a crucial role in implementing the concepts of multienergy complementarity and energy cascade usage in practical projects. Currently, research on integrated energy microgrids at a domestic level remains in the preliminary stage (Li et al., 2015).

Multivariate load forecasting is essential for planning, dispatching, and safety verification in integrated energy microgrids. Therefore, accurate multielement load forecasting is a prerequisite to achieving the safe and stable operation of integrated energy microgrids. In contrast to large-scale power systems, the demand-side load of integrated energy microgrids is relatively small and subject to various factors such as weather conditions, user behavior, and the high degree of randomness associated with renewable energy sources. This leads to a large volatility, resulting in greater difficulty in multielement load forecasting (Zhang et al., 2019). The improvement of the accuracy of multivariate load forecasting can provide a more reliable basis for the microgrid dispatching strategies and safety verification. This, in turn, enhances the microgrid's analytical and decision-making capabilities, reducing energy waste and improving energy utilization rates. Due to the influence of many internal and external factors, the multielement load presents the characteristics of randomness and nonlinearity. This complexity poses challenges for the mathematical model to accurately describe the relationship between the multielement load and the influencing factors, hindering ideal load forecasting results. Artificial intelligence technology, based on data-driven approaches, is excels in handling various nonlinear complex models and holds significant promise in multivariate load forecasting. The continuous advancement of energy measuring devices and the Internet makes it easier to access massive, multitype, and multienergy load data, providing the conditions and support needed to construct multiple load forecasting models using artificial intelligence technology.

Building a comprehensive energy industry will become an important strategic support for promoting China's energy revolution. The integrated energy microgrid has the potential to increase the consumption of renewable energy sources, enhance the usage efficiency of nonrenewable energy sources, mitigate environmental pollution, and alleviate the pressure on the energy supply. This technology is crucial for advancing energy efficiency and promoting new avenues toward sustainable development in modern society (Bie et al., 2017).

Multiload forecasting for integrated energy microgrids refers to the prediction of diverse energy demands from users. The primary approach involves analyzing historical data on users' energy consumption and potential factors that may influence their energy usage, investigating patterns in energy load changes, examining the relationship between load and influencing factors, constructing a model based on the collected information, and subsequently making precise multiload predictions through model calculation derived from learned data.

Precise multivariate load forecasting plays a pivotal role in the planning, dispatching, and operation of integrated energy microgrids. However, as the development of these microgrids is still in its early stages, there has been limited research on this subject. The multiload of integrated energy microgrids is influenced by numerous external factors, including weather conditions and daily routines. In contrast to traditional single-energy systems, the load of an integrated energy microgrid encompasses a diverse array of energy forms, which are plentiful and varied. Loads of various energy forms affect each other, creating a close coupling relationship. The complex relationships of the internal system make the change law of the multiple loads more complicated. Hence, in contrast to traditional single-energy load forecasting, multielement load forecasting necessitates not only the analysis of external factors and their impact on energy load but also the examination of interdependent relationships between different loads. This approach enhances the accuracy of multielement load forecasting. Deep learning, adept at handling complex nonlinear problems, has been successfully applied in power load forecasting. Hence, the deep learning method is selected to build a more complex multivariate load forecasting model.

Given that the integrated energy microgrid is still in its infancy, there are only a limited number of such grids in operation. Furthermore, they are not geographically concentrated, and their operational history is relatively short. Consequently, these microgrids may encounter challenges associated with inadequate data quantity or quality, as well as low data diversity. If each integrated energy microgrid only trains a local model with its data, the training effect of the model will not be ideal when facing these three problems. For example, the insufficient amount of microgrid data will lead to inadequate network training, and ultimately the network cannot fully learn the relationship between multiple loads as well as between loads and influencing factors, thus failing to achieve the desired prediction effect. If the data quality of the microgrid is poor, the network may pay too much attention to the wrong information during the model training, resulting in the distortion of the effect of the characteristics on the load reality. The network cannot learn the truest and most accurate information in this scenario. Additionally, if the data diversity within a microgrid is limited, it can be challenging to fully understand the unique characteristics of a small set of cases, such as equipment maintenance or legal holidays. To address these issues, a centralized learning mode has emerged. This approach involves each microgrid serving as a client that sends historical user data in the microgrid system to a central server for uniform processing. Although centralized learning eliminates data problems, under this mode each microgrid needs to send the actual load data of users in the system to the server. Information communication technology must be used for data transmission between the microgrid and the server. However, the use of communication technology introduces the risk of various network. Due to local policies, data privacy, system security, and other considerations, platforms such as WeChat may refuse to share user data and may not be able to participate in training. In addition to data privacy concerns, massive data transmission also imposes high requirements on communication conditions.

Similarly, the server must employ a local model to extract features from large-scale data, demanding high computing capacity of server equipment. To overcome the shortcomings of centralized learning, the federated learning mode has been introduced. Federated learning is a distributed training approach. In this mode, the microgrid does not need to transmit the user's actual load data to the server but instead conducts a small amount of local training before sending only the model parameters to the server. The server only needs to process the parameters without further training and achieves the purpose of training the model through multiple information exchanges. Compared with the actual load data, the model parameters required by federated learning reveal less, more fuzzy, and more abstract information. This significantly improves security compared with centralized learning. In addition, due to the small amount of data sent, the requirements for communication and server resources are significantly reduced. In summary, federated learning not only improves the various data problems faced by individual model training but also mitigates issues of data privacy, high communication requirements, and high server requirements involved in centralized training to a certain extent.

A deep learning-based multiple-load forecasting model is constructed to extract the coupling relationship between multiple loads and explore the impact of influencing factors on the load. Subsequently, the federated learning mode is introduced to enhance data quality and diversity while maintaining data privacy in the integrated energy micro network, thus thereby alleviating communication and computing pressure.

The significance of constructing a multivariate load forecasting model based on federated learning and deep learning as an essential component of the integrated energy microgrid can be summarized as follows:

1. Multivariate load forecasting can assist the dispatcher of an integrated energy microgrid in arranging the dispatching strategy better and achieving the optimal allocation of energy through the reasonable arrangement of equipment operation status (Cui & Wang, 2022; Shensi Cooling, 2018).

2. Multivariate load forecasting can help plan energy use, transform loads of different energy types, reduce energy waste, and improve the microgrid's economy, security, and reliability.

3. Multivariate load forecasting can be used to conduct safety verification, help monitor the equipment operation status, find abnormalities in time, improve the speed of handling abnormal conditions, and reduce system losses.

4. The introduction of federated learning can realize distributed training of multielement load forecasting models for multiple integrated energy micro networks and improve the poor training of forecasting models due to data quality and other problems.

5. Compared with the centralized learning mode, the model based on federated learning does not need to transmit the real load data to the server, which not only protects the privacy of customers but also relieves the communication pressure caused by massive data transmission and reduces the requirements for model training equipment.

6. Federated learning is known for its robustness, as it is capable of maintaining high prediction accuracy even in the presence of false data injection attacks. This makes it an ideal candidate to ensure the secure and stable operation of integrated energy microgrids.

6.2 Multivariate load characteristics

Multiple loads are influenced not only by social factors but also by various natural elements. The grid connection of renewable energy also increases the volatility of multiple loads. The energy coupling relationship in the integrated energy microgrid further increases load instability. The historical data of multielement load contains much information that can be mined. Through the analysis of multielement load data, a better understanding of load characteristics can be gained, facilitating more effective multielement load forecasting. The attributes of a multielement load are mainly four points: periodicity, randomness, coupling, and diversity.

6.2.1 Periodicity

The load demand is intricately linked to the economic operations of society. As these operations exhibit periodicity, so too does the load demand.

The load exhibits periodic characteristics on a daily, weekly, monthly, and yearly basis.

6.2.2 Randomness

The changing trend of multiple loads is determined by the demand of users, which is affected by both subjective and external objective factors. Under the comprehensive influence of various influencing factors, although the multielement load shows periodicity in the general trend, each cycle is also different and has an element of randomness.

6.2.3 Coupling

As the energy in the integrated energy microgrid has various conversion relationships, the load end uses energy conversion equipment, such as air conditioning, which converts electric energy into the cold load to fulfill cooling needs by consuming electric energy. Similarly, the hot water generated by burning gas is used to satisfy thermal demand. The loads also exhibit certain coupling.

6.2.4 Diversity

Different integrated energy microgrids will exhibit multiple load curves due to geographical, economic, local policies, social and cultural factors. Applying the multielement load forecasting model trained by one microgrid directly to another microgrid with different load laws may lead to an increase in the error of multielement load forecasting or even unavailability because of the diversity of multielement loads. As shown in Figs. 6.2−6.4, the two microgrids have not only different load capacities but also different curve shapes.

6.3 Research status of multienergy coincidence prediction

The multienergy load in integrated energy microgrids comprises various energy forms, and due to the interrelatedness of energy systems within the microgrid, renewable energy is characterized by greater uncertainty than conventional electric energy. Consequently, forecasting a multienergy load is more challenging than forecasting a single-energy load.

In recent years, as comprehensive energy development has gained momentum, several research studies have focused on multivariate load forecasting. One such study (Cui & Wang, 2022) proposes a multielement load collaborative forecasting model based on multiobjective stacking ensemble learning to address the issue of independent load forecasting, ignoring the coupling relationship between multiple loads. The maximum information coefficient is introduced to analyze the correlation between multiple loads and weather factors, while the load coupling morphological index is established to delve deeply into the coupling relationship between multiple loads. Finally, an example is provided to demonstrate the effectiveness and superiority of the proposed model. In another study by Liu et al. (2020) long-term and short-term memory networks are applied to multivariate load forecasting. Due to the significant volatility and randomness of the multielement load in integrated energy microgrids, accurate prediction is challenging. To address this issue, an ultra–short-term multielement load forecasting method using variable mode decomposition (VMD) and multimodel fusion was proposed in Ye et al. (2022). Initially, the VMD approach decomposes various load series into distinct intrinsic mode functions (IMF). Subsequently, each IMF is integrated with meteorological information to create distinct feature sets, which are then fed into support vector regression (SVR), short-term memory neural networks, and one-dimensional convolutional neural networks for prediction. The output of the three models is then combined through SVR fusion to obtain the final prediction value. Results demonstrate that the proposed multimodel fusion approach outperforms single-model prediction methods and exhibits superior prediction accuracy for power, cold, and thermal loads.

Building upon this foundation, Wu et al. (2021) proposed a combined cold, thermal, and power load forecasting approach based on variational modal decomposition and a deep belief network. On the other hand, Chen et al. (2022) explore the coupling relationships among various energy sources in comprehensive energy systems by using a residual neural network (RESNET) to extract long-term spatial coupling interaction features from multiple load data. Short-term memory neural networks are used to extract time-series features from multiple load data, while an attention mechanism is employed to target different multitask features with varying degrees of focus. A multitask learning model using a RESNET-LSTM network and an attention mechanism is constructed for the joint prediction of multiple loads. In (He et al., 2021), an improved domain adaptive neural network (DANN) load forecasting model was developed to uniformly model and forecast various loads. Meanwhile, the authors in (Yongli et al., 2022) analyze the coupling relationship between multiple loads in integrated energy systems using copula theory and establishe a multiple load forecasting model by employing the optimization algorithm of the bottle sea thermal group and the least square support vector machine. In (Chen Zhao et al., 2021), a comprehensive energy load forecasting method based on wavelet optimization and multitask learning is proposed, which can prevent getting trapped in local minima and aid in improving the forecasting speed.

The multitask learning approach can extract the coupling relationships between energy sources and enhance the model's generalization capabilities. In (Chen Hu et al., 2021), empirical mode decomposition is initially used to decompose the load, followed by kernel principal component analysis for dimensionality reduction of the data. The resulting components are then predicted using deep bidirectional long-short-term memory neural networks and multiple linear regression. The final predictions are obtained through the reconstruction of the results. Sun et al. (2021) introduced multitask learning based on long-short term memory neural networks to extract the coupling relationships between multiple loads and enhance prediction accuracy. Jianpeng et al. (2020) constructed a multivariate load forecasting model based on a generalized regression neural network, using kernel principal component analysis for dimensionality reduction of data and genetic algorithm optimization. The authors in (Zhu et al., 2019) incorporated a convolutional neural network into long-short-term memory neural networks for predicting the multiple loads in cogeneration systems. In Wang Wang et al. (2020), a gradient-boosting decision tree is combined with the encoder-decoder model based on long short-term memory neural networks. Tan et al. (2020) propose a model that combines multitask learning and least squares support vector machines to predict multielement loads. Based on this work, (Yan & Zhang, 2021) optimize the parameter settings of the model using particle swarm optimization (PSO). The authors in (Li et al., 2021) and (Wang Ma et al., 2020) use stacked autoencoders (SAEs) for predicting multivariate loads.

Research on single-energy load forecasting typically relies on mathematical models or artificial intelligence. Mathematical model-based forecasting methods typically only predict historical load data, neglecting the influence of other external factors. Artificial intelligence-based methods can, however, comprehensively consider both historical load data and a range of other influencing factors, allowing for more comprehensive information processing. Therefore, an artificial intelligence algorithm is suitable for multivariate load forecasting with complex internal coupling relationships.

 ## 6.4 Prediction process of an artificial intelligence method

6.4.1 Abnormal data processing

Electricity, cooling, and thermal load data exhibit randomness, volatility, and periodicity. In addition, historical load data and meteorological data may be inaccurately recorded or partially lost during storage. Directly training the model with such raw data can be hindered by numerous interfering factors that ultimately affect the model's effectiveness. In order to ensure data accuracy and completeness, missing values and outliers in data samples are filled in and replaced with average values of time points before and after the data point.

The 3σ principle, commonly used for abnormal data detection, calculates the average value of the sequence (μ) and standard deviation (σ). It determines whether the sample is an outlier according to the relationship between the sample and the mean and standard deviation. According to the 3σ principle, the data are expected to be distributed within the range (μ-3σ, $\mu + 3\sigma$), covering a probability of 0.9973. Therefore, samples beyond this range are considered outliers.

6.4.2 Select influencing factors

In addition to the interdependence between different energy sources, multivariate load forecasting requires the consideration of a range of factors, including environmental conditions, daily routines, and other relevant variables. The more comprehensive the theoretical framework guiding the analysis, the greater the accuracy achieved. However, excessive consideration of multiple factors may result in an abundance of data, which can impede effective network training. As a result, it is imperative to employ a rational quantitative method to discern the impact of external factors on each load, screen these factors, eliminate those with negligible effect on load forecasting, and retain those which exhibit strong correlation (Strive to build a world-class energy internet enterprise N State Grid News, 2019). Among the three primary correlation coefficients (Kendall, Pearson, and Spearman), the Kendall coefficient is appropriate for calculating classification variables. The Pearson coefficient is suitable for determining linear correlations between variables, while the Spearman rank correlation coefficient can describe nonlinear correlations between variables. Therefore, the Spearman rank correlation coefficient is selected as the evaluation index for correlation.

The Spearman rank correlation coefficient is used to indicate the strength of the correlation between two sequences, and its value ranges from -1 to $+1$. A value of zero indicates no correlation. Positive and negative values indicate positive and negative correlations, respectively. The larger the absolute value of the coefficient, the stronger the correlation between the two sequences. The mathematical expression is defined by:

$$\rho_{xy} = \frac{\sum_n (x_i - \overline{x})(y_i - \overline{y})}{\sqrt{\sum_n (x_i - \overline{x})^2 \sum_n (y_i - \overline{y})^2}} \tag{6.1}$$

where, ρ_{xy} represents the correlation degree of sequence X and Y, n is the total number of sequence data, x_i and y_i represents the value of the sequence X and Y, respectively, $\overline{x}$ and $\overline{y}$ correspond to the average value of the data.

Once the pertinent factors have been selected, the input data is finalized. It is essential to partition the data set into training, validation, and testing sets in order to enable the subsequent model training and evaluation. In this chapter, the three datasets are split with a ratio of 7:2:1.

6.4.3 Normalization

Given the dissimilarities in units and large ranges of values among various variables, the direct prediction may result in distorted feature extraction and ultimately alter their true effects. The processing of each sequence of input data is, therefore, necessary to make its value within a range. Normalization is a common processing method.

Data normalization is the process of scaling load data to be uniformly mapped to the interval [0, 1]. After normalization, variables are transformed into dimensionless values that can be more reasonably compared and weighted. Normalization helps prevent gradient explosion during training and improves the effectiveness of model training. The normalization formula is defined as:

$$X' = \frac{X - X_{min}}{X_{max} - X_{min}} \qquad (6.2)$$

where X the data before normalization processing, X' is the corresponding normalized data, X_{min} and X_{max} are the minimum and maximum values in the sequence, respectively. The advantage of normalization is that all "0" values can be kept unchanged for the series with more "0" values, which applies to the load data with multiple "0" values.

6.4.4 Select evaluation indicators

The error value is commonly used as the evaluation metric to assess the performance of multifunctional load forecasting models. In the context of multienergy load forecasting, the error refers to the discrepancy between the predicted and actual values. A smaller error indicates higher accuracy of the multienergy load forecasting model and a more effective forecasting outcome. In training the model, the error value is also fed back to the model as an important reference index so that the model can be continuously optimized in iteration to obtain better prediction results. In machine learning, different problems are suitable for different evaluation indicators, and selecting appropriate evaluation indicators impacts model training. In the context of multielement load forecasting for integrated energy microgrids, root mean square error, average absolute percentage error, and weighted average accuracy are commonly used as evaluation metrics to provide feedback on the forecasting model's performance.

6.4.4.1 RMSE

The calculation method of root mean square error is defined as

$$RMSE = \sqrt{\frac{1}{N}\sum_{i=1}^{N}\left(y_i - \hat{y}_1\right)^2} \qquad (6.3)$$

where N is the number of sequence data, y_i is the actual value of each load and $\hat{y}_i$ is the predicted value of each load.

As the value of RMSE increases, so too does the disparity between predicted and actual load values, resulting in lower prediction accuracy. RMSE error values are commensurate with the dimensionality of actual load values and share a similar order of magnitude with such values, rendering them an intuitive measure of prediction error.

6.4.4.2 MAPE

The calculation method of average absolute percentage error is defined by:

$$MAPE = \frac{100\%}{N} \sum_{i=1}^{N} \left| \frac{y_i - \hat{y}_i}{y_i} \right| \qquad (6.4)$$

where N is the number of sequence data, y_i is the actual value of each load and $\hat{y}_i$ is the predicted value of each load.

As stated by Eq. (6.4), lower MAPE values indicate smaller disparities between predicted and actual load values, thereby resulting in higher prediction accuracy. MAPE has no dimension and describes the proportion of error value in the original sequence as a percentage, which more intuitively reflects the prediction effect of the prediction model. A lower RMSE and MAPE indicate higher accuracy for the prediction model.

6.4.4.3 WMA

Because the integrated energy microgrid includes not only electric loads but also other energy loads, calculating the RMSE and MAPE indexes for each individual energy load may not be intuitive to reflect the performance of the overall multivariate load forecasting model for the entire integrated energy microgrid. The multivariate load prediction model is assessed as a whole to evaluate the prediction effect of the model more intuitively. In addition to RMSE and MAPE, the weighted average accuracy is selected to weight the prediction accuracy of each prediction task based on its importance. This results in the overall evaluation index of the prediction model. The WMA can be determined as follows:

$$MA = 1 - MAPE \qquad (6.5)$$

$$WMA = \alpha_1 MA_1 + \alpha_2 MA_2 \qquad (6.6)$$

where MA (mean accuracy) is the average accuracy while α_1 and α_2 represent the weights of different prediction tasks.

6.5 Standard prediction models and their mechanisms

The internal energy relationship of the integrated energy microgrid is complex, with a strong coupling relationship between multiple loads. Deep learning has been successfully applied in the energy field. To efficiently and accurately extract the characteristics of the load and the complex coupling relationship between loads, this chapter shall initially use deep learning to build a multivariate load forecasting model.

6.5.1 Convolutional Neural Network (CNN)

Kunihiko Fukushima was inspired by studying the cat optic nerve and proposed the basic idea of convolution and pooling as early as 1980. On this basis, Yann LeCun proposed the concept of the receptive field in 1989. In 1998, the concept of convolution was first proposed by Yann LeCun. However, the convolutional neural network did not develop rapidly due to the lack of computer computing power at that time. It wasn't until 2012, when AlexNet won by an absolute advantage in a competition, that the convolutional neural network, which is an important part of AlexNet, quickly attracted a lot of attention. Convolutional neural networks have found widespread applications in diverse domains to date.

The CNN is a deep feedforward neural network that essentially functions as a data feature extractor. Due to the sharing of internal parameters within the network, CNNs are effective at processing high-dimensional data and possess automatic feature selection capabilities. A typical convolutional neural network typically consists of an input layer, convolutional layers, pooling layers, fully connected layers, and an output layer. Among the different layers of a CNN, the input layer receives the most basic form of data, while the convolutional layer extracts relevant features from the input data. The pooling layer reduces the dimensionality of the data and optimizes network structure by selecting the most representative features. The fully connected layer integrates feature information extracted by the pooling layer, performing classification or regression calculations. Ultimately, the output layer produces the final calculation result.

A typical CNN structure is shown in Fig. 6.1 while the details of each layer are described in the following subsections.

6.5.1.1 Convolutional layer

The fundamental role of the convolutional layer is to execute convolutions using a convolution kernel, which acts like a filter. This kernel raverses the input data, carries out element-wise multiplication at each position, and aggregates these results to generate a feature matrix. Fig. 6.2 shows the 2D convolution process of a convolution kernel with 3 * 3 receptive fields and one sliding step.

In this figure, the origin of the 8 in the upper left corner of the convolution feature map is $1 * 1 + 1 * 0 + 0 * 1 + 0 * 0 + 1 * 1 + 1 * 0 + 3 * 1 + 0 * 0 + 2 * 1 = 8$. The red box corresponding to the convolution kernel in the input matrix is called the "receptive field." The larger the receptive field is, the more comprehensive the observed information will be, and the extracted features will also tend to be global. The smaller

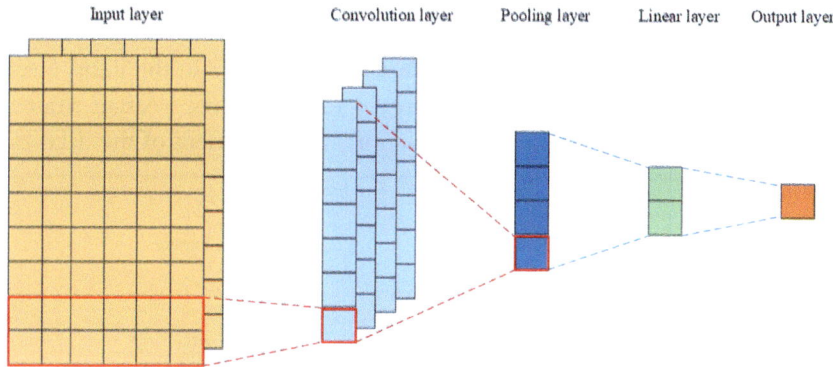

Figure 6.1 Structure of CNN.

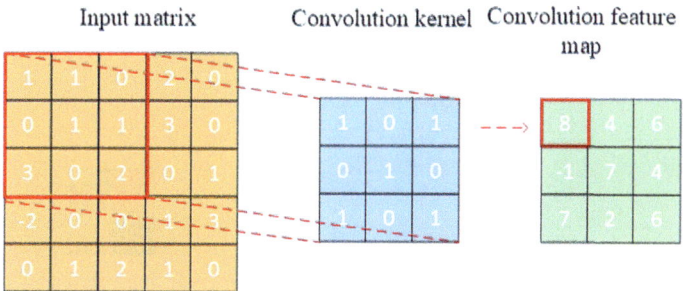

Figure 6.2 Principle of the convolution kernel.

the receptive field is, the more detailed the extracted features will be. The sliding stride refers to the distance that the convolution kernel moves. In Fig. 6.2, the convolution kernel slides from left to right and from top to bottom, one step at a time, until the entire input matrix is scanned and finally gets the convolutional feature map. The sliding stride affects the size of the convolution feature map. A smaller sliding stride results in a larger feature map, while a larger sliding stride leads to a smaller feature map. Taking Fig. 6.3 as an example, when the sliding stride is 1, the convolution kernel moves one step at a time, and the final convolution feature map is 3*3 in size. When the sliding stride is 2, the convolution kernel moves two-step at a time, and the final convolutional feature map is 2*2 in size.

One can observe that when the sliding step is small, the extracted features are more specific, and more information is retained. When the sliding step is large, the repeated data analysis can be reduced, and a smaller amount of data can be obtained. Typically, a smaller sliding stride is selected to extract the information as comprehensively as possible in the initial layers of the network. As the network progresses to later layers, a larger stride is selected to achieve dimensionality reduction and refinement of the data.

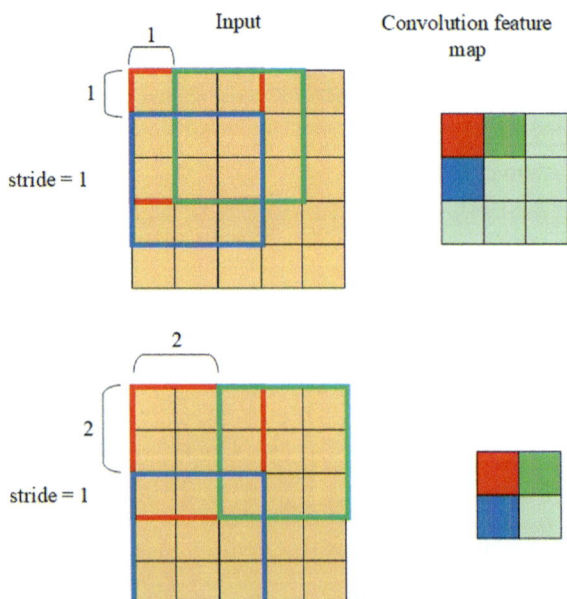

Figure 6.3 Sliding step.

No padding   Padding

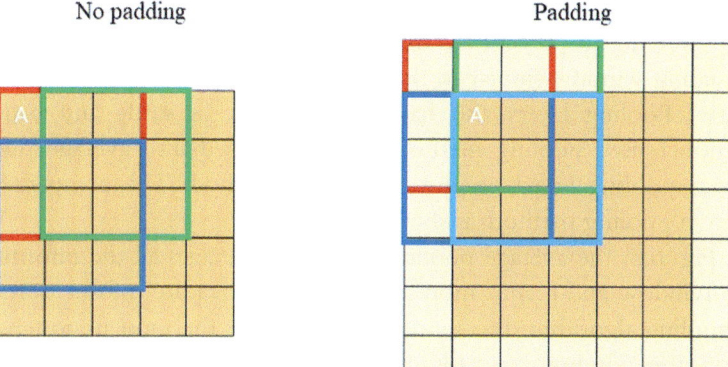

Figure 6.4 Padding.

Sometimes, in order to ensure that the feature map is not too small or to enable the convolution kernel to slide from the beginning to the end at a fixed stride, padding is added to the periphery of the input matrix. This padding, shown in the light yellow area on the right of Fig. 6.4, involves the addition of zeros to prevent noise interference and preserve feature extraction. An additional benefit of introducing padding is that it strengthens the extraction of edge information, as shown in Fig. 6.4A. Without padding, the convolution kernel can only sense the edge once, while the central area is sensed multiple times. By introducing the padding illustrated in Fig. 6.4A, the information in this region participates in the calculation four times. Thus, padding serves to reduce the weakening of edge information to a certain degree.

Depending on the direction of movement of the convolution kernel, CNN commonly uses 1D convolution and 2D convolution. The convolution kernel of a 2D convolutional neural network traverses both horizontally and vertically, making it suitable for image processing. On the other hand, a 1D convolutional neural network operates strictly in the vertical direction and is typically used for language or data processing. In this chapter, a 1D convolutional neural network has been adopted as the model architecture, with the kernel sliding only along the temporal axis.

6.5.1.2 Pooling layer

Following the convolutional layer, a pooling layer is typically incorporated to decrease the dimensionality of the data. This layer has the ability to compress data while preserving feature information as much as possible, thereby helping to prevent overfitting to some degree. The pooling layer has a sliding window similar to the convolution kernel, called the pooling window. Pooling layers also have the concept of stride and padding. Commonly used pooling methods can be divided into average pooling and max pooling based on the calculation method. The comparison of these two pooling methods is shown in Fig. 6.5.

In Fig. 6.5, the average pooling algorithm averages all the information in the window and retains more comprehensive feature information. The max pooling algorithm directly selects and retains the most important features in the window, discards other information, and retains the most effective features to the greatest extent. In practice, max pooling is more commonly used compared to average pooling.

6.5.1.3 Fully connected layer

In the last part of the convolutional network, the fully connected layer is usually used to integrate the previously learned features, analyze the relationship between scattered features and samples, complete the classification or fitting work, and then pass the results to the output layer.

CNN is capable of effectively extracting the latent features of multiple loads, as well as capturing the interdependent relationships between loads and their associations with other relevant factors. Considering that the multivariate load has a significant periodicity and CNN cannot extract the temporal characteristics, the LSTM layer is subsequently added to enhance its performance.

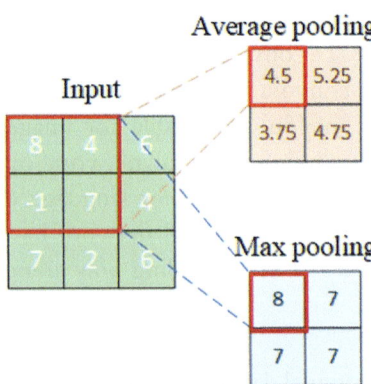

Figure 6.5 Pooling layer.

6.5.2 LSTM

As early as 1990, Jeffrey L. Elman published the article "Finding Structure in Time," which discussed the problem of how to extract time features in time series data and form a Recurrent Neural Network (RNN). As the most basic framework, RNNs serve as the foundation for the subsequent development of time-series neural networks. The structure of an RNN is depicted in Fig. 6.6. In the figure, h_{t-1} and h_t are the output at time t-1 and time t, respectively, while x_t is the input at time t. The hyperbolic tangent activation function, tanh, is employed, and the internal formula is as follows:

$$\tan hx = \frac{e^x - e^{-x}}{e^x + e^{-x}} \tag{6.7}$$

The original RNN has been widely employed across various domains and has effectively addressed a multitude of challenges. Nonetheless, there exist notable deficiencies in the architectures. RNNs are highly susceptible to gradient vanishing and exploding during training. These issues can arise due to factors such as the choice of training rules, activation functions, and other factors. Long-term memory storage is furthermore required, especially when learning data intervals are distant, leading to extensive training periods. Given the aforementioned circumstances, the Long

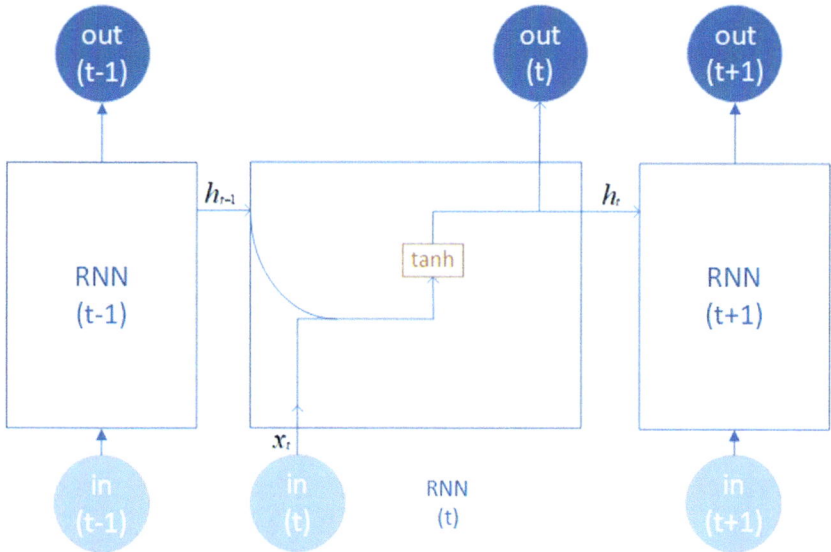

Figure 6.6 Structure of RNN.

Short-Term Memory Network (LSTM) paradigm was initially introduced by Sepp Hochreiter and Jürgen Schmidhuber in 1997. LSTM is an enhanced version of RNN that incorporates a forgetting gate, input gate, and output gate to facilitate information selection. LSTM improves upon its predecessor by selectively retaining crucial features while appropriately discarding nonessential data. The tripartite gating mechanism facilitates the dynamic processing of input data by weighing the contributions from each gating unit. This helps mitigate issues such as gradient vanishing and exploding. By assigning weights to the information retained in memory, LSTM can preserve pertinent information for prolonged periods, effectively addressing the challenge of long-term dependency.

Fig. 6.7 shows the basic structural unit of LSTM. In this figure, C_{t-1} is the memory state at time t-1, C_t is the memory state at time t, the "$\times$" and "$+$" in the circle in the figure represent vector dot product and addition, respectively while σ is the sigmoid activation function, defined as:

$$\sigma(x) = 1/(1 + \exp(-x)) \tag{6.8}$$

f_t, i_t, $\tilde{C}_t$, O_t are the internal calculation results of the three gating units, and the specific working principles shall now be defined.

6.5.2.1 Forget gate
The forget gate, shown in Fig. 6.8, selectively discards information, aiming to forget the information with less impact while retaining important

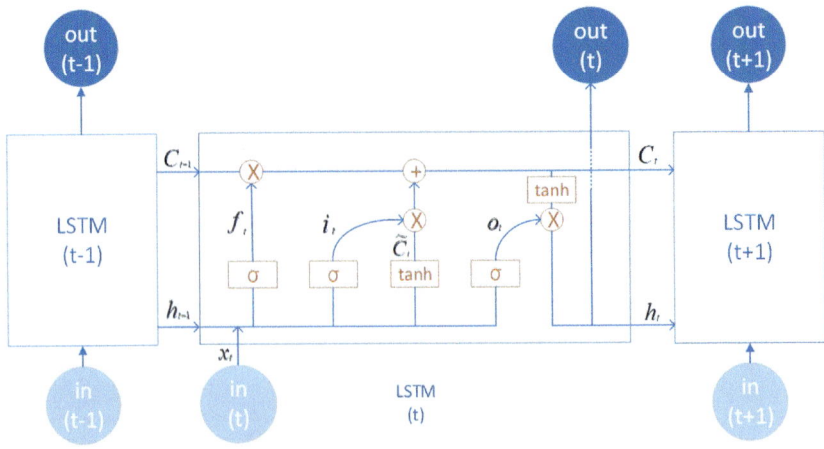

Figure 6.7 Structure of LSTM.

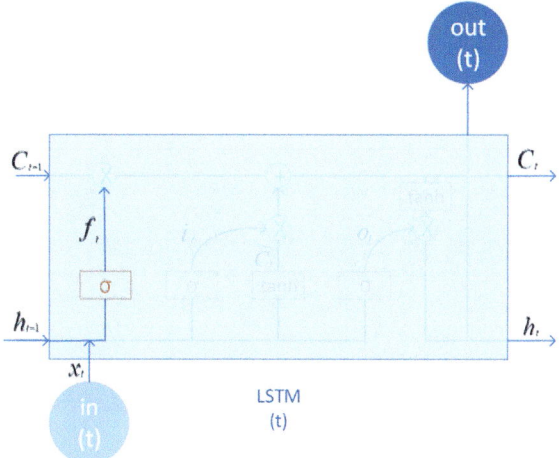

Figure 6.8 Forget gate.

information. This gate receives the output information from the previous layer of LSTM and the input information at the current moment. It processes this information through the sigmoid activation function, producing an output value in the range of 0 to 1. This output value represents the degree of forgetting. The smaller the value, the higher the degree of forgetting, while the larger the value, the lower the degree of forgetting. When the value of "0" means that the information is forgotten, whereas a value of "1" means that the information is fully retained.

The formula can be defined as follows:

$$f_t = \sigma(W_f \cdot [h_{t-1}, x_t] + b_f) \tag{6.9}$$

where W_f and b_f are the weight matrix and bias vector of the forget gate, respectively.

6.5.2.2 Input gate

The input gate is shown in Fig. 6.9 and the marked part in Fig. 6.8. The function of the input gate is to select the data to be input, and it comprises of two parts. Upon receiving h_{t-1} and x_t, one part calculates the forgetting degree of the information through the sigmoid function to obtain a vector i_t. The other part is activated through the tanh function to create an alternative vector $\tilde{c}_t$. By multiplying with i_t and $\tilde{c}_t$, the information will be sorted according to the importance, waiting for the subsequent input.

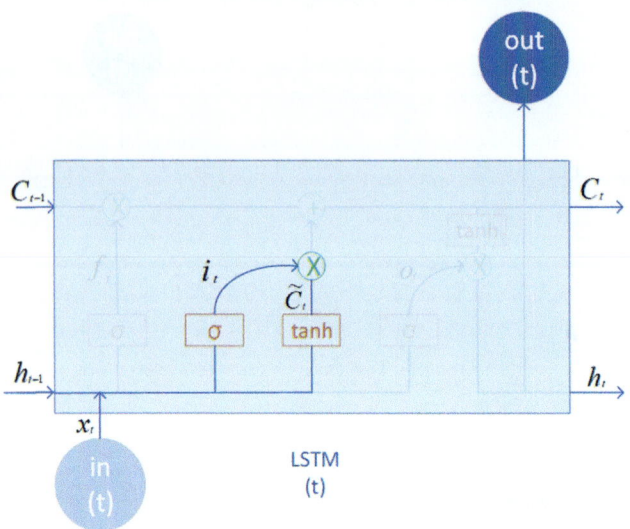

Figure 6.9 Input gate.

The input gate can be mathematically expressed by:

$$i_t = \sigma(W_i \cdot [h_{t-1}, x_t] + b_i) \tag{6.10}$$

$$\tilde{C}_t = \tanh(W_C \cdot [h_{t-1}, x_t] + b_C) \tag{6.11}$$

where W_i and b_i, W_C and b_C are the corresponding weight matrix and bias vector, respectively.

After obtaining the above results, the memory state C_{t-1} can already be updated. Firstly, the memory state C_{t-1} is multiplied by the output of the forget gate to forget information. Secondly, the multiplied result is added to the calculation result of the input gate to obtain a new memory state C_t. This can be expressed by:

$$C_t = f_t * C_{t-1} + i_t * \tilde{C}_t \tag{6.12}$$

6.5.2.3 Output gate

The output gate, shown in Fig. 6.10, determines the portion of the output. Initially the information that needs to be forgotten and retained is determined by the sigmoid function, and the calculation result is stored as

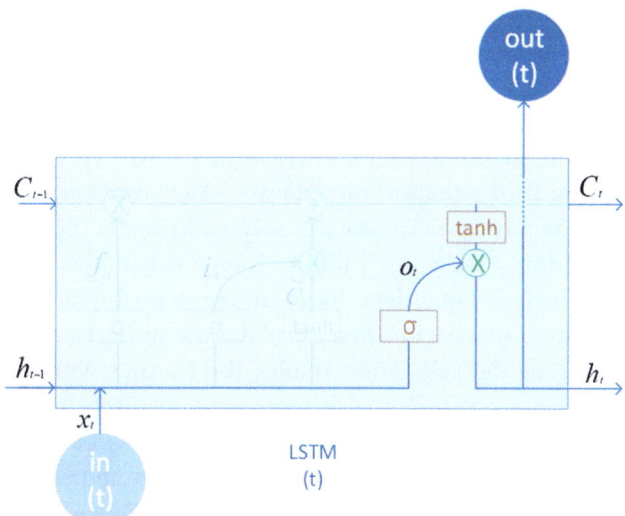

Figure 6.10 Output gate.

O_t. Subsequently, the tanh activation function is used to process the updated memory state C_t. Additionally, O_t, is multiplied by the activated C_t to ensure that the output information is consistent with the decision result. The relevant calculation formula for the output gate is as follows:

$$O_t = \sigma(W_o[h_{t-1}, x_t] + b_o) \qquad (6.13)$$

$$h_t = O_t \cdot \tanh(C_t) \qquad (6.14)$$

where W_o and b_o are the weight matrix and bias vector of the output gate, respectively.

LSTM has been widely used to deal with timing-related problems. Multivariate loads have strong temporal characteristics, and LSTM can analyze the temporal characteristics of data well. The introduction of the LSTM layer compensates for the limitation of CNN in extracting time series features.

6.5.3 Attention mechanism

In the simulation of the information processing mechanism of the human brain, the attention mechanism is a process that allocates limited information processing capabilities according to feature importance. The attention mechanism was first proposed by the Google Mind team in 2014, who applied it to the RNN model for image classification. In 2017, the Google Machine Translation team used self-attention mechanisms extensively in a study on text learning, bringing widespread attention to the use of attention mechanisms.

Attention mechanisms can be categorized into two types: soft attention mechanisms and hard attention mechanisms. After receiving the information, the soft attention mechanism will assign corresponding weights to all feature information. Finally, each feature is retained and has a corresponding weight, which is equivalent to obtaining a probability distribution. The hard attention mechanism first calculates the importance of each feature. According to the calculation results, the features with strong influence are selectively retained, the features with relatively weak influence are directly discarded, and the retained features are not given weights. Soft and hard attention mechanisms usually rely on an encoder-decoder framework to analyze the information between the encoder and the decoder. As an enhancement to the attention mechanism, the self-attention mechanism transcends the constraints of predefined structures, mitigates reliance on external information sources, and prioritizes the extraction of salient features intrinsic to the data for learning more abstract representations. This chapter selects the most widely used self-attention mechanism without framework constraints.

The attention mechanism is a model framework known as Query-Key-Value. Its fundamental purpose is to compute the similarity or correlation between Query and Key in order to assign a weight to the corresponding value. This process can be broken down into three steps: initially, the weight coefficient is calculated using the Query and Key. Secondly, the weight coefficient is normalized and transformed into an attention probability distribution with a sum of 1, which highlights the weight of important features. Finally, the attention value is obtained by multiplying the value by the normalized weight coefficient. The architecture of the attention mechanism is depicted in Fig. 6.11.

When inputting to the attention mechanism, it is assumed that the input data is a matrix, and three vectors of Q, K, and V are created through the linear transformation layer. Among them, the weight matrix is initially initialized randomly and is gradually updated and optimized in the subsequent training process. The attention score is computed as the dot product between the query vector and the transposed key vector of the corresponding sample, normalized by a scaling factor. For a given

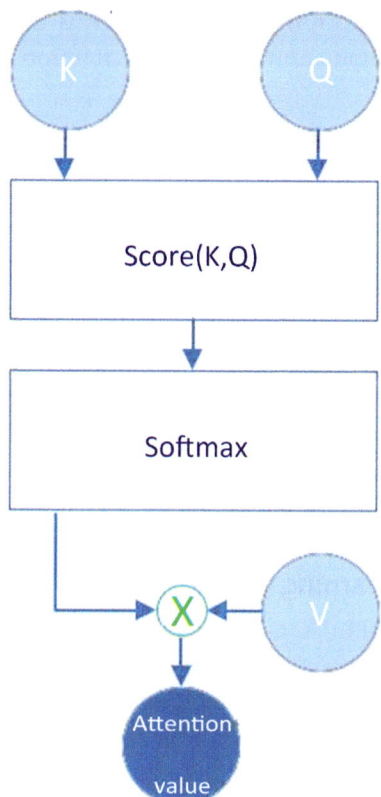

Figure 6.11 Attention mechanism.

node pair (a,b) in the hidden layer of the network, the three related vectors are presented in Eq. (6.15), and the score is evaluated according to:

$$Q_b = W^Q \cdot x_b, K_a = W^K \cdot x_a, V_a = W^V \cdot x_a \qquad (6.15)$$

$$score = Q_b \cdot K_a^T \qquad (6.16)$$

where $W^Q \in R^{\frac{d \times d}{K}}$, $W^K \in R^{\frac{d \times d}{K}}$, $W^V \in R^{\frac{d \times d}{K}}$ are the corresponding weight matrices, respectively.

In order to prevent the inner product from being too large, these calculated scores are divided by the square root $\sqrt{K_{\mathrm{dim}}}$ of the key vector dimension to obtain a stable gradient and normalized by softmax. The normalized value is multiplied by the V vector to obtain the attention value. The computation formula for Softmax is given by the following equation, and the computation formula for attention value is as follows:

$$Softmax(s_i) = \frac{e^{s_i}}{\sum_{j=1}^{n} e^{s_j}} \tag{6.17}$$

$$Attention(Q, K, V) = softmax\frac{score}{\sqrt{K_{dim}}} V \tag{6.18}$$

where S_i represents the ith value in the sequence and n is the number of features.

After comprehensively considering the global information, the self-attention mechanism learns the dependencies between the information and assigns weights to different features. The incorporation of the attention mechanism enables the model to selectively focus on significant information, thereby enhancing computational efficiency.

6.5.4 Federated learning

Federated Learning (FL) is a paradigm of machine learning where the majority of the training procedure is executed through distributed training on the client side. Federated learning was first proposed by Google. Its design concept is to realize the collaborative learning of a shared prediction model on mobile phones. The training data does not need to be sent to the server but is only kept on the mobile phone, which separates the training part of the model from the need for storing data. Not only does it train better models, but it also uses lower power consumption while protecting customer data privacy. In addition to furnishing updates to the shared model, federated learning harbors another advantage it can instantly deliver an enhanced model tailored to user habits for a personalized experience. Federated learning is suitable for dealing with situations where data privacy is more sensitive, and the dataset is large and unevenly distributed.

Federated learning can be categorized into three types based on the relationship between clients, namely horizontal federated learning, vertical federated learning, and federated transfer learning, as illustrated in Fig. 6.12. In horizontal federated learning, there is more feature overlap between customers and less user overlap. The essence is to combine samples, expanding the number of samples to increase data diversity. In vertical federated learning, the features between customers are mostly

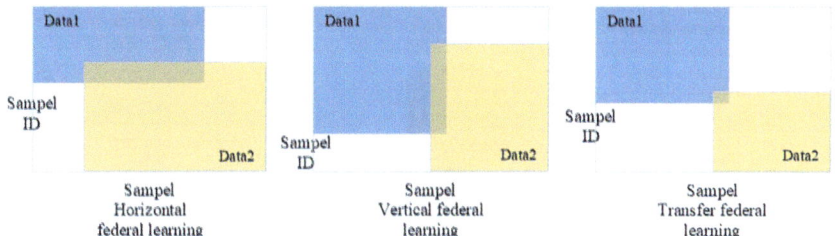

Figure 6.12 Federated learning classification.

different, but there is a lot of user overlap. This involves combining features to analyze customers more comprehensively and in more directions. The features and users of federated transfer learning have less overlap between customers and users, and it is essentially a kind of transfer learning. Considering that the multivariate load forecasting microgrids carried out in this chapter have the same characteristics and different geographical locations, which means users are different, the horizontal federated learning model is finally selected.

Federated learning is an effective approach for jointly training a model in a distributed manner while ensuring customer privacy protection, data security, and regulatory compliance. Recently, this technique has been applied to various domains, such as banking, transportation, and communication. In the energy sector, research on federated learning has primarily focused on smart grids.

Federated learning can be divided into two main parts: the server side and the client side. The federated learning process typically consists of several rounds, with each round, involving multiple steps. Specifically, in each round, the client sends encrypted gradients to the server; the server then collects the parameters of the client, aggregates them, and sends back new parameters, which the client uses to update its model.

Fig. 6.13 shows the architecture of federated learning. In federated learning, an optimization problem is mainly solved by the server:

$$minf(x) = \frac{1}{m}\sum_{i=1}^{m} F_i(x) \tag{6.19}$$

where $F_i(x)$ is the loss function of the ith client and m is the number of clients. ServerOpt in Fig. 6.13 refers to the self-adaptive optimization method solved.

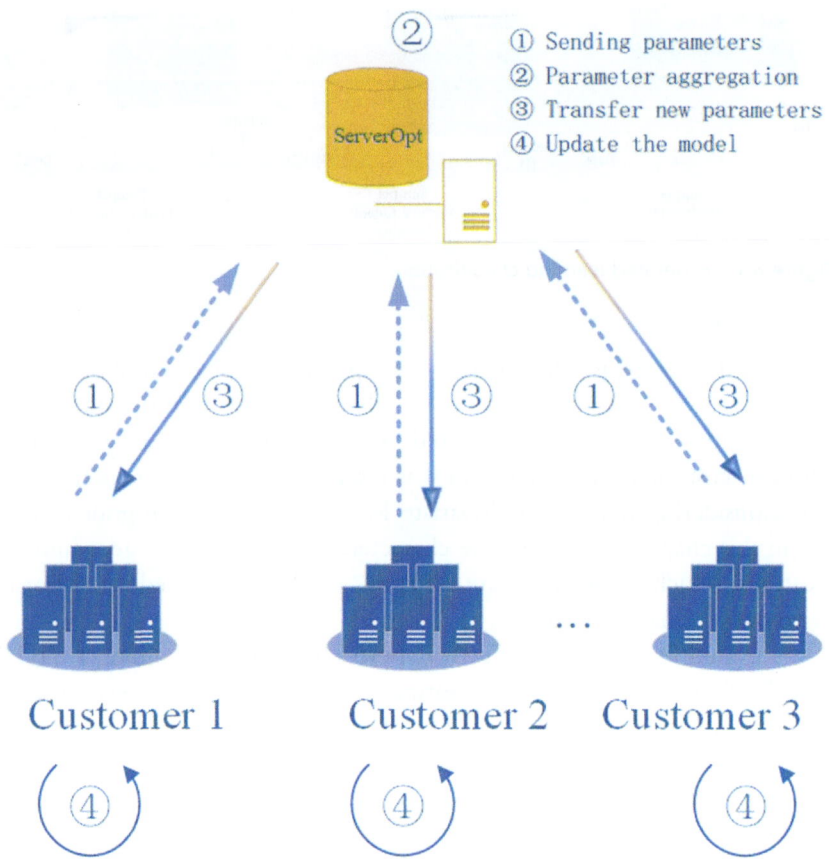

Figure 6.13 Federated learning architecture.

Algorithm 6.1 is the pseudocode of the federated model operation:

Algorithm 6.1. Federated Model Operation
1. Initialize model parameters, sampling scale
2. **for** = 1,2... **do**
3. Select the proportion f of customers from the set S of all customers
4. Server sends to selected client
5. Selected client for local training with iteration number E
6. The selected client gets the new model x_k^t and sends the model parameter difference to the server

7. The server computes a pseudogradient and updates the parameters via SERVEROPT:

$$\Delta_t = \frac{1}{|S|}\sum_{i \in S} \Delta_k^t, x_{t+1} = SERVEROPT\left(x_t, \eta_s, \Delta_t\right)$$

where t represents the t-th interaction, S is the client set, x_t represents the server model after the t-th interaction, x_{t+1} represents the server model in the $t + 1$th round, x_k^t represents the kth client model after local training in the t-th round and η_s is the learning rate of the server model.

Algorithm 6.1 is a generalization of federated learning models that incorporates four different aggregation strategies: FedAvg, FedAdagrad, FedYogi, and FedAdam. These correspond to the cases where SERVEROPT uses the optimization methods SGD, Adagrad, Yogi, and Adam, respectively.

In the FedAvg-based federated learning model, the server updates the global model parameters to the mean value of the parameters transmitted by individual clients. The parameters are updated as follows:

$$x_{t+1} = \frac{1}{|S|}\sum_{i \in S} x_i^t \tag{6.120}$$

The federated model is based on FedAdagrad, FedYogi, and FedAdam policies; the parameters are updated as follows:

$$x_{t+1} = x_t + \eta_s \frac{m_t}{\sqrt{v_t} + \tau} \tag{6.21}$$

$$m_t = \beta_1 m_{t-1} + (1 - \beta_1)\Delta_t \tag{6.22}$$

Among them, the corresponding to FedAdagrad, FedYogi, and FedAdam are different, respectively (6.23), (6.24), and (6.25):

$$v_t = v_{t-1} + \Delta_t^2 \tag{6.23}$$

$$v_t = v_{t-1} - (1 - \beta_2)\Delta_t^2 sign(v_{t-1} - \Delta_t^2) \tag{6.24}$$

$$v_t = \beta_2 v_{t-1} + (1 - \beta_2)\Delta_t^w \tag{6.25}$$

In the formula, τ is the adaptive coefficient, and the smaller the value is, the higher the adaptive degree is; β_1 and β_2 are the attenuation coefficients.

After the server aggregates the parameters sent by the client using different aggregation strategies, it redistributes the parameters to the client for updating.

Federated learning can improve the problems faced by individual training and centralized learning. Individual training means that each integrated energy microgrid only uses the data of this microgrid to locally train the prediction model. Each microgrid conducts model training separately, which faces poor data diversity and may have poor data quality, such as missing data and many outliers. These issues reduce the generalization and fault tolerance of the predictive model, promted the emergence of centralized learning. Centralized learning collects data from all microgrids together, and a server uses this data for model training. This mode improves the inadequacy of individual training to a certain extent, but the final trained model may not be suitable for all customers and faces data privacy leakage. To address the former limitation, personalization becomes a viable approach. Personalization can be achieved by retraining the model locally for a small number of iterations using the user's data exclusively. For data privacy concerns, some campuses cannot conduct centralized training due to local policies, campus regulations, user privacy, and other issues, and federated learning faces fewer privacy risks than centralized learning. In centralized learning mode, even though the client is anonymous when sending data, the identity of the client can still be discovered and attacked through reverse engineering. In the federated learning mode, each client only needs to send model parameters, revealing little information. In addition, the parameter aggregation process is short, and the aggregated parameters are discarded after they are sent and are not stored on the server. Although the data still needs to be sent and recorded, the parametric data reveals much less detail than the original payload data.

The federated learning approach can not only enhance the model's generalization performance and increase data diversity but also provide effective privacy preservation mechanisms for client data.

6.5.5 Model building

Based on the above three deep learning networks and mechanisms, this chapter constructs a CNN-Attention–LSTM model for multivariate load prediction, including one layer of CNN, two layers of LSTM, and one layer of attention mechanism. The model employs a convolutional neural network (CNN) layer to analyze input data and extract features from it. The first layer of long short-term memory (LSTM) is employed for extracting time-series features in electricity, cold, thermal loads, and meteorological data. The integration of the attention mechanism is used for computing feature importance and assigning corresponding weights. The second layer of LSTM is leveraged for processing the extracted information and synthesizing the received weight matrix to execute multivariate load forecasting tasks.

The overall structure of the CNN-Attention-LSTM model is shown in Fig. 6.14. Input data encompasses electricity, cold, thermal load data, meteorological data, and day type. CNN denotes the CNN layer, LSTM1 signifies the first LSTM layer, and the attention mechanism computes the feature weight matrix. LSTM2 represents the second LSTM layer, with the forecast result being the predicted value of electricity, cold, and thermal loads.

The process of building a model can be broadly divided into the following five steps:

1. Cleaning the original dataset, filling in missing values, and replacing outliers.
2. Calculating the correlation coefficients and selecting influential factors with strong correlations to include in the full data set.
3. Dividing the full dataset into training, validation, and testing sets.
4. Normalizing each dataset to improve the effectiveness of model training.
5. Training the network using the training set, optimizing internal parameters using the validation set, and finally, obtaining the prediction model. The efficacy of the model is tested using the test set.

The model-building process is shown in Fig. 6.15:

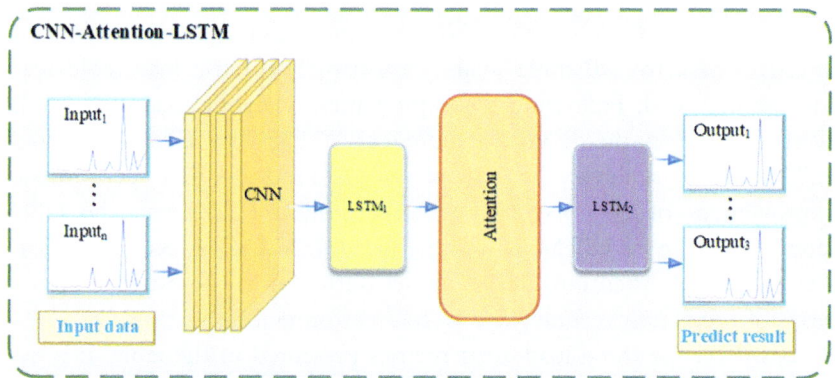

Figure 6.14 Structure of CNN-Attention-LSTM.

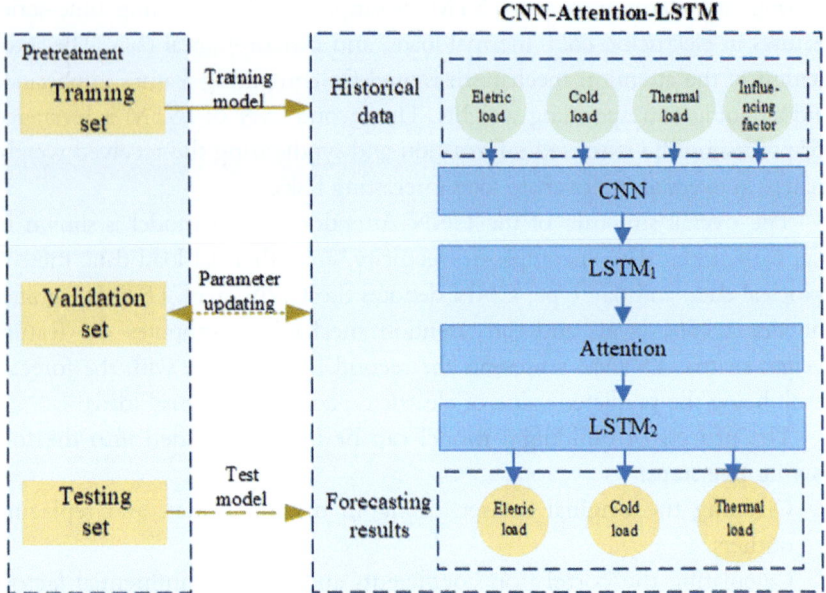

Figure 6.15 Model building process.

6.6 Example analysis

6.6.1 CNN-Attention-LSTM prediction results comparison

The trained model is tested using the test set, and the prediction error and WMA are shown in Table 6.1, In the table, C-L refers to the CNN-LSTM model, and C-A-L refers to the CNN-Attention-LSTM model.

As shown in Table 6.1 the CNN-Attention-LSTM model exhibits the lowest error across all three load types, namely electric load, cold load, and thermal load. Following a comprehensive weighting calculation, the WMA of RF is the lowest, standing at 77.78. The CNN-Attention-LSTM model achieves the highest WMA index, with a performance improvement of 2.90% over the CNN model, 2.12% over the DNN model, 1.00% over LSTM, and 0.8% over CNN-LSTM. Results indicate that the CNN-Attention-LSTM model outperforms the other models in terms of prediction accuracy and overall performance.

Based on the three load error metrics presented in the table, it is evident that the electric load exhibits the highest degree of controllability

Table 6.1 Forecasting error and accuracy of models.

Model	Electrical load		Cold load		Thermal load		WMA (%)
	MAPE (%)	RMSE (kW)	MAPE (%)	RMSE (Tons)	MAPE (%)	RMSE (mmBTU)	
RF	13.75	3266.7	12.95	2112.45	35.72	3.99	77.78
CNN	4.53	1025.11	9.49	571.53	7.79	0.84	92.18
DNN	3.98	890.31	9.37	570.01	6.24	0.69	92.96
LSTM	3.31	793.25	7.38	376.99	5.77	0.66	94.08
C-L	3.01	749.77	7.41	358.45	5.38	0.61	94.28
C-A-L	2.70	691.85	5.99	326.44	4.95	0.55	95.08

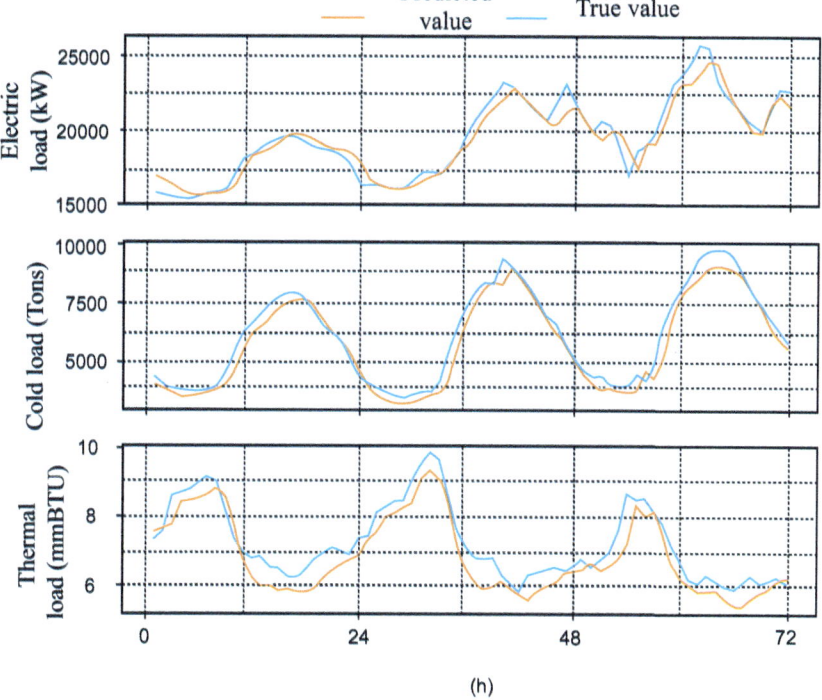

Figure 6.16 Forecasting results of the CNN-Attention-LSTM model.

and the least prediction error. The error of cold load and thermal load is slightly higher than that of electric load due to weak controllability.

Fig. 6.16 illustrates a comparison between the forecasted and actual values of the CNN-Attention–LSTM model for electric load, cold load, and thermal load.

6.6.2 Influence of coupling relationship on results

In a conventional energy network, the electric, refrigeration, and thermal systems typically operate in isolation, with each energy load forecasted independently. Integrated energy microgrids, have interconnected systems, and the coupling relationship between energy sources must be taken into account when making predictions. To examine the impact of coupling relationships on multiload forecasting, the CNN–Attention–LSTM model was employed in this chapter to simulate joint forecasting for multiple loads, as well as independent forecasting for single loads, and compared the prediction results. Multivariate load forecasting takes into account the coupling relationship between various energy systems by using electric, refrigeration, and thermal loads as inputs and outputs for model training and forecasting. In contrast, single-load forecasting employs only one type of load as input data and prediction result. Moreover, the network architecture remains unchanged for the two approaches. The multivariate load forecasting model is illustrated in Fig. 6.17, while the single-load forecasting diagram is depicted in Fig. 6.18.

Table 6.2 shows that the electric load prediction MAPE under multiple load prediction is 1.56% lower than that of single-load prediction, the cold load prediction MAPE is reduced by 1.29%, the thermal load

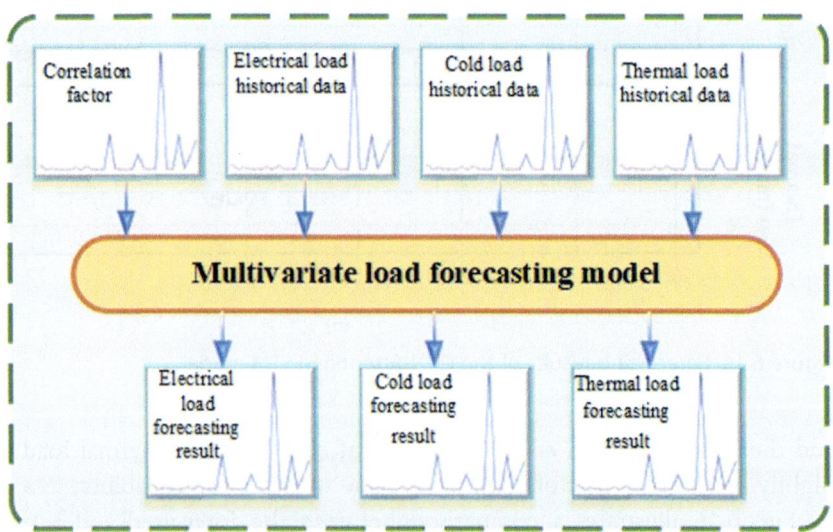

Figure 6.17 Schematic diagram of the load forecasting model.

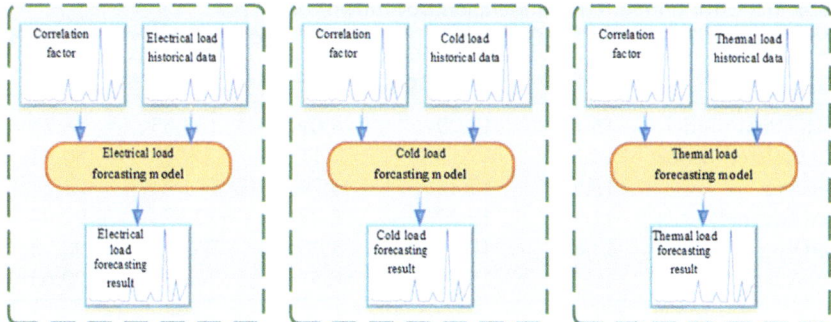

Figure 6.18 Schematic diagram of the single-energy load forecasting model.

Table 6.2 Multienergy and single-energy load forecasting results.

	Single-load		Multiple loads	
	MAPE(%)	RMSE	MAPE(%)	RMSE
Electric load	4.26	1051.26	2.70	691.85
Cold load	7.28	396.78	5.99	326.44
Thermal load	6.23	0.75	4.95	0.55
WMA	93.74		95.08	

prediction MAPE is reduced by 1.28%, and the WMA is increased by 1.34%. Observably, taking into account the interdependence between energy sources, multivariate load forecasting attains greater precision than individual load forecasting. Within an integrated energy microgrid, accounting for the interrelationships between energy systems is advantageous for enhancing the accuracy of multiload forecasting.

6.6.3 Comparison of the results of normal federal learning

The model's error metric and weighted moving average (WMA) prediction under standard operation is simulated to evaluate the performance of each model, as summarized in Table 6.3, Table 6.4, Table 6.5, and Table 6.6.

From Tables 6.3 to 6.6 one can observe that, no matter which campus, the prediction error of all models is smaller than that of individual models, of which the centralized model and FedAdagrad have the highest prediction accuracy.

In the context of the downtown campus, the Weighted Mean Absolute (WMA) error of the centralized model is 89.94%, which surpasses that of the local model by 1.21%. The WMA errors of the four

Table 6.3 Error during regular operation(Downtown).

Model	Cold load		Electric load		WMA(%)
	MAPE(%)	RMSE(Tons)	MAPE(%)	RMSE(kW)	
Individual model	15.13	13.79	8.69	103.65	88.73
Centralized Model	14.50	17.19	7.11	109.45	89.94
FedAvg	12.50	12.42	5.76	72.91	91.54
FedAdagrad	11.74	12.22	5.78	72.12	91.84
FedYogi	13.65	14.29	5.93	76.57	90.98
FedAdam	15.42	15.54	5.31	72.58	90.65

Table 6.4 Error during regular operation(Polytechnic).

Model	Cold load		Electric load		WMA(%)
	MAPE(%)	RMSE(Tons)	MAPE(%)	RMSE(kW)	
Individual model	27.35	37.61	10.61	229.39	81.02
Centralized Model	23.66	27.65	6.26	162.85	85.04
FedAvg	26.78	30.24	7.58	161.52	82.82
FedAdagrad	23.91	28.35	6.40	141.68	84.85
FedYogi	27.70	34.20	6.72	151.62	82.79
FedAdam	24.85	28.59	6.43	146.70	84.36

Table 6.5 Error during regular operation(Tempe).

Model	Cold load		Electric load		WMA(%)
	MAPE(%)	RMSE(Tons)	MAPE(%)	RMSE(kW)	
Individual model	11.31	682.65	4.03	917.23	90.87
Centralized Model	6.79	400.49	2.93	719.94	94.36
FedAvg	9.10	531.36	3.60	820.49	92.55
FedAdagrad	7.87	425.15	3.74	821.08	93.37
FedYogi	9.08	468.76	3.36	775.60	92.63
FedAdam	8.95	468.22	3.60	819.17	92.65

Table 6.6 Error during regular operation(West).

Model	Cold load		Electric load		WMA(%)
	MAPE(%)	RMSE(Tons)	MAPE(%)	RMSE(kW)	
Individual model	25.35	189.27	17.76	293.80	76.93
Centralized Model	27.51	175.43	10.60	189.71	77.56
FedAvg	21.03	156.02	11.73	211.20	81.76
FedAdagrad	19.51	123.20	12.46	205.27	82.60
FedYogi	19.87	122.33	13.81	234.64	81.95
FedAdam	22.02	159.05	12.04	217.89	80.97

federal models are 1.92% to 3.11% higher than that of the regional model, indicating superior performance compared to the centralized model. Among them, FedAdagrad exhibits the best prediction accuracy, with a WMA error of 91.84%, 3.11% higher than the single model and 1.9% higher than the centralized model.

Regarding the Polytechnic campus, the centralized model achieved a Weighted Mean Absolute (WMA) error of 85.04%, exceeding that of the local model by 4.02%. In comparison to the single model, the four federal models also enhance prediction accuracy to varying degrees. Notably, FedAdagrad obtained a WMA error of 84.85%, surpassing the single model by 3.83% and differing from the centralized model by only 0.19%.

Concerning the Tempe campus, the centralized model yields a Weighted Mean Absolute (WMA) error of 94.36%, while FedAdagrad achieves 93.37%, surpassing the single model by 3.49% and 2.50%, respectively. Other federal models also exhibit superior performance compared to individual models in terms of WMA error.

In terms of the West Campus, FedAdagrad exhibits the highest predictive performance, with a WMA error of 82.60%, surpassing the single and centralized models by 5.67% and 5.04%, respectively. Compared to the single model, the prediction accuracy of the other three federal models is also significantly enhanced, ranging from 4.04% to 5.02%, which is superior to that of the centralized model.

Based on the aforementioned data, it is evident that under standard operation, the federated learning-based model can attain comparable or superior accuracy compared to a centralized model. The FedAdagrad-based model exhibits the highest degree of accuracy of the models considered. Furthermore, the performance gains achieved by federated models are particularly pronounced when dealing with schools characterized by low data quality, such as West Campus.

Using the Downtown campus as a case study, Fig. 6.19 illustrates the predicted and actual value curves of FedAdagrad.

In addition to regular operation, there is a risk of false data injection due to the inclusion of data transmission in federal learning. The stability of the multivariate load forecasting model based on national education is, therefore, tested by simulating false data injection attacks.

6.6.3.1 Forecast results under FDIA

This part of the experiment simulates the process of transmitting model parameters from the school district to the server, and the parameters are

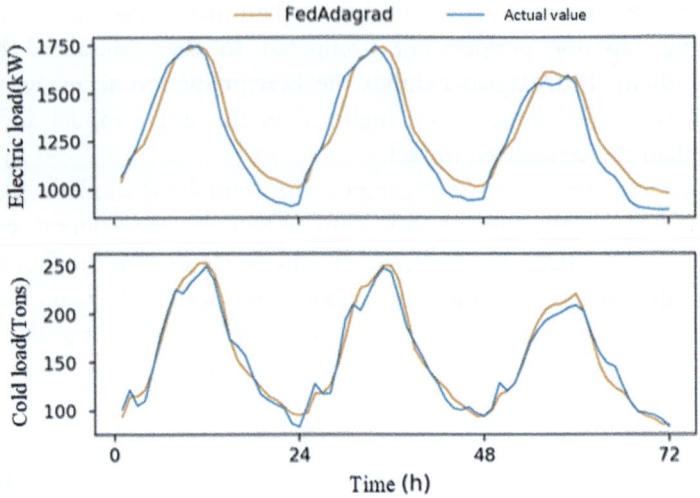

Figure 6.19 Forecast curve of Downtown.

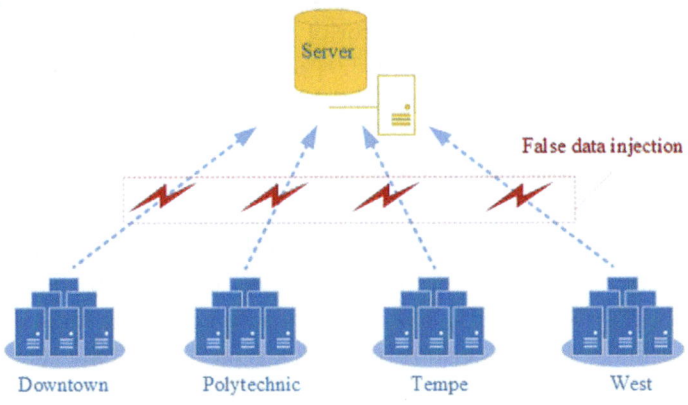

Figure 6.20 Schematic diagram of the fake data injection attack.

randomly added noise. Reference (Mode et al., 2019.), the noise range is set from 0.01% to 0.05%. A false data injection attack is depicted in Fig. 6.20.

In the case of an attack, the prediction errors for FedAvg, FedAdagrad, FedYogi, and FedAdam are listed in Table 6.7, Table 6.8, Table 6.9, and Table 6.10.

From Tables 6.7 to 6.10, we can see that the WMA of FedAdagrad is still the highest of the four federal models, i.e., the best predictor, regardless of which school district it encounters with FDIA.

Table 6.7 Errors of the federal model under FDIA (Downtown).

Model	Cold load		Electric load		WMA(%)
	MAPE(%)	RMSE(Tons)	MAPE(%)	RMSE(kW)	
FedAvg	15.35	16.44	4.69	58.43	91.05
FedAdagrad	13.94	15.13	5.13	66.58	91.35
FedYogi	15.08	15.65	6.25	83.62	90.22
FedAdam	15.29	14.76	6.48	98.88	89.99

Table 6.8 Errors of the federal model under FDIA (Polytechnic).

Model	Cold load		Electric load		WMA(%)
	MAPE(%)	RMSE(Tons)	MAPE(%)	RMSE(kW)	
FedAvg	27.62	31.29	8.08	173.69	82.15
FedAdagrad	26.13	0.26	6.52	143.44	83.68
FedYogi	30.63	42.83	7.69	187.41	80.84
FedAdam	31.23	31.02	6.71	165.41	81.03

Table 6.9 Errors of the federal model under FDIA (Tempe).

Model	Cold load		Electric load		WMA(%)
	MAPE(%)	RMSE(Tons)	MAPE(%)	RMSE(kW)	
FedAvg	10.67	567.61	3.71	917.69	91.42
FedAdagrad	10.26	527.35	3.92	859.08	91.64
FedYogi	11.41	571.13	3.31	810.36	91.02
FedAdam	11.27	617.29	3.69	870.25	91.01

Table 6.10 Errors of the federal model under FDIA (West).

Model	Cold load		Electric load		WMA(%)
	MAPE(%)	RMSE(Tons)	MAPE(%)	RMSE(kW)	
FedAvg	22.40	156.28	13.22	215.08	80.36
FedAdagrad	20.91	147.21	11.35	197.21	81.96
FedYogi	23.39	173.17	9.87	176.56	80.66
FedAdam	22.16	136.31	12.57	217.80	80.72

Tables 6.3 to 6.6 and Tables 6.7 to 6.10 are compared, and WMA under FDIA is subtracted from the WMA of each model under normal operating conditions, resulting in a WMA decrease value to analyze the impact of FDIA on federal models, and summarized in Table 6.11.

From Table 6.11, one can observe that in the four school districts, the WMA of FedAvg decreased by 0.49% to 1.40%, FedAdagrad decreased

Table 6.11 WMA difference between the federation model under regular operation and FDIA.

Model	WMA Difference (%)			
	Downtown	Polytechnic	Tempe	West
FedAvg	0.49	0.67	1.13	1.40
FedAdagrad	0.49	1.17	1.73	0.64
FedYogi	0.76	1.95	1.61	1.29
FedAdam	0.66	3.33	1.64	0.25

Table 6.12 WMA differences between the federated models under FDIA and individual models.

Model	WMA (%)			
	Downtown	Polytechnic	Tempe	West
FedAvg	2.32	1.13	0.55	3.43
FedAdagrad	2.62	2.66	0.77	5.03
FedYogi	1.49	–0.18	0.15	3.73
FedAdam	1.26	0.01	0.14	3.79

by 0.49% to 1.73%, FedYogi decreased by 0.76% to 1.95%, and FedAdam decreased by 0.25% to 3.33%. Under false data injection attacks, the prediction accuracy of each federated model exhibits varying levels of degradation compared to normal operation. With the exception of FedAdam, the other models experience a relative performance decline of less than 2% and demonstrate good robustness. By comparing Tables 6.1 and 6.2, which present the difference between cold load and thermal load in various school districts, it can be inferred that FDIA has a greater impact on thermal load than electric load. WMA is less dramatic in the districts with larger power load capacity, that is, those with the more significant weight of power load prediction task in WMA. For example, in the downtown campus, the consequences of electric and cold load forecasting jobs are 0.6 and 0.4, respectively, and WMA is reduced by 0.49% to 0.76% after FDIA.

On the contrary, WMA is more affected in schools with larger cold load capacity. For example, on the Tempe campus, the weights of electric and challenging load forecasting tasks are 0.3 and 0.7, respectively. After FDIA, WMA decreased by 1.13% to 1.73%, which is more affected than Downtown.

The WMA difference between the Federal model under FDIA and the individual model is shown in Table 6.12.

The difference refers to Tables 6.7 to 6.10 minus the WMA of individual models in Tables 6.3 to 6.6. From Table 6.12, one can observe that other federal models, except FedYogi which has a 0.18% decrease in WMA for Polytechnic school districts, still have higher accuracy than the individual models. FedAdagrad remains the tallest model in WMA of all federal models. Hence, even after suffering from FDIA, the national model remains stable and achieves better predictions than individual models.

To more visually reflect the effects of the FedAdagrad model, visualize the WMA of the individual model, the centralized model, the regular operation, and the FedAdagrad model under FDIA, as shown in Fig. 6.21.

From Fig. 6.21, it can be seen that the FedAdagrad model is less affected by FDIA and can maintain similar accuracy to regular operation. Moreover, the FedAdagrad model always has a higher WMA than the individual model, whether running normally or under attack. The FedAdagrad model is capable of achieving and possibly surpassing, the prediction performance of the centralized model. Notably, in school districts characterized by low data quality, the performance gains are particularly pronounced for those with poor predictive results under the single-model approach.

6.6.4 Training time comparison

In addition to the model's prediction accuracy, the time required to train each model is also counted in Table 6.13. This part expands the data into

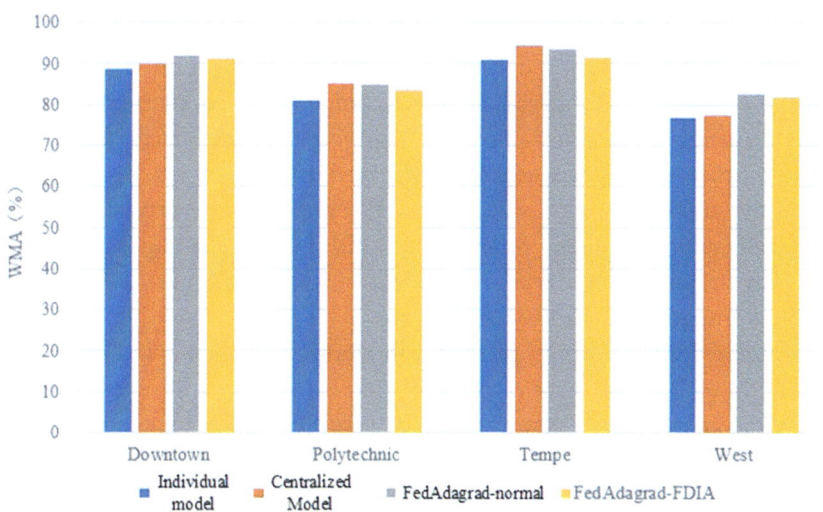

Figure 6.21 WMA comparison of the FedAdagrad model with other models.

Table 6.13 Comparison of time evaluation of different models.

Model	Number of interactions	Number of correction zones	Number of iterations	Time (s)
Individual model	–	1	100	140.21
Centralized Model	–	8	100	350.42
FedAvg	20	8	5	229.04
FedAdagrad	20	8	5	214.09
FedYogi	20	8	5	210.35
FedAdam	20	8	5	238.99

eight school districts based on the historical load data of Downtown, Polytechnic, Tempe, and West campuses to make the comparison more apparent. It simulates how eight school districts jointly train the multivariate load prediction model. For each model to run under the same hardware conditions, only 20% of the GPU memory is used for training each model here.

The term "centralized model time" in this context strictly refers to the duration required to train the model, excluding any data transmission time. By contrast, the time required by the federated model encompasses multiple stages, including data transfer, local training, and server parameter aggregation. The number of interactions is a unique parameter in federal learning, so individual and centralized models are represented by '-.'

From Table 6.13, it can be concluded that individual models take the shortest time, 140.21 s, as unique models not only have relatively small amounts of data but also do not require data transmission. Due to the large volume of data involved, centralized models are constrained to operate on a single device and are heavily impacted by the hardware conditions of that device, leading to prolonged training times. In contrast, while the amount of data processed by federated models may also be sizable, federated learning represents a distributed training framework whereby training can occur concurrently across multiple devices with minimal impact from hardware constraints.

In the federated learning paradigm, the server selects only four clients for training in each round, with each client contributing only half of its local data to the training process. Consequently, the federated model requires less training time compared to the centralized model. Although the FedAdagrad model requires 73.8 s more training time than the individual model, its accuracy can be improved by 2.50% to 5.67%, which is of great economic significance.

6.7 Conclusions and outlook

6.7.1 Conclusion

Accurate multivariate load prediction is the precondition for integrating energy microgrids' safe and stable operation. The combined energy microgrid contains a variety of energy systems, and the designs are coupled with each other, so the internal energy conversion relationship is complex. Traditional power load forecasting methods can not be fully applied to multivariate load forecasting because there is only one form of energy in the system, and there is no energy coupling relationship. By analyzing the research status, this chapter selects in-depth learning to build the basic model, introduces the Federal Learning Improvement Training mode, and finally makes the CNN-Attention-LSTM model in the Federal Learning mode. The work of this article is summarized as follows:

1. A literature review was performed to comprehend the research status of integrated energy microgrid and multivariate load prediction. The internal structure of integrated energy microgrid, the characteristics of the multivariate load, and the classification of multivariate load prediction are further analyzed.

2. Processing the abnormal data and analyzing the relationship between loads and load and influencing factors by calculating the Spearman rank correlation coefficient. Highly correlated data are then selected as input. The dataset is normalized after partitioning the dataset.

3. The CNN-Attention-LSTM model is constructed, and its predictive accuracy is evaluated relative to other models through illustrative simulations. By comparing the performance of multivariate load forecasting with that of single-load forecasting, the beneficial impact of accounting for coupling relationships in multivariate load forecasting is confirmed.

4. This study proposes the CNN-Attention-LSTM model as the global and Federal Learning model. Four Federation models with distinct Federation strategies, namely FedAvg, FedAdagrad, FedYogi, and FedAdam, are simulated to compare their prediction results against the individual and centralized models, thereby evaluating each model's performance. The results of the experiment indicate that the predictive accuracy of the four federal models is greater than that of individual

models and comparable to or even higher than that of centralized models. Additionally, when dealing with microns that possess low-quality data, the federal model exhibits a noteworthy advantage in predictive accuracy compared to the centralized model.

5. False data injection attacks were simulated to examine the stability and reliability of the federal model. Simulation results demonstrate that the federal model remains stable and reliable in the face of such attacks. FedAdagrad is always the most accurate predictor of the four federal models, whether in regular operation or under attack by spurious data injection.

6. An analysis of the computational resource requirements for each model was conducted by comparing their respective training times. The results indicate that the federal model is more privacy-preserving than the centralized model while also achieving similar or even better prediction outcomes, shorter training times, and lower hardware requirements.

6.7.2 Expectation

In this chapter, the multivariate load forecasting of integrated energy microgrids has been explored and studied, and some progress has been made, but there are still many shortcomings. To further improve the accuracy of multivariate load forecasting, the following improvements are needed:

1. Increase the type of energy. The integrated energy microgrid simulated in this paper contains only three energy sources: electricity, cold, and thermal. As some microgrids lack thermal data, the coupling relationship between thermal energy and other energy sources cannot be reflected in the training of the federal model. Follow-up attempts to add points such as "gas."

2. Consider more influencing factors. The simulation in this paper only collects meteorological data and day type. Many factors affect the multivariate load. Other factors, such as energy price, charging, and discharging of electric vehicles, will be explored and studied in the future.

3. Explore more accurate multivariate load forecasting models. This paper builds the CNN-Attention-LSTM model. While it has demonstrated improved prediction results compared to certain models, the

prediction accuracy of some microgrids remains suboptimal. Consider looking for newer and more appropriate models, or try to modify the prediction results.

4. Explore new federal policies to simulate more types of attacks. This paper builds four federal models, which only simulate a false data injection attack. Subsequent studies consider trying other national strategies or learning models to investigate different types of attacks and corresponding defense mechanisms.

References

Bie, C., Wang, X., & Hu, Y. (2017). Summary and prospect of energy internet planning research. *Chinese Journal of Electrical Engineering*, *37*(22), 6445−6462.

Chengshan, W., Dan, W., & Yue, Z. (2015). Intelligent distribution system architecture analysis and technical challenges. *Power System Automation*, *39*(09), 2−9.

Chen, J., Hu, Z., & Chen, W. (2021). Load forecasting of integrated energy system based on quadratic mode decomposition combined with DBiLSTM-MLR. *Power System Automation*, *45*(13), 86−94.

Chen, W., Ying, W., Tao, Z., Zemei, D., & Kaifeng, Z. (2022). Multi-Energy Load Forecasting in Integrated Energy System Based on ResNet-LSTM Network and Attention Mechanism. *Transactions of China Electrotechnical Society*, *37*(7), 1789−17999.

Chen, G., Zhao, P., & Shan, J. (2021). Comprehensive energy load forecasting based on wavelet optimization and multi-task learning. *Journal of Liaoning University of Engineering and Technology (Natural Science Edition)*, *40*(02), 163−169.

Cui, S., & Wang, X. (2022). Multivariate load forecasting of integrated energy system based on maximum information coefficient and multi-objective stacking integrated learning. *Power Automation Equipment*, *05*, 1−9.

He, G., Jin, L., & Li, K. (2021). Multi-energy load forecasting of integrated energy system based on improved DANN. *Power Engineering Technology*, *40*(06), 25−33.

Jia, H., Wang, D., & Xu, X. (2015). Research on several issues of regional integrated energy system. *Power System Automation*, *39*(07), 198207.

Jianpeng, M., Wenjie, G., & Zhisheng, Z. (2020). Based on Copula theory and KPCA-GRNN, a short-term forecasting model for multiple loads of regional comprehensive energy system. *New Technology of Electrical Energy*, *03*, 24−31.

Jianzhong, W. (2016). The driving force and current situation of the development of European integrated energy system. *Power System Automation*, *40*(05), 1−7.

Liu, E., Wang, Y., & Huang, Y. (2020) *Short-term forecast of multi-load of electrical heating and cooling in regional integrated energy system based on deep LSTM RNN*, 2020 IEEE 4th Conference on Energy Internet and Energy System Integration (EI2 2020), Wuhan, China, October 2020, 2994−2998. Available from https://doi.org/10.1109/EI250167.2020.9347300.

Li, J. (2018). Research on the development strategy of China's electricity market under the energy internet. *Electrical Engineering*, *10*, 64−69.

Li Q., Tang X., Luo Y., Liu W., Chen Q., & Hu J. (2021) *Integrated Energy System Load Forecast Based on Information Entropy and Stacked Auto-Encoders*, 2021 IEEE/IAS Industrial and Commercial Power System Asia, (I & CPS Asia), Chengdu, China, January 2021, pp. 385−390. Available from https://doi.org/10.1109/ICPSAsia52756.2021.9621631.

Li, Y., Wu, M., & Zhou, H. (2015). Discussion on several issues of regional multi-energy system based on the all-energy flow model. *Power Grid Technology*, *39*(08), 2230−2237.

Mode, G.R., Calyam, P., & Hoque, K.A. (2019). False data injection attacks in internet of things and deep learning enabled predictive analytics. arXiv preprint arXiv:1910.01716. Available from https://doi.org/10.48550/arXiv.1910.01716.

Shensi Cooling, heating and power load prediction of the park using distributed energy. 2018.

Strive to build a world-class energy internet enterprise [N] State Grid News. 2019.

Sun, Q., Wang, X., & Zhang, Y. (2021). Multiple load forecasting of integrated energy system based on LSTM and multi-task learning. *Power System Automation*, *45*(05), 63−70.

Tan, Z., De, G., Li, M., Lin, H., Yang, S., Huang, L., & Tan, Q. (2020). Combined electricity-heat-cooling-gas load forecasting model for integrated energy system based on multi-task learning and least square support vector machine. *Journal of Cleaner Production*, *248*, 119252. Available from https://doi.org/10.1016/j.jclepro.2019.119252.

Wang Y., Ma K., Li X., Liang Y., Hu Y., Li J., & Liu H. (2020). *Multi-type Load Forecasting of IES Based on Load Correlation and Stacked Auto-Encode Extreme Learning Machine*, 10th International Conference on Power and Energy Systems, ICPES 2020 585−589, Chengdu, China, 25 December 2020. Available from https://doi.org/10.1109/ICPES51309.2020.9349738.

Wang, S., Wang, S., Chen, H., & Gu, Q. (2020). Multi-energy load forecasting for regional integrated energy systems considering temporal dynamic and coupling characteristics. *Energy*, *195*116964. Available from https://doi.org/10.1016/j.energy.2020.116964.

Wang, Y., Zhang, N., & Guan, Y. (2020). Inheritance and expansion of current research topics on energy internet and smart grid. *Power System Automation*, *44*(04), 1−8.

Wu, W., Wu, J., & Lei, Z. (2021). Deep belief network method of cross and cross optimization for CCHP user cooling and heating load forecasting. *China Southern Power Grid Technology.*, *15*(12), 1−10.

Xiaping, Z. (2019). Comprehensive energy system optimization planning based on energy hub. *South Energy Construction*, *6*(04), 6−12.

Xing, Z. (2006). Review and prospect of national policies to promote energy efficiency and renewable energy. *Solar*, *03*, 12−14.

Yan Y., Zhang Z. (2021). *Cooling, Heating and Electrical Load Forecasting Method for Integrated Energy System based on SVR Model*, Proceedings of the 2021 6th Asia Conference on Power and Electrical Engineering (ACPEE), 1753−1758. Chongqing, China, April 2021. Available from https://doi.org/10.1109/ACPEE51499.2021.9436990.

Ye, J., Cao, J., Yang, L., & Luo, F. (2022). Ultra-short term load forecasting of user-level integrated energy system based on variational mode decomposition and multi-model fusion. *Power System Technology*, *46*(07), 2610−2622. Available from https://doi.org/10.13335/j.1000-3673.pst.2021.2566.

Yongli, W., Minhan, Z., Suhang, Y., et al. (2022). Multi-energy coupling mechanism based multi-load collaborative forecasting model of integrated energy system. *Journal of North China Electric Power University*, *49*(2), 118−126.

Yu, X., Xu, X., & Chen, S. (2016). Overview of integrated energy system and energy internet. *Journal of Electrical Technology*, *31*(01), 1−13.

Zhang, Y., Ai, Q., Lin, L., Yuan, S., & Li, Z. (2019). etc A regional-level ultra-short-term load forecasting method based on deep long-term and short-term memory network. *Power grid technology*, *43*(06), 1884−1892.

Zeng, M., Yang, Y., & Liu, D. (2016). Energy Internet "source-network-load-storage" coordinated and optimized operation mode and key technologies. *Power Grid Technology*, *40*(01), 114−124.

Zhu, R., Guo, W., & Gong, X. (2019). Short-term load forecasting for CCHP systems considering the correlation between heating, gas and electrical loads based on deep learning. *Energies*, *12*(17), 3308. Available from https://doi.org/10.3390/en12173308.

Intelligent learning approaches for demand-side controller for BIPV-integrated buildings

Zhengxuan Liu[1], Linfeng Zhang[2] and Shaojun Wang[3]
[1]Faculty of Architecture and the Built Environment, Delft University of Technology, Delft, The Netherlands
[2]Department of Underground Engineering, School of Transportation, Southeast University, Nanjing, P.R. China
[3]China Construction Fourth Engineering Division, Corp. Ltd, Guangdong, P.R. China

7.1 Introduction

Residential and commercial buildings extensively utilize energy for diverse purposes, ranging from heating and cooling to lighting and powering electronic devices. Energy consumption in buildings is a significant contributor to the overall energy consumption in vast majority of countries (Liu, Queena et al., 2022; Xie et al., 2022; Zhang et al., 2021). Renewable energy presents a promising solution for meeting the ever-increasing demand for building energy and reducing building carbon emissions (Jalil-Vega et al., 2020; Liu et al., 2018; Shen et al., 2022; Zhou & Liu, 2023; Zhou et al., 2021), especially to achieve carbon neutrality strategic target for the building sector (Liu et al., 2023). Common renewable energy sources include solar energy (Vassiliades et al., 2022; Wang, Liu et al., 2021; Zhou, Cao et al., 2020), geothermal energy (Liu et al., 2017, 2018, 2019a,b,c, 2022; Tang et al., 2020), wind energy (Bansal, 2022; Dupré la Tour, 2023), ocean energy (Liu et al., 2023; Liu, Zhou et al., 2023), biomass energy (Amjith & Bavanish, 2022; Chaudhary & Kumar, 2022; Okafor et al., 2022), and other advanced clean energy systems (Chen et al., 2022; He et al., 2022, He, Zhou, Wang et al., 2021, He, Zhou, Yuan et al., 2021; Liu, Xie et al., 2023). Among these renewable energy sources, solar photovoltaic (PV) energy systems are particularly attractive, as they are virtually unlimited in availability worldwide and can

Intelligent Learning Approaches for Renewable and Sustainable Energy
DOI: https://doi.org/10.1016/B978-0-443-15806-3.00007-3

be easily integrated with existing building roofs and facades (Li et al., 2020; Md Khairi et al., 2022; Zheng & Zhou, 2023). According to statistical data from the International Energy Agency (IEA), solar PV electricity is projected to account for 50% of the total renewable energy mix by 2024 (IEA). In fact, over the past few decades, PV systems in buildings have demonstrated the highest growth rate in installed capacity worldwide, with their share increasing each year (Ballesteros-Gallardo et al., 2021; You et al., 2023). PV systems are primarily integrated into different building components as four main types of BIPV: PV facades, PV roofs, PV windows, and PV sunshades (Zhang et al., 2018). Notably, building integrated photovoltaics (BIPV) energy systems are increasingly recognized and valued by the industry as an efficient, environmentally friendly, and cost-effective energy solution.

The current global energy crisis has made energy management of BIPV systems a hot topic in current research (Zhou, 2022g; Zhou & Lund, 2022; Zhou & Zhou, 2023). Demand-side management response and control are the most common solutions for increasing building energy flexibility and enhancing the smart grid's resilience to energy supply fluctuations (Zhou & Zhou, 2023; Zhou, 2022c, 2023). With the rapid development of AI, machine learning (ML) techniques have been widely applied in building energy systems, including accurate prediction of building energy demand (Guo et al., 2018; Sun et al., 2020, 2022b; Zhou, 2022b) and demand-side management (Liu, Zhang et al., 2023; Pallonetto et al., 2019; Song et al., 2022). It is crucial to use efficient and advanced demand-side controllers based on ML algorithms to accurately predict the demand for heating, cooling, and lighting loads, as well as the electricity generated by BIPV, in order to improve building energy management and energy flexibility (Liu, Sun et al., 2022; Sun et al., 2022a; Sun, Haghighat et al., 2022; Zhou, Zheng, Liu et al., 2020).

Due to the benefits of using AI learning algorithms for BIPV, some review studies have been conducted to examine their most advanced techniques and methods. Multiple review articles have been published between 2003 and 2023, obtained through keyword searches of Scopus and ScienceDirect databases for "AI" or "ML" and "BIPV." Table 7.1 lists the existing review studies on the application of AI algorithms to BIPV. These reviews mainly focus on (1) integrated energy system modeling, (2) fault diagnosis, (3) building load prediction, (4) optimization design, and (5) flexibility. It is worth noting that, as shown in Table 7.1, most of the reviewed literature only considers one of the many possible applications

Table 7.1 The main contributions of previous review works on intelligent learning approaches for BIPV.

Reference	Year	Topic	Main findings/contributions
Alabi et al. (2022)	2022	ML, integrated energy systems, integrated optimization techniques for prediction and decision-making.	This study provides an overview of the structure of integrated energy systems, as well as the methods and technologies used in their modeling. It explores the application of ML in the research of integrated energy systems and the trend toward integrating ML and optimization techniques to achieve optimal and feasible planning of integrated energy systems.
Sohani et al. (2022)	2022	ML, fault detection, solar energy, and smart energy.	This study provides a comprehensive review of ML techniques applied to PV systems. It includes an introduction to traditional modeling methods for PV systems from the perspectives of electrical and thermal aspects, a discussion of the applications of ML in the analysis of PV systems, and a review of the use of ML algorithms for predicting performance and detecting faults.
Yahya et al. (2022)	2022	Digital image processing, ML, deep learning, PVs.	This study presents a literature review on the application of image pattern recognition for the detection of PV modules primarily using spectral data. The first section provides an overview of image acquisition guidelines and the main detectable defect patterns under each spectrum. The second section introduces various image preprocessing steps used to prepare inspection-ready datasets. The third section investigates defect detection and classification through digital image processing and machine/deep learning techniques.
Waqar Akram et al. (2022)	2022	AI and deep learning; PV cells; Module failures and fire risks.	This study focuses on the fire risks and fault detection/measurement associated with PV modules, as well as fault detection in PV modules based on computer/machine vision or AI. All types of faults occurring in PV components, including recently reported on-site faults, are discussed.

(Continued)

Table 7.1 (Continued)

Reference	Year	Topic	Main findings/contributions
Gassar and Cha (2021)	2021	Solar PV potential; Geographic information system (GIS)–based analysis; ML	This study aims to comprehensively review the methods for estimating the solar PV potential of rooftops based on GIS. The study categorizes GIS–based methods into sampling, geostatistics, modeling, and ML. The application, advantages, and limitations of each method are reviewed and discussed.
Youssef et al. (2017)	2017	AI; PV systems; fault diagnosis	This study investigates the application of AI technology in PV research by summarizing over 100 research articles. The role of AI algorithms in PV system modeling, scaling, control, fault diagnosis, and output estimation is explored.
Kow et al. (2016)	2016	PV grid-connected systems, AI, and power quality.	This study aims to investigate the negative impacts of grid-connected PV systems on the power grid and study the performance of AI and traditional methods in mitigating power quality events.
Mellit et al. (2009)	2009	AI; PV systems; optimal design	This study analyzes various AI technologies used to determine PV systems, including independent PV systems, grid-connected PV systems, and PV–wind hybrid systems.
Mellit and Kalogirou (2008)	2008	AI; PV systems; prediction and optimization	This study provides an overview of the understanding of the operation of AI systems in PV system design by discussing some issues in the application of PV systems. It demonstrates the potential of AI as a tool for designing PV systems.
Seyedmahmoudian et al. (2016)	2016	PV systems; AI; maximum power point tracking	This study reviews the control strategies of maximum power point tracking methods based on AI technology and analyzes their effectiveness and feasibility, as well as their limitations and advantages.

of intelligent learning methods in PV systems, such as performance prediction, fault detection, or optimization.

Although there has been an extensive review of applying ML algorithms to PV systems, existing reviews have paid little attention to the demand-side control of intelligent learning methods in BIPV systems, particularly considering the analysis of intelligent learning methods involving demand-side control of BIPV systems, and the summary and discussion of related cases. This chapter focuses on the systematic analysis and summary of intelligent learning methods for demand-side control of BIPV systems, including a summary of existing research, current challenges, and analysis of relevant cases. This aims to provide a broader insight for researchers, policymakers, and investors in this field, to promote the application of ML methods in PV systems, and to provide valuable comments on current trends, main obstacles, and future prospects and opportunities.

7.2 Literature review

Rapid developments in ML technology have provided enormous opportunities for building demand-side control and intelligent energy management strategies (Zhou, 2022a, 2022e). It can intelligently control indoor environmental parameters, such as temperature, humidity, and air quality, by using sensor data and user feedback to improve indoor comfort and health while reducing unnecessary energy consumption (Chen et al., 2023; Tien et al., 2022; Zhou, 2021). ML applied to building-integrated PV systems can be divided into supervised learning, unsupervised learning, and reinforcement learning. The following sections summarize and analyze the research topics related to this classification.

7.2.1 Supervised learning

Supervised learning methods can learn from historical data and use existing features and labels to predict future demand-side control strategies (Ahmad et al., 2018; Naganathan et al., 2016). For example, algorithms, such as support vector machine (SVM) and artificial neural network (ANN), can be used to predict building energy demand and PV generation, thereby determining the optimal demand-side control strategies for building-integrated PV systems (Zhou, 2022d).

Efficient and accurate advanced demand-side controllers are crucial for enhancing the energy flexibility provided by buildings. However, existing current research rarely proposes model development and mechanisms for demand-side control. Zhou and Zheng (2020) used supervised ML methods to deal with the complexity of building demand prediction and developed an advanced demand-side controller, as shown in Fig. 7.1. Multiple linear regression, support vector regression, and backpropagation neural networks were employed. This study introduced the configuration of artificial neural networks and performed sensitivity analysis on the learning performance for different training times. The energy flexibility of complex

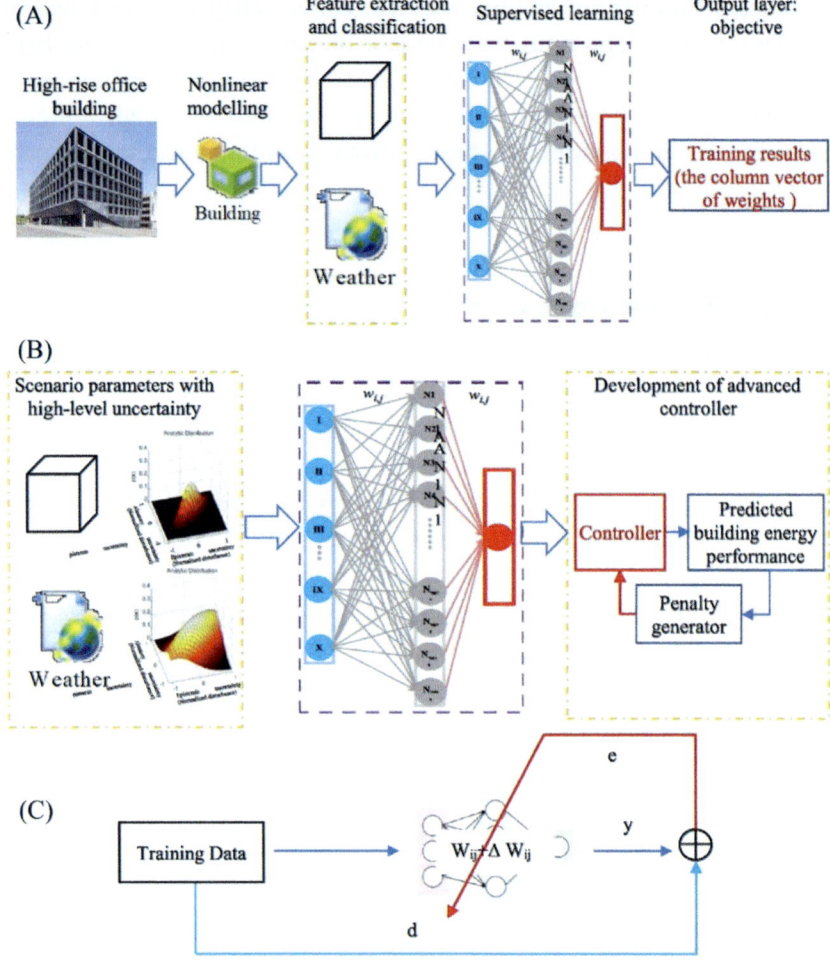

Figure 7.1 Overview of ML for developing advanced controllers: (A) deterministic case; (B) case with uncertainties: system control penalty based on supervised learning; (C) dynamic training process (Zhou & Zheng, 2020).

building energy systems was quantitatively characterized through a series of quantifiable indicators. To use the energy flexibility of demand-side management, four different controllers were developed and compared, including rule-based controllers, predictive controllers, iterative feedback controllers, and hybrid controllers. Case studies demonstrated the technical feasibility of the developed hybrid controller in enhancing energy flexibility. With the implementation of the developed hybrid controller, the peak power input to the grid could be reduced from 500.3 kW to 195 kW, a decrease of 61%.

Accurate model training usually requires a large amount of labeled data. However, in practice many application scenarios lack sufficient locally labeled data, which limits the application of supervised learning (Kabir et al., 2021). A common solution to the lack of local training data is to train models on publicly available datasets. Using open datasets, poses a problem as the differences between these datasets and the target application scenario may lead to performance degradation. This discrepancy arises since open datasets may come from different domains and have different data distributions, features, and label definitions, which are different from the data in the target application scenario. Therefore, models trained on open datasets may not generalize well to the target application scenario. To address this issue, Liu, Chen et al. (2022) proposed a novel self-supervised learning method, which generates pseudo-labeled data by using the structure and information of the data itself for training, thereby avoiding the reliance on additional manually annotated data. The advantages of self-supervised learning include good data efficiency and generalization ability, which can be widely applied in various application scenarios. Fig. 7.2A and B illustrate the different labeling processes between supervised learning and self-supervised learning for the net load decomposition problem. Fig. 7.5C shows the overall framework of the proposed self-supervised learning method. Additionally, an end-to-end network architecture was proposed as the basic estimation model. Based on the linear embedding of PV power generation, the proposed end-to-end architecture can be directly trained with PV power generation labels, which simplifies the training process and improves estimation performance.

7.2.2 Unsupervised learning

While supervised learning algorithms are commonly used for electric power quality event prediction, the disadvantage of supervised learning is the

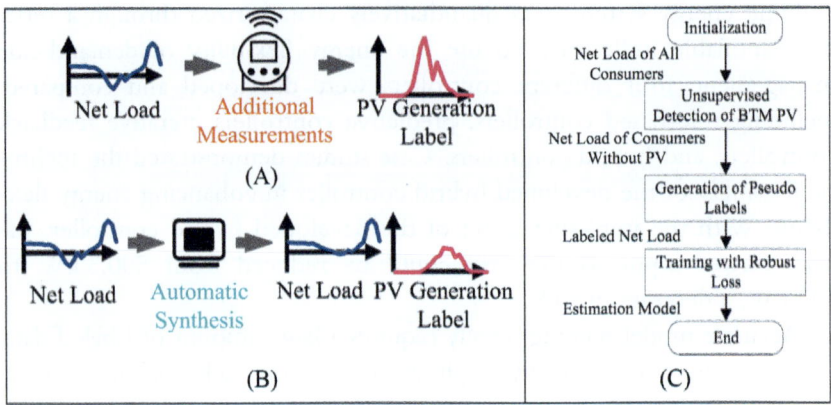

Figure 7.2 (A) The labeling process of supervised learning and (B) self-supervised learning; (C) the overall framework of the proposed self-supervised learning method (Liu Chen et al., 2022).

expensive and time-consuming data labeling process (Hebboul et al., 2015). Unsupervised learning methods can automatically discover patterns and correlations from data, extract features and regularities, and cluster data without prior knowledge (Homod et al., 2022; Jiménez-Fernández et al., 2022).

The lack of accurate understanding of PV systems can have adverse effects on the operation of the distribution network in terms of voltage regulation, frequency control, and feedback power flow. It is necessary to effectively identify all PV households on the distribution network to avoid these problems. Most research on load pattern classification focuses on using raw time series data with unsupervised ML algorithms. Using raw data may result in the model being unable to effectively distinguish between different classes. In addition, unreported PV installations can have adverse effects on voltage control and frequency regulation, leading to reverse power flow, hence the need to list these locations. Hu et al. (2021) introduced a load pattern classification method based on smart meter data, with the flowchart shown in Fig. 7.3. Unlike existing research, this chapter developed a set of interpretable and discriminative global and peak period features to extract physical information of load patterns and then used a feature-based two-step clustering method for classification. A quantitative composition analysis of the clustered groups was developed to describe the energy use variability and composition changes of each household. The proposed PV household identification method can help grid owners identify PV installations in specific areas for voltage control and ancillary services. In the long run, effective load

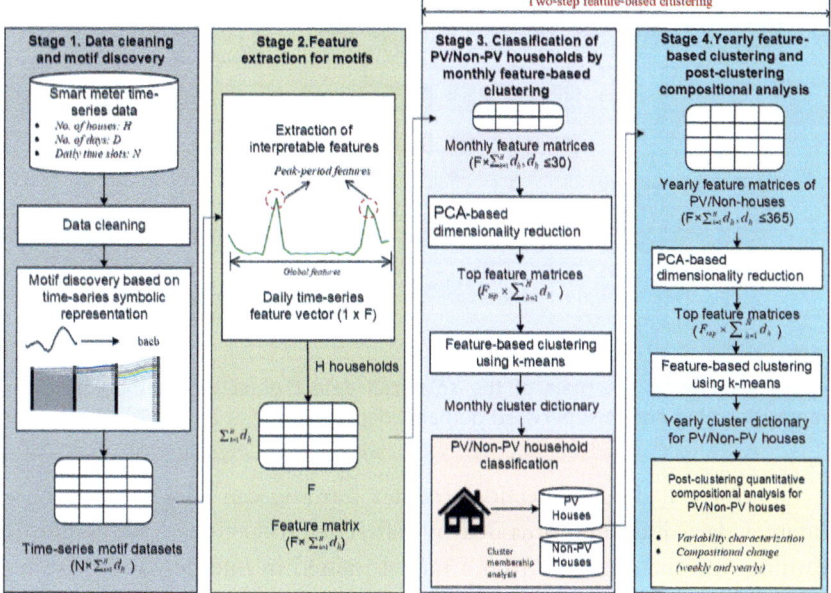

Figure 7.3 Flowchart of the developed overall analysis method for smart meter data (Hu et al., 2021).

pattern classification and feature analysis after clustering can help better understand the demand characteristics, network usage, and load growth changes at the network level for different customer groups.

Kow et al. (2018) introduced an incremental unsupervised neural network algorithm, namely a memory self-organizing incremental neural network (M-SOINN), for predicting the output power and detecting power fluctuation events in PV microgrid systems. M-SOINN uses clustering techniques to form a data graph and identify the most similar patterns for predicting PV output power and detecting power fluctuation events. The system achieved a detection rate of up to 92.69% under real-life environmental data, outperforming traditional self-organizing map (SOM), k-nearest neighbor (KNN), focused time delay neural network (FTDNN), and nonlinear autoregressive with exogenous inputs (NARX) networks. Integrating M-SOINN into a power management system mitigated power fluctuation events in the PV grid, resulting in a reduction of 79.62% of power fluctuation events, energy loss of 2.16%, and battery state-of-charge remaining between 30% and 100%. The proposed system outperformed rule-based controllers and slope controllers by 44.02% and 27.57%, respectively, in mitigating power fluctuation events.

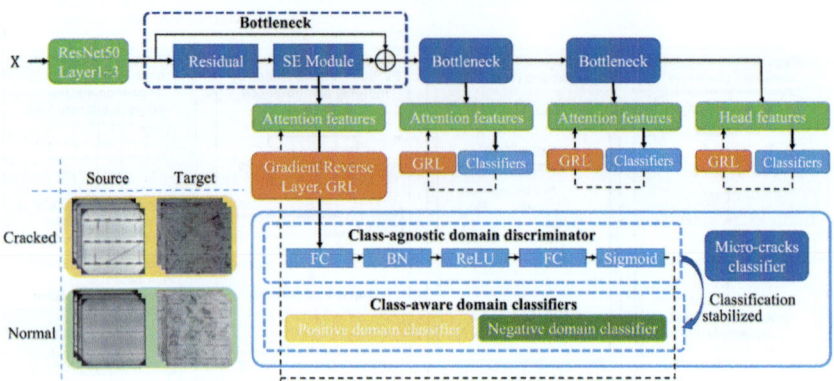

Figure 7.4 Overall flowchart of the solar cell defect detection method based on transfer learning and unsupervised domain adaptation (Xie et al., 2023).

In solar cell defect detection, transfer learning can address the issue of different data distributions caused by different production technologies. By learning universal features from a model trained in one domain, these features can be used in another domain, reducing the need for labeled data and improving model performance (Xie et al., 2023). Unsupervised domain adaptation is a commonly used transfer learning method that can use unlabeled data in one domain for model training. In addition, techniques such as adversarial learning and attention mechanism can improve the model's transferability and performance. Xie et al. (2023) proposed a transfer learning and unsupervised domain adaptation-based method for solar cell defect detection, as shown in Fig. 7.4. Due to the differences between production schemes and solar cell types, a trained model may not achieve good performance on new datasets. To address this issue, the authors used transfer learning and adversarial learning methods, embedding a domain discriminator into the classifier so that the feature extractor can learn domain-invariant features, making the model more adaptable to new data. In the scenario of defect detection in electroluminescence, the authors also proposed a method based on attention-based transfer learning and class-aware domain discriminator to further improve the model's performance. Experimental results show that the proposed method increased the F1-score by 0.2631, with recall and precision rates of 84.70% and 90.15%, respectively, demonstrating the effectiveness of the proposed method.

7.2.3 Reinforcement learning

Reinforcement learning aims to achieve optimal decision-making strategies through trial-and-error learning. In reinforcement learning, an algorithm interacts with an environment, continuously trying different actions and observing the environment's feedback to gradually learn how to maximize long-term rewards (Zhang et al., 2022). RL is one of the most active research fields in AI in the coming years and has been applied to various recent research areas (Wang & Hong, 2020), such as IoT (Min et al., 2019) and energy management (Cao & Xiong, 2017). RL has received significant attention and research in the field of building-integrated PV systems. It can use building data and simulation environments for simulation training, learning the optimal control strategy through continuous trial and error (Avila et al., 2020; Jung et al., 2021).

Using renewable energy to power buildings has always been a complex and difficult-to-control problem. Wang et al. (2022) proposed the use of reinforcement learning control methods to optimize renewable energy in green buildings, minimizing energy consumption and power loss. The algorithm was tested using SVR and a mixed SVR-Wavelet algorithm with minimal RSME parameters. The results showed that the reinforcement algorithm reduced cooling and heating energy underwater and air conditions by 15% and 26%, respectively. In addition, the inappropriate amount was reduced by 73% and 84%, and high humidity time was reduced by 66% and 75%.

Reinforcement learning has made significant progress in the field of building control as an advanced control algorithm. Due to the uncertainty of renewable energy and battery safety issues, reinforcement learning faces many challenges in off-grid operation research. In recent years, deep reinforcement learning (DRL) has become an important development in the field of AI. DRL has achieved excellent results in the field of smart buildings due to the combination of the powerful nonlinear fitting ability of deep neural networks and the excellent decision-making ability of reinforcement learning (Berghout et al., 2022; Mason & Grijalva, 2019). In traditional sequence optimization problems, the typical approach involves optimizing the output results by a given set of input data. In DRL, the agent obtains a high-dimensional observation vector through interaction with the environment and selects actions that maximize cumulative rewards. This consideration of future multistep scenarios enables the agent to make optimal decisions.

In the field of smart buildings, DRL has been widely used in the operational optimization of building energy systems. By using DRL, we can achieve optimal decisions based on the current indoor environmental conditions and the real-time operation of building energy systems, such as adjusting the settings of HVAC systems, controlling lighting systems, etc.,

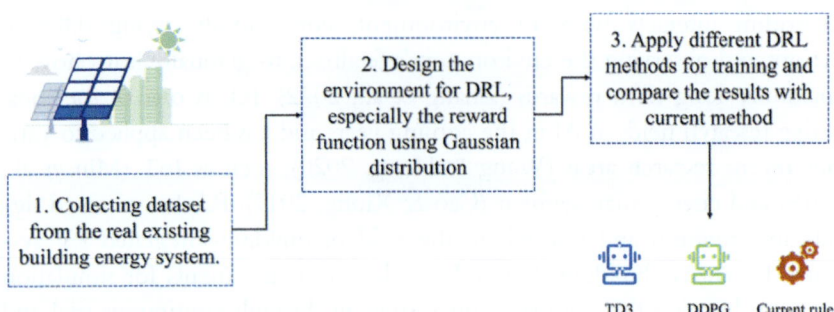

Figure 7.5 Workflow for operating optimization of off-grid renewable building energy systems using deep reinforcement learning.

to achieve optimal energy efficiency and comfort (Lissa et al., 2021; Liu, Ren et al., 2022).

Gao et al. (2022) proposed a new deep reinforcement learning algorithm for optimizing off-grid operation and battery safety issues of renewable building energy systems, as shown in Fig. 7.5. This study validated the effectiveness of the proposed deep reinforcement learning algorithm in actual scenarios through offline reinforcement learning of existing measurement data sets of actual buildings in Japan. The research results showed that the proposed deep reinforcement learning algorithm can better achieve the two optimization objectives of off-grid operation and battery safety under normal and extreme conditions. This provides a new solution for future off-grid operation coupling of renewable energy and building systems.

AI has the potential to address the energy optimization problem in residential demand response, particularly in resolving the scheduling of various types of appliances. Chu et al. (2022) proposed an energy scheduling optimization framework for a home energy management system that integrates PV and energy storage design to achieve optimized scheduling of household appliances. The schematic diagram of the framework is shown in Fig. 7.6. A model-free energy scheduling method based on actor-critic using Kronecker-factored trust region (ACKTR) is introduced to search for the optimal device scheduling strategy in complex and variable environments. The method has high sampling efficiency and can handle discrete and continuous control actions to jointly optimize the scheduling strategy of various types of appliances. The numerical results of the case study verify the effectiveness of the method and demonstrate that it can significantly reduce costs on typical test days, achieving an energy-saving effect of 25.37%.

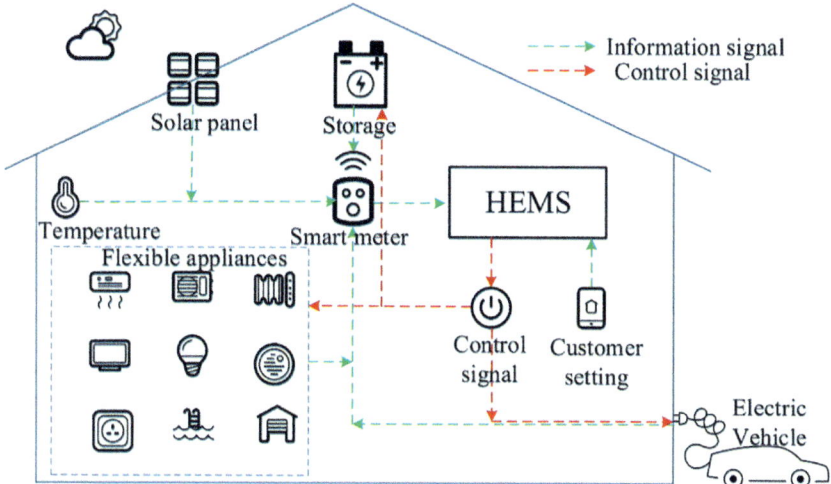

Figure 7.6 Smart home energy management structure using a household energy management system (Chu et al., 2022).

The reinforcement learning-based system optimization scheduling and reliability management-enhanced digital twin application is an emerging technology that can be used for optimizing and managing complex systems (Liu, Liu et al., 2023; Yuan & Xie, 2023). The digital twin is a virtual reality-based concept that uses simulation models of physical systems to predict their performance and behavior and helps design better systems (Alanne & Sierla, 2022). Zhou et al. (2023) proposed a method for energy management in smart homes by integrating PV, energy storage systems, and bidirectional energy flow, as shown in Fig. 7.7. The focus of this study is to address the energy management problem in smart homes under the conditions of distributed PV and energy storage systems. By optimizing the scheduling of home appliances through a Q-value-supported reinforcement learning method, the electricity cost is reduced, and demand response measures are implemented. Additionally, this study considers the limitations of user satisfaction with PV and energy storage systems to improve the user experience of the system. Through digital twin simulation, the study shows that the proposed home appliance scheduling recommendations are executed well, demand response measures are implemented, and electricity costs and system uncertainties are reduced. This study was aimed at providing an effective solution for energy management in smart homes and promoting the development of sustainable energy.

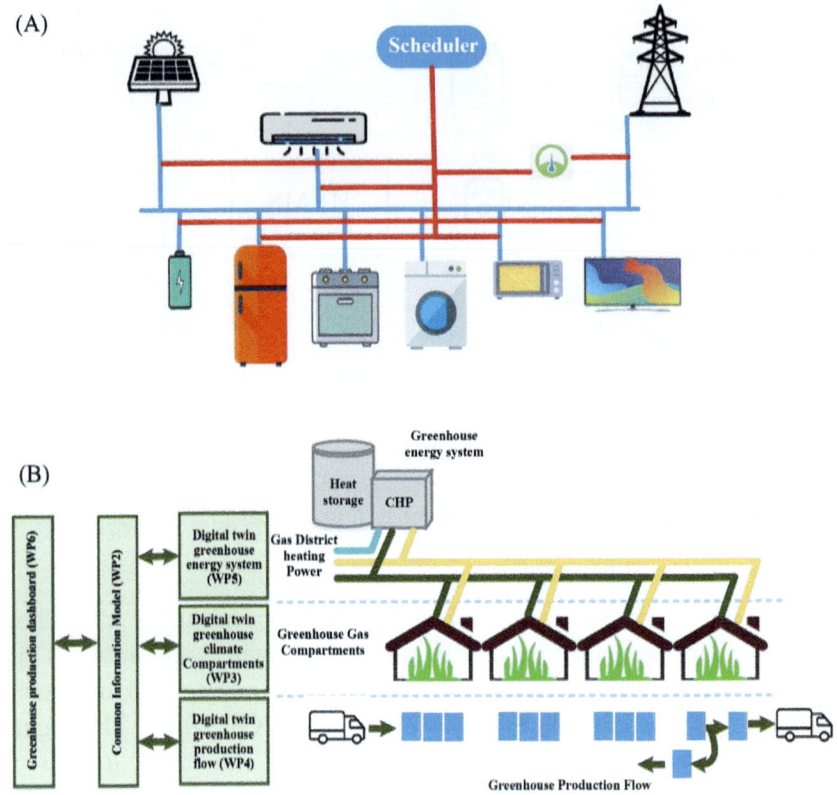

Figure 7.7 (A) Optimal home appliance scheduling structure with integrated PV and energy storage system; (B) Organized application of a digital twin in a greenhouse project (Zhou Yang et al., 2023).

Traditional rule-based control algorithms do not provide integrated closed-loop control, while the latest advances in deep reinforcement learning offer a choice for optimizing distributed energy systems and building load operation (Li et al., 2023; Zhang et al., 2023). The deep deterministic policy gradient algorithm was trained on synthetic data and evaluated in a physical test building. For example, Touzani et al. (2021) introduced a method that uses deep reinforcement learning (DRL) to integrate control of HVAC and battery storage systems in the presence of on-site PV generation using the deep deterministic policy gradient algorithm. Results show that the DRL-based controller can save up to 39.6% in cost while maintaining similar thermal comfort compared to the baseline controller and outperforming it in terms of energy efficiency, load shifting, and load shedding. The advantage of this approach is that it can handle the complexity of distributed energy systems, improve energy efficiency, provide grid services, and reduce setup effort. The use of deep reinforcement learning can provide better performance and efficiency for building energy management systems and is expected to become an important technology for future sustainable energy development.

7.3 Challenges

Intelligent learning methods have been widely applied to BIPV systems, but there are challenges and issues that need to be addressed in practical applications (Ahmad et al., 2022; Ifaei et al., 2023; Manandhar et al., 2023; Su et al., 2023). These challenges can mainly be divided into three categories: technical, policy, and social.

7.3.1 Technical considerations

When applying intelligent learning methods to control the demand-side controllers of BIPV systems, a large amount of data needs to be collected for training and validating the model. The sources of this data mainly include PV components, related electrical equipment, sensors, and monitoring devices (Khan et al., 2023; Balakumar et al., 2023b). Since BIPV systems are typically complex and contain multiple components and subsystems, data collection will, however, be a significant challenge. The performance of the demand-side controller for BIPV systems is affected by multiple factors, so the data collection process will face a large amount of data, such as indoor and outdoor temperature, humidity, lighting intensity, and energy consumption (Shen et al., 2023). This data may come from multiple sensors and monitoring devices and is stored in different formats and frequencies. It is, therefore, necessary to consider the issue of data source dispersion and ensure that data can be collected and integrated from different devices and systems. For the collected data, it is necessary to ensure its accuracy and reliability (Adhya et al., 2022; Huang et al., 2023). For example, when collecting temperature data, it may be necessary to consider the accuracy and stability of the temperature sensor, as well as the reliability of data transmission and storage. In addition, the collected data may need to be cleaned, standardized, and formatted. This may require the use of appropriate data processing techniques and algorithms to handle the collected data.

Developing effective algorithms for the demand-side controller of BIPV is a complex task. BIPV systems consist of multiple components, including PV modules, inverters, batteries, and building control systems, among others. There are complex interactions between these components, and the system performance varies with changes in the environment and usage. The development of algorithms suitable for BIPV systems needs to consider multiple factors, including energy production and consumption models, weather forecast models, building load models, and user behavior models. In addition, the algorithms developed for BIPV systems must be able to find a balance between real-time performance and accuracy. The algorithm needs to quickly respond to changes in energy production and consumption to achieve the best energy usage efficiency while also accurately predicting future energy demand and production situations to make optimal control decisions (Nwokolo et al., 2023; Pombo et al., 2022). Additionally, it's worth mentioning that most of the ML techniques mentioned above are essentially black-box models, where involving parameters that are difficult to interpret physically. In cases where a physical interpretation is required, the implementation of physics-informed ML presents a promising approach, such as, using grey-box modeling (Zhang et al., 2023a, 2023b).

7.3.2 Policy considerations

The development and deployment of intelligent learning methods face several policy challenges that may affect their effectiveness and adoption. These challenges primarily include a lack of relevant incentive policies, limited technical standards, data privacy, and security (Amini et al., 2020; Hu et al., 2023; Sharma & Shekhar, 2020). The lack of supportive policies is one of the main challenges in promoting the adoption of BIPV systems and demand-side controllers' intelligent learning methods. In many countries, governments lack policies to encourage and support BIPV systems and demand-side controllers, such as tax incentives, subsidies, or streamlined approval procedures. These policy measures can reduce the costs of BIPV systems and demand-side controllers, thereby promoting their deployment and application. However, the effective deployment and application of these technologies are limited by the lack of technical standards. Technical standards are critical to ensuring the interoperability, safety, and reliability of the system, which is a key factor in ensuring the normal operation of BIPV systems and demand-side controllers. The lack of technical standards can lead to compatibility issues, which can hinder the integration of different components of BIPV systems and demand-side controllers. Therefore, in the absence of technical standards, there may be compatibility or inconsistency issues between different components, which can lead to a decrease in system performance or even the inability to function properly. This situation not only increases the cost of system implementation and maintenance but also may lead to energy waste and security risks. The intelligent learning methods of BIPV systems and demand-side controllers require the collection and analysis of large amounts of data, including personal data. Therefore, policymakers need to consider how to ensure the privacy and security of this data. Policymakers need to establish data privacy regulations, encourage enterprises and organizations to take necessary measures to ensure data security and allow the collection and use of data for research and development purposes to promote the development of intelligent learning methods for BIPV systems and demand-side controllers.

7.3.3 Social considerations

Intelligent learning methods for BIPV and demand-side controllers face several social challenges. One of the challenges is the lack of awareness and understanding of the benefits and operation of BIPV and demand-side controllers among building residents and homeowners. Research shows that many building residents are not aware of the existence and benefits of BIPV, which may limit its adoption and use. This lack of awareness can also lead to resistance to demand-side management strategies because residents may not understand the necessity of adjusting energy use. Policymakers and the industry, need to strengthen the promotion and publicity of BIPV technology, enabling more people to understand the potential advantages of BIPV and its contribution to energy saving and emission reduction. Another social challenge is the lack of user participation in the implementation of demand-side management strategies. The participation of building residents and homeowners includes understanding, support, use, and feedback of the system. The success of demand-side controllers' intelligent learning methods depends on the active participation of building residents and homeowners; however, research shows that due to a lack of incentives, insufficient communication, and a lack of control over the system, many residents are not actively involved in implementing energy management strategies (Li et al., 2019; Wang Yin et al., 2021; Zhong et al., 2021).

7.4 Case study

The case study on intelligent learning approaches for demand-side control in BIPV-integrated buildings lies in its potential to provide valuable insights into the practical performance and effectiveness of these technologies (Balakumar et al., 2023a, 2023b; Sharda et al., 2021). In recent years, there has been an increasing interest in applying intelligent learning methods to demand-side control in BIPV systems for both research and practical purposes. The combination of demand-side management and home automation has given rise to a new concept known as active demand-side management (ADSM). ADSM allows for the modification of demand profiles to alleviate pressure on the power system, maximize consumption when resources are available, and minimize congestion. It can be implemented based on various criteria, such as energy prices, self-consumption maximization, or limiting the maximum power between households.

Matallanas et al. (2012) developed a distributed ANN-based ADSM controller for a real solar house, as illustrated in Fig. 7.8. They described the development of a distributed control system for demand-side management in residential areas. The distributed control system consisted of multiple artificial neural networks located on different devices in the houses providing local PV generation. The electrical system under investigation included local PV generation, energy storage systems, grid connections, and home automation systems. Both the scheduler and coordinator modules involved in the distributed control system were implemented using neural networks. The developed control system enhanced the building's energy flexibility by scheduling tasks as required by users and maximizing local generation. The results demonstrated that

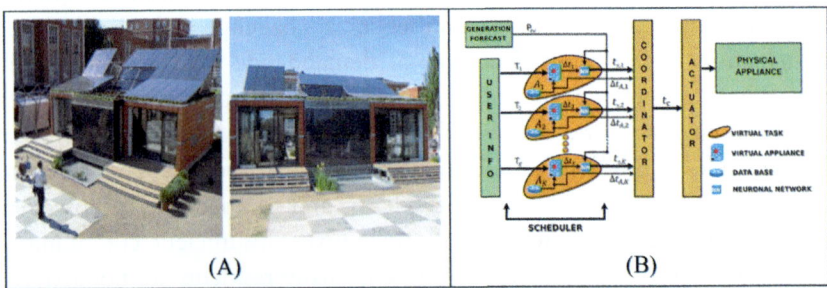

Figure 7.8 (A) Real solar house based on ADSM system; (B) Control structure of ADSM system (Matallanas et al., 2012).

the implemented ADSM system, which adopted a load-shifting strategy, improved the overall energy efficiency of the system and maximized the self-consumption of local energy.

Accurate energy demand and supply forecasting models can effectively improve the energy efficiency of buildings. In order to accurately predict energy demand and renewable energy production in buildings, Cordeiro-Costas et al. (2023) proposed a method to improve building performance through an intelligent energy management system. The main objective is to obtain high-precision PV electricity generation and consumption forecasting models using deep learning (DL) technology. The architecture of the proposed system is shown in Fig. 7.9. This method is based on the prediction of distributed energy and deep learning consumption and selects the best action in each period. The combination of these technologies allows for energy management that considers both current and future demands. This study takes the School of Mining and Energy Engineering at the University of Vigo as an example. Through the combination of AI technology, the building with an intelligent energy management system studied in this research increased its self-sufficiency ratio, resulting in a 22.65% increase in self-consumption and a 17.85% reduction in grid dependence; therefore, this intelligent energy management system signifies an improvement in building energy efficiency, with a reduction of 8.56 kg of carbon dioxide equivalent in carbon footprint and an annual saving of approximately 25,000 euros in electricity costs.

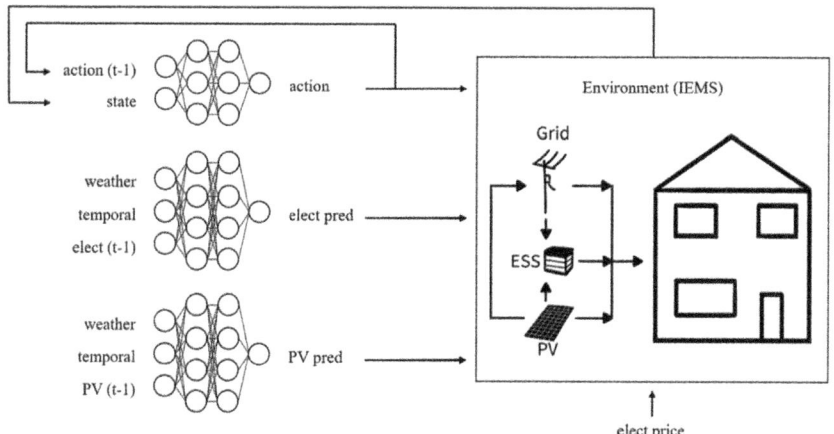

Figure 7.9 Topology diagram of the studied smart energy management system (Cordeiro-Costas et al., 2023).

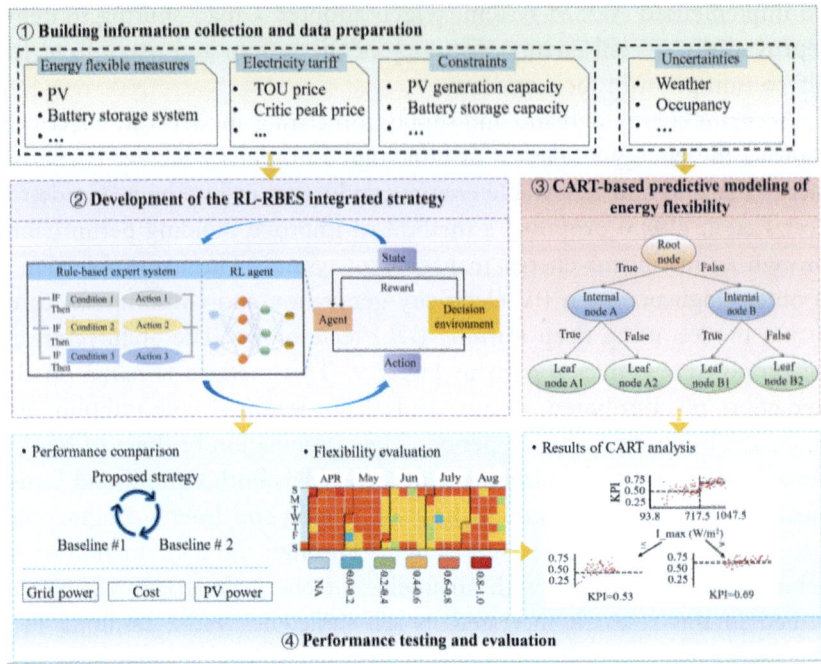

(A)

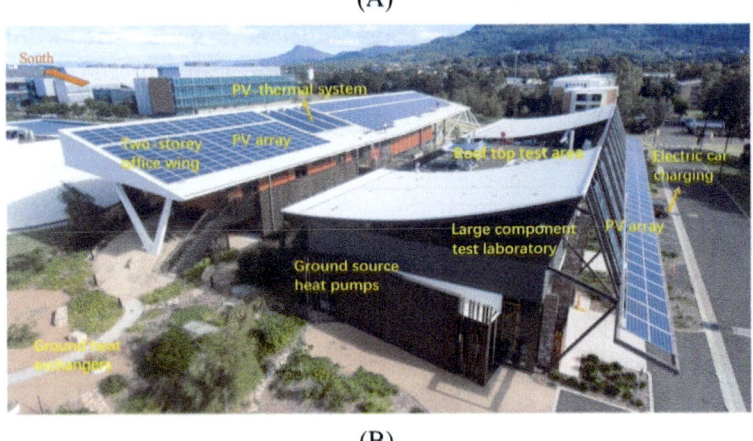

(B)

Figure 7.10 (A) Overview of the framework developed to enhance building energy flexibility; (B) Net zero energy office building at the University of Wollongong, Australia (Zhou Du et al., 2023).

The application of RL in building control faces challenges, among which the need for a large amount of data for training is one (Wang & Hong, 2020). This is because RL algorithms rely on trial-and-error learning in the control environment to obtain the optimal strategy (Fu et al., 2022; Zhou, Zheng, Zhang et al., 2020; Zhou, 2022a). To alleviate this problem, some researchers have used deep RL algorithms to reduce training time and improve algorithm efficiency (Frikha et al., 2021; Vázquez-Canteli & Nagy, 2019). In addition, to ensure the safety and robustness of the RL controller, optimization and improvement of the algorithm are required to prevent a negative impact on the building system operation. Zhou, Du et al. (2023) developed a new framework that combines RL, rule-based expert systems, and decision tree models to improve energy flexibility and battery storage systems in integrated PV buildings, as illustrated in Fig. 7.10A. In this framework, a comprehensive strategy using both RL agents and rule-based expert systems is employed to optimize battery charging and the grid power input of the battery to maximize self-consumption of PV power and minimize electricity costs by discharging at times with low electricity prices (Song, 2023a, 2023b). The framework was tested in a net-zero energy office building at the University of Wollongong in Australia, as shown in Fig. 7.10B. Results show that the integration of RL and rule-based expert systems can reduce electricity costs and consumption by 7.0% and 10.6%, respectively, compared with using only rule-based expert systems, while increasing self-consumption of PV power by 9.2%.

Through an analysis of the real-world performance of existing intelligent learning methods, case studies can provide valuable insights into how to improve and optimize these technologies to better meet the specific requirements of building environments. This can serve as a reference for developing more advanced and effective intelligent learning methods for building energy management.

References

Adhya, D., Chatterjee, S., & Chakraborty, A. K. (2022). Performance assessment of selective machine learning techniques for improved PV array fault diagnosis. *Sustainable Energy, Grids and Networks, 29,* 100582.

Ahmad, T., Chen, H., Huang, R., Yabin, G., Wang, J., Shair, J., Azeem Akram, H. M., Hassnain Mohsan, S. A., & Kazim, M. (2018). Supervised based machine learning models for short, medium and long-term energy prediction in distinct building environment. *Energy, 158,* 17−32.

Ahmad, T., Madonski, R., Zhang, D., Huang, C., & Mujeeb, A. (2022). Data-driven probabilistic machine learning in sustainable smart energy/smart energy systems: Key developments, challenges, and future research opportunities in the context of smart grid paradigm. *Renewable and Sustainable Energy Reviews, 160,* 112128.

Alabi, T. M., Aghimien, E. I., Agbajor, F. D., Yang, Z., Lu, L., Adeoye, A. R., & Gopaluni, B. (2022). A review on the integrated optimization techniques and machine learning approaches for modeling, prediction, and decision making on integrated energy systems. *Renewable Energy, 194*, 822–849.

Alanne, K., & Sierla, S. (2022). An overview of machine learning applications for smart buildings. *Sustainable Cities and Society, 76*, 103445.

Amini, M. H., Imteaj, A., & Pardalos, P. M. (2020). Interdependent networks: A data science perspective. *Patterns, 1*, 100003.

Amjith, L. R., & Bavanish, B. (2022). A review on biomass and wind as renewable energy for sustainable environment. *Chemosphere, 293*, 133579.

Avila, L., Paula, De, Trimboli, M., & Carlucho, I., M. (2020). Deep reinforcement learning approach for MPPT control of partially shaded PV systems in Smart Grids. *Applied Soft Computing, 97*, 106711.

Balakumar, P., Vinopraba, T., & Chandrasekaran, K. (2023a). Deep learning based real time Demand Side Management controller for smart building integrated with renewable energy and Energy Storage System. *Journal of Energy Storage, 58*, 106412.

Balakumar, P., Vinopraba, T., & Chandrasekaran, K. (2023b). Machine learning based demand response scheme for IoT enabled PV integrated smart building. *Sustainable Cities and Society, 89*, 104260.

Ballesteros-Gallardo, J. A., Arcos-Vargas, A., & Núñez, F. (2021). Optimal design model for a residential PV storage system an application to the Spanish Case. *Sustainability, 13*, 575.

Bansal, A. K. (2022). Sizing and forecasting techniques in photovoltaic-wind based hybrid renewable energy system: A review. *Journal of Cleaner Production, 369*, 133376.

Berghout, T., Benbouzid, M., & Muyeen, S. M. (2022). Machine learning for cybersecurity in smart grids: A comprehensive review-based study on methods, solutions, and prospects. *International Journal of Critical Infrastructure Protection, 38*, 100547.

Cao, J., & Xiong, R. (2017). Reinforcement learning-based real-time energy management for plug-in hybrid electric vehicle with hybrid energy storage system. *Energy Procedia, 142*, 1896–1901.

Chaudhary, B., & Kumar, V. (2022). Emerging technological frameworks for the sustainable agriculture and environmental management. *Sustainable Horizons, 3*, 100026.

Chen, Z., Wei, W., Song, L., & Ni, B.-J. (2022). Hybrid water electrolysis: A new sustainable avenue for energy-saving hydrogen production. *Sustainable Horizons, 1*, 100002.

Chen, Z., Xiao, F., Guo, F., & Yan, J. (2023). Interpretable machine learning for building energy management: A state-of-the-art review. *Advances in Applied Energy, 9*, 100123.

Chu, Y., Wei, Z., Sun, G., Zang, H., Chen, S., & Zhou, Y. (2022). Optimal home energy management strategy: A reinforcement learning method with actor-critic using Kronecker-factored trust region. *Electric Power Systems Research, 212*, 108617.

Cordeiro-Costas, M., Villanueva, D., Eguía-Oller, P., & Granada-Álvarez, E. (2023). Intelligent energy storage management trade-off system applied to Deep Learning predictions. *Journal of Energy Storage, 61*, 106784.

Dupré la Tour, M.-A. (2023). Photovoltaic and wind energy potential in Europe — A systematic review. *Renewable and Sustainable Energy Reviews, 179*, 113189.

Frikha, M. S., Gammar, S. M., Lahmadi, A., & Andrey, L. (2021). Reinforcement and deep reinforcement learning for wireless Internet of Things: A survey. *Computer Communications, 178*, 98–113.

Fu, Q., Han, Z., Chen, J., Lu, Y., Wu, H., & Wang, Y. (2022). Applications of reinforcement learning for building energy efficiency control: A review. *Journal of Building Engineering, 50*, 104165.

Gao, Y., Matsunami, Y., Miyata, S., & Akashi, Y. (2022). Operational optimization for off-grid renewable building energy system using deep reinforcement learning. *Applied Energy, 325*, 119783.

Gassar, A. A. A., & Cha, S. H. (2021). Review of geographic information systems-based rooftop solar photovoltaic potential estimation approaches at urban scales. *Applied Energy, 291,* 116817.

Guo, Y., Wang, J., Chen, H., Li, G., Liu, J., Xu, C., Huang, R., & Huang, Y. (2018). Machine learning-based thermal response time ahead energy demand prediction for building heating systems. *Applied Energy, 221,* 16—27.

He, Y., Zhou, Y., Liu, J., Liu, Z., & Zhang, G. (2022). An inter-city energy migration framework for regional energy balance through daily commuting fuel-cell vehicles. *Applied Energy, 324,* 119714.

He, Y., Zhou, Y., Wang, Z., Liu, J., Liu, Z., & Zhang, G. (2021a). Quantification on fuel cell degradation and techno-economic analysis of a hydrogen-based grid-interactive residential energy sharing network with fuel-cell-powered vehicles. *Applied Energy, 303,* 117444.

He, Y., Zhou, Y., Yuan, J., Liu, Z., Wang, Z., & Zhang, G. (2021b). Transformation towards a carbon-neutral residential community with hydrogen economy and advanced energy management strategies. *Energy Conversion and Management, 249,* 114834.

Hebboul, A., Hachouf, F., & Boulemnadjel, A. (2015). A new incremental neural network for simultaneous clustering and classification. *Neurocomputing, 169,* 89—99.

Homod, R. Z., Togun, H., Kadhim Hussein, A., Noraldeen Al-Mousawi, F., Yaseen, Z. M., Al-Kouz, W., Abd, H. J., Alawi, O. A., Goodarzi, M., & Hussein, O. A. (2022). Dynamics analysis of a novel hybrid deep clustering for unsupervised learning by reinforcement of multi-agent to energy saving in intelligent buildings. *Applied Energy, 313,* 118863.

Hu, H., Xu, J., Liu, M., & Lim, M. K. (2023). Vaccine supply chain management: An intelligent system utilizing blockchain, IoT and machine learning. *Journal of Business Research, 156,* 113480.

Hu, M., Ge, D., Telford, R., Stephen, B., & Wallom, D. C. H. (2021). Classification and characterization of intra-day load curves of PV and non-PV households using interpretable feature extraction and feature-based clustering. *Sustainable Cities and Society, 75,* 103380.

Huang, J., Koroteev, D. D., & Rynkovskaya, M. (2023). Machine learning-based demand response in PV-based smart home considering energy management in digital twin. *Solar Energy, 252,* 8—19.

Ifaei, P., Nazari-Heris, M., Tayerani Charmchi, A. S., Asadi, S., & Yoo, C. (2023). Sustainable energies and machine learning: An organized review of recent applications and challenges. *Energy, 266,* 126432.

Jalil-Vega, F., Kerdan, García, & Hawkes, A.D., I. (2020). Spatially-resolved urban energy systems model to study decarbonisation pathways for energy services in cities. *Applied Energy, 262,* 114445.

Jiménez-Fernández, E., Sánchez, A., & Sánchez Pérez, E. A. (2022). Unsupervised machine learning approach for building composite indicators with fuzzy metrics. *Expert Systems with Applications, 200,* 116927.

Jung, S., Jeoung, J., Kang, H., & Hong, T. (2021). Optimal planning of a rooftop PV system using GIS-based reinforcement learning. *Applied Energy, 298,* 117239.

Kabir, F., Yu, N., Yao, W., Yang, R., & Zhang, Y. (2021). Joint estimation of behind-the-meter solar generation in a community. *IEEE Transactions on Sustainable Energy, 12,* 682—694.

Khan, K., Rashid, S., Mansoor, M., Khan, A., Raza, H., Zafar, M. H., & Akhtar, N. (2023). Data-driven green energy extraction: Machine learning-based MPPT control with efficient fault detection method for the hybrid PV-TEG system. *Energy Reports, 9,* 3604—3623.

Kow, K. W., Wong, Y. W., Rajkumar, R., & Isa, D. (2018). An intelligent real-time power management system with active learning prediction engine for PV grid-tied systems. *Journal of Cleaner Production*, *205*, 252–265.

Kow, K. W., Wong, Y. W., Rajkumar., Rajparthiban, K., & Rajkumar, R. K. (2016). A review on performance of artificial intelligence and conventional method in mitigating PV grid-tied related power quality events. *Renewable and Sustainable Energy Reviews*, *56*, 334–346.

Li, G., Xuan, Q., Akram, M. W., Golizadeh Akhlaghi, Y., Liu, H., & Shittu, S. (2020). Building integrated solar concentrating systems: A review. *Applied Energy*, *260*, 114288.

Li, H. X., Horan, P., Luther, M. B., & Ahmed, T. M. F. (2019). Informed decision making of battery storage for solar-PV homes using smart meter data. *Energy and Buildings*, *198*, 491–502.

Li, Y., Wang, Z., Xu, W., Gao, W., Xu, Y., & Xiao, F. (2023). Modeling and energy dynamic control for a ZEH via hybrid model-based deep reinforcement learning. *Energy*, *277*, 127627.

Lissa, P., Deane, C., Schukat, M., Seri, F., Keane, M., & Barrett, E. (2021). Deep reinforcement learning for home energy management system control. *Energy and AI*, *3*, 100043.

Liu, C. C., Chen, H., Shi, J., & Chen, L. (2022). Self-supervised learning method for consumer-level behind-the-meter PV estimation. *Applied Energy*, *326*, 119961.

Liu, H., Liu, Q., Rao, C., Wang, F., Alsokhiry, F., Shvetsov, A. V., & Mohamed, M. A. (2023). An effective energy management layout-based reinforcement learning for household demand response in digital twin simulation. *Solar Energy*, *258*, 95–105.

Liu, J., Yu, Z., Liu, Z., Qin, D., Zhou, J., & Zhang, G. (2017). Performance analysis of Earth-air heat exchangers in hot summer and cold winter areas. *Procedia Engineering*, *205*, 1672–1677.

Liu, Z., Queena, K., Henk, V., & Zhang, G. (2022). Review on shallow geothermal promoting energy efficiency of existing buildings in Europe. *IOP Conference Series: Earth and Environmental Science*, *1085*, 012026.

Liu, X., Ren, M., Yang, Z., Yan, G., Guo, Y., Cheng, L., & Wu, C. (2022). A multi-step predictive deep reinforcement learning algorithm for HVAC control systems in smart buildings. *Energy*, *259*, 124857.

Liu, Z., Sun, Y., Xing, C., Liu, J., He, Y., Zhou, Y., & Zhang, G. (2022). Artificial intelligence powered large-scale renewable integrations in multi-energy systems for carbon neutrality transition: Challenges and future perspectives. *Energy and AI*, *10*, 100195.

Liu, Z., Xie, M., Zhou, Y., He, Y., Zhang, L., Zhang, G., & Chen, D. (2023). A state-of-the-art review on shallow geothermal ventilation systems with thermal performance enhancement system classifications, advanced technologies and applications. *Energy and Built Environment*, *4*, 148–168.

Liu, Z., Yu, C., Queena, K., Huang, R., You, K., Henk, V., & Zhang, G. (2023). Incentive initiatives on energy-efficient renovation of existing buildings towards carbon–neutral blueprints in China: Advancements, challenges and prospects. *Energy and Buildings*, *296*, 113343.

Liu, Z., Yu, Z., Yang, T., Letizia, R., Sun, P., Li, S., Zhang, G., & Mohamed, E. (2019a). Numerical modeling and parametric study of a vertical earth-to-air heat exchanger system. *Energy*, *172*, 220–231.

Liu, Z., Yu, Z., Yang, T., Li, S., Mohamed, E., Letizia, R., Qin, D., & Zhang, G. (2019b). Experimental investigation of a vertical earth-to-air heat exchanger system. *Energy Conversion and Management*, *183*, 241–251.

Liu, Z., Yu, Z., Yang, Y., Mohamed, E., Letizia, R., Sun, Y., Sun, P., Li, H., & Zhang, G. (2019c). Experimental and numerical study of a vertical earth-to-air heat exchanger system integrated with annular phase change material. *Energy Conversion and Management, 186,* 433−449.

Liu, Z., Yu, Z., Yang, T., Qin, D., Li, S., Zhang, G., Haghighat, F., & Joybari, M. M. (2018). A review on macro-encapsulated phase change material for building envelope applications. *Building and Environment, 144,* 281−294.

Liu, Z., Zeng, C., Zhou, Y., & Xing, C. (2022). 9 - The main utilization forms and current developmental status of geothermal energy for building cooling/heating in developing countries. In Y. Noorollahi, M. N. Naseer, & M. M. Siddiqi (Eds.), *Utilization of thermal potential of abandoned wells* (pp. 159−190). Academic Press.

Liu, Z., Zhang, X., Sun, Y., & Zhou, Y. (2023). Advanced controls on energy reliability, flexibility and occupant-centric control for smart and energy-efficient buildings. *Energy and Buildings, 297,* 113436.

Liu, Z., Zhou, Y., Yan, J., & Marcos, T. (2023). Frontier ocean thermal/power and solar PV systems for transformation towards net-zero communities. *Energy, 284,* 128362.

Manandhar, P., Rafiq, H., & Rodriguez-Ubinas, E. (2023). Current status, challenges, and prospects of data-driven urban energy modeling: A review of machine learning methods. *Energy Reports, 9,* 2757−2776.

Mason, K., & Grijalva, S. (2019). A review of reinforcement learning for autonomous building energy management. *Computers & Electrical Engineering, 78,* 300−312.

Matallanas, E., Castillo-Cagigal, M., Gutiérrez, A., Monasterio-Huelin, F., Caamaño-Martín, E., Masa, D., & Jiménez-Leube, J. (2012). Neural network controller for Active Demand-Side Management with PV energy in the residential sector. *Applied Energy, 91,* 90−97.

Md Khairi, N. H., Akimoto, Y., & Okajima, K. (2022). Suitability of rooftop solar photovoltaic at educational building towards energy sustainability in Malaysia. *Sustainable Horizons, 4,* 100032.

Mellit, A., & Kalogirou, S. A. (2008). Artificial intelligence techniques for photovoltaic applications: A review. *Progress in Energy and Combustion Science, 34,* 574−632.

Mellit, A., Kalogirou, S. A., Hontoria, L., & Shaari, S. (2009). Artificial intelligence techniques for sizing photovoltaic systems: A review. *Renewable and Sustainable Energy Reviews, 13,* 406−419.

Min, M., Xiao, L., Chen, Y., Cheng, P., Wu, D., & Zhuang, W. (2019). Learning-based computation offloading for IoT devices with energy harvesting. *IEEE Transactions on Vehicular Technology, 68,* 1930−1941.

Naganathan, H., Chong, W. O., & Chen, X. (2016). Building energy modeling (BEM) using clustering algorithms and semi-supervised machine learning approaches. *Automation in Construction, 72,* 187−194.

Nwokolo, S. C., Obiwulu, A. U., & Ogbulezie, J. C. (2023). Machine learning and analytical model hybridization to assess the impact of climate change on solar PV energy production. *Physics and Chemistry of the Earth, Parts A/B/C, 130,* 103389.

Okafor, C. C., Nzekwe, C. A., Ajaero, C. C., Ibekwe, J. C., & Otunomo, F. A. (2022). Biomass utilization for energy production in Nigeria: A review. *Cleaner Energy Systems, 3,* 100043.

Pallonetto, F., De Rosa, M., Milano, M., & Finn, D.P., F. (2019). Demand response algorithms for smart-grid ready residential buildings using machine learning models. *Applied Energy, 239,* 1265−1282.

Pombo, D. V., Bacher, P., Ziras, C., Bindner, H. W., Spataru, S. V., & Sørensen, P. E. (2022). Benchmarking physics-informed machine learning-based short term PV-power forecasting tools. *Energy Reports, 8,* 6512−6520.

Seyedmahmoudian, M., Horan, B., Soon, T. K., Rahmani, R., Than, Oo, A. M., Mekhilef, S., & Stojcevski, A. (2016). State of the art artificial intelligence-based MPPT techniques for mitigating partial shading effects on PV systems — A review. *Renewable and Sustainable Energy Reviews, 64,* 435—455.

Sharda, S., Sharma, K., & Singh, M. (2021). A real-time automated scheduling algorithm with PV integration for smart home prosumers. *Journal of Building Engineering, 44,* 102828.

Sharma, A., & Shekhar, H. (2020). Intelligent learning based opinion mining model for governmental decision making. *Procedia Computer Science, 173,* 216—224.

Shen, G., Xing, R., Zhou, Y., Jiao, X., Luo, Z., Xiong, R., Huang, W., Tian, Y., Chen, Y., Du, W., Shen, H., Cheng, H., Zhu, D., & Tao, S. (2022). Revisiting the proportion of clean household energy users in rural China by accounting for energy stacking. *Sustainable Horizons, 1,* 100010.

Shen, Z., Xu, W., Li, W., Shi, Y., & Gao, F. (2023). Digital twin application for attach detection and mitigation of PV-based smart systems using fast and accurate hybrid machine learning algorithm. *Solar Energy, 250,* 377—387.

Sohani, A., Sayyaadi, H., Cornaro, C., Shahverdian, M. H., Pierro, M., Moser, D., Karimi, N., Doranehgard, M. H., & Li, L. K. B. (2022). Using machine learning in photovoltaics to create smarter and cleaner energy generation systems: A comprehensive review. *Journal of Cleaner Production, 364,* 132701.

Song, A., & Zhou, Y. (2023a). Advanced cycling ageing-driven circular economy with E-mobility-based energy sharing and lithium battery cascade utilisation in a district community. *Journal of Cleaner Production, 415,* 137797. Available from https://doi.org/10.1016/j.jclepro.2023.137797.

Song, A., & Zhou, Y. (2023b). A hierarchical control with thermal and electrical synergies on battery cycling ageing and energy flexibility in a multi-energy sharing network. *Renewable Energy, 212,* 1020—1037.

Song, G., Ai, Z., Liu, Z., & Zhang, G. (2022). A systematic literature review on smart and personalized ventilation using CO2 concentration monitoring and control. *Energy Reports, 8,* 7523—7536.

Su, Z., Xing, L., Ali, H. E., Alkhalifah, T., Alturise, F., Khadimallah, M. A., & Assilzadeh, H. (2023). Latest insights on separation and storage of carbon compounds in buildings towards sustainable environment: Recent innovations, challenges, future perspectives and application of machine learning. *Chemosphere, 329,* 138573.

Sun, Y., Fung, B. C. M., & Haghighat, F. (2022a). The generalizability of pre-processing techniques on the accuracy and fairness of data-driven building models: A case study. *Energy and Buildings, 268,* 112204.

Sun, Y., Fung, B. C. M., & Haghighat, F. (2022b). In-processing fairness improvement methods for regression data-driven building models: Achieving uniform energy prediction. *Energy and Buildings, 277,* 112565.

Sun, Y., Haghighat, F., & Fung, B. C. M. (2020). A review of the-state-of-the-art in data-driven approaches for building energy prediction. *Energy and Buildings, 221,* 110022.

Sun, Y., Haghighat, F., & Fung, B. C. M. (2022). Trade-off between accuracy and fairness of data-driven building and indoor environment models: A comparative study of pre-processing methods. *Energy, 239,* 122273.

Tang, L., Liu, Z., Zhou, Y., Qin, D., & Zhang, G. (2020). Study on a dynamic numerical model of an underground air tunnel system for cooling applications—Experimental validation and multidimensional parametrical analysis. *Energies, 13,* 1236.

Tien, P. W., Wei, S., Darkwa, J., Wood, C., & Calautit, J. K. (2022). Machine learning and deep learning methods for enhancing building energy efficiency and indoor environmental quality — A review. *Energy and AI, 10,* 100198.

Touzani, S., Prakash, A. K., Wang, Z., Agarwal, S., Pritoni, M., Kiran, M., Brown, R., & Granderson, J. (2021). Controlling distributed energy resources via deep reinforcement learning for load flexibility and energy efficiency. *Applied Energy*, *304*, 117733.

Vassiliades, C., Agathokleous, R., Barone, G., Forzano, C., Giuzio, G. F., Palombo, A., Buonomano, A., & Kalogirou, S. (2022). Building integration of active solar energy systems: A review of geometrical and architectural characteristics. *Renewable and Sustainable Energy Reviews*, *164*, 112482.

Vázquez-Canteli, J. R., & Nagy, Z. (2019). Reinforcement learning for demand response: A review of algorithms and modeling techniques. *Applied Energy*, *235*, 1072−1089.

Wang, G., Yin, J., Hossain, M. S., & Muhammad, G. (2021). Incentive mechanism for collaborative distributed learning in Artificial Intelligence of Things. *Future Generation Computer Systems*, *125*, 376−384.

Wang, L., Zhang, G., Yin, X., Zhang, H., & Ghalandari, M. (2022). Optimal control of renewable energy in buildings using the machine learning method. *Sustainable Energy Technologies and Assessments*, *53*, 102534.

Wang, P., Liu, Z., & Zhang, L. (2021). Sustainability of compact cities: A review of Inter-Building Effect on building energy and solar energy use. *Sustainable Cities and Society*, *72*, 103035.

Wang, Z., & Hong, T. (2020). Reinforcement learning for building controls: The opportunities and challenges. *Applied Energy*, *269*, 115036.

Waqar Akram, M., Li, G., Jin, Y., & Chen, X. (2022). Failures of photovoltaic modules and their detection: A review. *Applied Energy*, *313*, 118822.

Xie, X., Lai, G., You, M., Liang, J., & Leng, B. (2023). Effective transfer learning of defect detection for photovoltaic module cells in electroluminescence images. *Solar Energy*, *250*, 312−323.

Xie, M., Wang, Y., Liu, Z., & Zhang, G. (2022). Effect of the location pattern of rural residential buildings on natural ventilation in mountainous terrain of central China. *Journal of Cleaner Production*, *340*, 130837.

Yahya, Z., Imane, S., Hicham, H., Ghassane, A., & Bouchini-Idrissi Safia, E. (2022). Applied imagery pattern recognition for photovoltaic modules' inspection: A review on methods, challenges and future development. *Sustainable Energy Technologies and Assessments*, *52*, 102071.

You, K., Yu, Y., Cai, W., & Liu, Z. (2023). The change in temporal trend and spatial distribution of CO_2 emissions of China's public and commercial buildings. *Building and Environment*, *229*, 109956.

Youssef, A., El-Telbany, M., & Zekry, A. (2017). The role of artificial intelligence in photo-voltaic systems design and control: A review. *Renewable and Sustainable Energy Reviews*, *78*, 72−79.

Yuan, G., & Xie, F. (2023). Digital Twin-Based economic assessment of solar energy in smart microgrids using reinforcement learning technique. *Solar Energy*, *250*, 398−408.

Zhang, F., Guo, H., Liu, Z., & Zhang, G. (2021). A critical review of the research about radiant cooling systems in China. *Energy and Buildings*, *235*, 110756.

Zhang, B., Hu, W., Xu, X., Li, T., Zhang, Z., & Chen, Z. (2022). Physical-model-free intelligent energy management for a grid-connected hybrid wind-microturbine-PV-EV energy system via deep reinforcement learning approach. *Renewable Energy*, *200*, 433−448.

Zhang, B., Hu, W., Xu, X., Zhang, Z., & Chen, Z. (2023). Hybrid data-driven method for low-carbon economic energy management strategy in electricity-gas coupled energy systems based on transformer network and deep reinforcement learning. *Energy*, *273*, 127183.

Zhang, X., Lau, S., Lau, S., & Zhao, Y. (2018). Photovoltaic integrated shading devices (PVSDs): A review. *Solar Energy*, *170*, 947−968.

Zhang, X., Saelens, D., & Roels, S. (2023a). Data-driven estimation of time-dependent solar gain coefficient in a two-zone building with synthetic occupants: Two B-splines integrated grey-box modeling approaches. *Building and Environment*, 110311.

Zhang, X., Saelens, D., & Roels, S. (2023b). Impact of solar gain estimation on heat loss coefficient determination using in-situ data: Comparing co-heating test with B-splines integrated grey-box modelling. *Building and Environment*, 110417.

Zheng, X., & Zhou, Y. (2023). A three-dimensional unsteady numerical model on a novel aerogel-based PV/T-PCM system with dynamic heat-transfer mechanism and solar energy harvesting analysis. *Applied Energy*, *338*, 120899.

Zhong, S., Wang, X., Zhao, J., Li, W., Li, H., Wang, Y., Deng, S., & Zhu, J. (2021). Deep reinforcement learning framework for dynamic pricing demand response of regenerative electric heating. *Applied Energy*, *288*, 116623.

Zhou, Y., & Lund, P. D. (2022). Peer-to-peer energy sharing and trading of renewable energy in smart communities — trading pricing models, decision-making and agent-based collaboration. *Renewable Energy*, *207*, 177−193.

Zhou, J., Yang, M., Zhan, Y., & Xu, L. (2023). Digital twin application for reinforcement learning based optimal scheduling and reliability management enhancement of systems. *Solar Energy*, *252*, 29−38.

Zhou, L., & Zhou, Y. (2023). Study on thermo-electric-hydrogen conversion mechanisms and synergistic operation on hydrogen fuel cell and electrochemical battery in energy flexible buildings. *Energy Conversion and Management*, *277*, 116610.

Zhou, X., Du, H., Sun, Y., Ren, H., Cui, P., & Ma, Z. (2023). A new framework integrating reinforcement learning, a rule-based expert system, and decision tree analysis to improve building energy flexibility. *Journal of Building Engineering*, *71*, 106536.

Zhou, Y. (2021). Artificial neural network-based smart aerogel glazing in low-energy buildings: A state-of-the-art review. *iScience*, *24*, 103420.

Zhou, Y. (2022a). Advances of machine learning in multi-energy district communities-mechanisms, applications and perspectives. *Energy and AI*, *10*, 100187.

Zhou, Y. (2022b). Artificial intelligence in renewable systems for transformation towards intelligent buildings. *Energy and AI*, *10*, 100182.

Zhou, Y. (2022c). Demand response flexibility with synergies on passive PCM walls, BIPVs, and active air-conditioning system in a subtropical climate. *Renewable Energy*, *199*, 204−225.

Zhou, Y. (2022d). A multi-stage supervised learning optimisation approach on an aerogel glazing system with stochastic uncertainty. *Energy*, *258*, 124815.

Zhou, Y. (2022e). A regression learner-based approach for battery cycling ageing prediction—advances in energy management strategy and techno-economic analysis. *Energy*, *256*, 124668.

Zhou, Y. (2022f). Ocean energy applications for coastal communities with artificial intelligencea state-of-the-art review. *Energy and AI*, *10*, 100189.

Zhou, Y. (2022g). Incentivising multi-stakeholders' proactivity and market vitality for spatio-temporal microgrids in Guangzhou-Shenzhen-Hong Kong Bay Area. *Applied Energy*, *328*, 120196.

Zhou, Y. (2023). A dynamic self-learning grid-responsive strategy for battery sharing economy—multi-objective optimisation and posteriori multi-criteria decision making. *Energy*, *266*, 126397.

Zhou, Y., Cao, S., Hensen, J. L. M., & Hasan, A. (2020). Heuristic battery-protective strategy for energy management of an interactive renewables−buildings−vehicles energy sharing network with high energy flexibility. *Energy Conversion and Management*, *214*, 112891.

Zhou, Y., & Liu, Z. (2023). A cross-scale 'material-component-system' framework for transition towards zero-carbon buildings and districts with low, medium and high-temperature phase change materials. *Sustainable Cities and Society*, *89*, 104378.

Zhou, Y., Liu, Z., & Zheng, S. (2021). 15 - Influence of novel PCM-based strategies on building cooling performance. In F. Pacheco-Torgal, L. Czarnecki, A. L. Pisello, L. F. Cabeza, & C.-G. Granqvist (Eds.), *Eco-efficient materials for reducing cooling needs in buildings and construction* (pp. 329–353). Woodhead Publishing.

Zhou, Y., & Zheng, S. (2020). Machine-learning based hybrid demand-side controller for high-rise office buildings with high energy flexibilities. *Applied Energy, 262,* 114416.

Zhou, Y., Zheng, S., Liu, Z., Wen, T., Ding, Z., Yan, J., & Zhang, G. (2020). Passive and active phase change materials integrated building energy systems with advanced machine-learning based climate-adaptive designs, intelligent operations, uncertainty-based analysis and optimisations: A state-of-the-art review. *Renewable and Sustainable Energy Reviews, 130,* 109889.

Zhou, Y., Zheng, S., & Zhang, G. (2020). A state-of-the-art-review on phase change materials integrated cooling systems for deterministic parametrical analysis, stochastic uncertainty-based design, single and multi-objective optimisations with machine learning applications. *Energy and Buildings, 220,* 110013.

Zhou, L., & Zhou, Y. (2023). Study on thermo-electric-hydrogen conversion mechanisms and synergistic operation on hydrogen fuel cell and electrochemical battery in energy flexible buildings. *Energy Conversion and Management, 277,* 116610.

Intelligent learning methods for optimizing integrated energy systems

Intelligent learning methods for optimizing integrated energy systems (Predictive and prescriptive approaches to optimizing integrated energy systems that take into account uncertainty)

Zheng Liqin[1,2], Bai Xiaoqing[1], Wang Puming[1] and Shi Xiaoqing[1]
[1]Key Laboratory of Power System Optimization and Energy Saving Technology, Guangxi University, Nanning, P.R. China
[2]State Grid Xiamen Electric Power Supply Company, Xiamen, P.R. China

8.1 Introduction

8.1.1 Background

Integrated energy systems (IES) can interact with various energy and improve energy usage. The optimization of IES is conducive to improving energy efficiency and achieving energy transformation (Cheng et al., 2019). The increased uncertainties of renewable energy and multiple electric (EV) and hydrogen vehicles (HVs) will, however, significantly impact the IES's energy dispatch. Therefore, optimization problems with uncertainties have become the focus of research. Nowadays, the analytics to derive optimal results for decision problems with uncertainties are mainly divided into prediction and prescription. The prediction is to express what uncertainty will happen by predictive techniques, and the prescription is to show how feasible results can be achieved by optimization techniques. However, major studies on analyzing these problems predict uncertainty first and then optimize the solutions according to the prediction. Although the predictive technique is constantly improving and ultrashort-term interval prediction is becoming more and more reliable (Tu et al., 2022), separating prediction and optimization may lead to a suboptimal decision. The impact of prediction errors on optimization problems remains unknown. Ben-Tal et al. (2009) indicated that a small prediction error can seriously worsen the objective function value. Specifically, Dragoon and Milligan (2003) showed the impact of prediction error for wind production on incremental reserve requirements and imbalance costs. The concerns about uncertainty regions in optimization techniques like robust optimization are crucial, and overbudget regions cause the solutions to be more conservative and consume computing resources. Therefore, the intelligent learning approaches, integrated prediction and prescription are significant. Furthermore, the selection of predictive and optimization techniques affects the performance of the solutions.

8.1.2 Relevant literature about intelligent learning approaches

Optimization of IES involves two key approaches: predicting uncertainty using predictive techniques and addressing operational issues through optimization techniques. These complementary methods contribute to the overall optimization of IES. Some traditional predictive techniques are backpropagation neural network (BPNN), elastic-net regression (ENR), general regression neural network (GRNN), k-nearest neighbors (knn), partial least squares regression (PLSR), group method of data handling (GMDH), and support vector regression (SVR). To improve the performance of prediction, Moradzadeh et al. (2022) proposed a hybrid model called group support vector regression (GSVR). This model applies a combination of GMDH and SVR models to forecast the cooling load (CL) and heating load (HL). Moreover, a combined prediction model of electricity, heat, cooling, and gas loads based on the multitask learning and least square support vector machine was constructed by using the weight-sharing mechanism (Tan et al., 2020). To further consider the uncertainty of integrated demand response (IDR), Li et al. (2022) proposed a two-level optimal distribution model considering renewable energy, load, and IDR. The traditional source-load uncertainty probability model was improved by combining nonparametric density estimation and Copula theory.

Solving operation problems of IES needs complex interaction among the internal subjects of the IES and the flexible participation of the external multienergy market. The two-layer game model constructed by (Yang et al., 2021) optimized the comprehensive energy system of the park internally and externally. The Stackelberg game model between the upper energy network and the park system was used to optimize the comprehensive energy system of the park. The user of the lower level, the gas supply system, and the IES were used to optimize the internal IES. Sivanantham and Gopalakrishnan (2020) established a Stackelberg game between the service provider and the users, which reduced the user's electricity bill by considering the game optimization of the demand response. On the other hand, the problem of matching supply and demand is also a challenge for the IES since the IES is a comprehensive relationship between various energy sources and demands. Liu et al. (2019) put forward a new type of supply and demand matching strategy. Under the same conditions, the changes in supply and demand were considered at the same time to achieve optimal economic matching.

The basic operation problem is multienergy flow optimization, which seriously affects the safety and economy of IES. The multenergy flow problem is mainly solved by the Newton-Raphson method in terms of the solution methods. Considering the characteristics of the multisystem coupling of the IES, they can be divided into unified solution method (Zeng et al., 2016; Liu & Mancarella, 2016) and decomposition solution method (Wei et al., 2017; Li et al., 2016). The basic idea of the unified solution method is as follows: Based on the power flow calculation of the AC power system, the variables of the gas system and the thermal system are regarded as extended variables, which are then integrated with the variables of the power system for a unified solution. The basic idea of the decomposition solution method is as follows: In the iterative process, the equations of different systems are solved separately. When solving the power system equations, the nodes coupled with the natural gas system and the thermal system are equivalent to PV nodes or PQ nodes. When solving the natural gas system or thermal system equation, the nodes coupled with other systems are equivalent to source nodes or load nodes. Qin et al. (2019) further proposed a generalized quasi-dynamic model and a decomposition-iteration solving method for electric-heat coupling IES to achieve accurate calculation of multienergy flow. The above literature focused on optimizing IES with predictive techniques and optimization techniques separately.

8.1.3 Challenges of optimizing IES using intelligent learning methods

When optimizing IES with the intelligent learning approaches combined with the predictive analytics method and prescriptive analytics method, three challenges must be addressed:

1. The selection of the predictive analytics method: The structure of the predictive analytics method should not be too complex and performs interpretability well. The predicted process could be expressed with historical data, and predictions are made based on the mean or mode of past observations similar to the one at hand.
2. The prescriptive analytics method: The uncertainty of the IES optimization problem should be described accurately and solved fast with the prescriptive analytics method. The solutions approximate the optimal based infinite number of samples.
3. The combination way of the two methods: The predictive analytics process can be included in the prescriptive analytics method well, and the intelligent learning approaches combined with the predictive analytics method and prescriptive analytics method are tractable.

8.2 Literature on intelligent learning methods

8.2.1 Predictive analytics method

In the predictive method, common techniques include predictive techniques based on the probability distribution function of renewable energy output (Seo et al., 2019), neural networks (Hua et al., 2022), and nonparametric regressions (Bertsimas & Kallus, 2020). The predictive techniques based on mathematical statistics, such as probability distribution function and nonparametric regressions, offer interpretability. Moreover, nonparametric regressions do not assume a probability distribution function, simplifying the modeling process.

8.2.1.1 Mathematical statistics

The autoregressive model (AR model) is the earliest predictive analysis method that uses mathematical statistics to analyze time series. Developed from linear regression in regression analysis, it is utilized to deal with the linear prediction of time series. It uses its own variable sequence for prediction, which needs a few variables but requires that the variables must be autocorrelated. The autocorrelation coefficient should be more than 0.5, or the prediction results will be extremely inaccurate (Box et al., 1994). Motivated by the AR model, the moving average model (MA model), autoregressive moving average model (ARMA model), and autoregressive integrated moving average model

(ARIMA model) are established one after another. For stationary time series, fitting can be achieved using the general ARMA model and AR model, MA model, or combined ARMA model. When there are more than 50 observations, the ARMA model is generally adopted (Hossain et al., 2020). On the other hand, the ARIMA model is generally adopted for nonstationary time series (Adebiyi et al., 2014). However, the application of these methods based on mathematical statistics is limited to a small number of stationary data.

8.2.1.2 Machine learning

A strain of nonparametric machine learning (ML) methods based on local learning is developed. Knn is the most fundamental technique, defining the prediction as a locally constant function depending on which k data points are closest to each other (Triguero et al., 2019). A comparable method is kernel regression (KR), which stands out for being extremely adaptable to theoretical study but is less commonly used in practice (Agarwal et al., 2006). Local regression is a more popular local learning regression method than KR. The recursive partitioning methods in the form of trees and classification and regression trees (CART) are even more popular (Gey & Nedelec, 2005). Most notably, random forest (RF) ensembles of trees are known to be exceedingly adaptable and perform well across a wide range of prediction challenges (Speiser et al., 2019). The latter combines multiple such averages, whereas the former averages locally over a partition created using the data (the leaves of a tree).

8.2.2 Prescriptive analytical method

In the prescription method, stochastic optimization (SO) and robust optimization (RO) are popular optimization techniques under uncertainty. The SO assumes the probability distribution of the uncertainty from historical data, which is critical to be similar to real distribution (Aaslid et al., 2022). The RO generally uses an ambiguity set to obtain solutions based on the worst-case analysis (Jin et al., 2021).

8.2.2.1 Stochastic optimization

In the process of solving the uncertainty problem, SO is generally used for the uncertainty problem with substantial data. The advantages of SO are proactive and intuitive, and the predictions can be directly obtained from the probability information. However, it also has some shortcomings: (1) although SO can extract the distribution model of high-frequency uncertainty parameters from historical data, it is difficult to describe the low-frequency uncertainty, and (2) the increase of uncertain parameters will lead to an exponential increase in the number of scenarios of SO, and then a large number of scenarios will aggravate the difficulty of solving the optimization problem.

8.2.2.2 Robust optimization

To enhance the traditional uncertainty optimization methods, RO can model uncertain parameters using an uncertainty set and transform it into a form without uncertain variables. In that case, solutions will be optimal without being affected by uncertain parameters. In a robust model, the optimal solutions may be overly conservative with an underperforming set. The robust uncertainty sets usually contain box, polyhedral, and ellipsoidal sets (Zhao et al., 2021; Malekpour & Pahwa, 2017; Vatani et al., 2018). Previously, multiple pieces of literature have confirmed that the RO models with box sets and polyhedral sets contain many irrelevant uncertainties and aggravate conservativeness. In comparison, the ellipsoidal set performs less conservatively and preserves the correlation of uncertainties (Wu et al., 2017). Furthermore, many studies have also committed to motivating the standard ellipsoidal set. A Newton's iteration method using ellipsoidal boundaries was proposed to ensure both satisfactory accuracy and efficiency in handling uncertainty (Qiu & Jiang, 2021). However, the boundary for large uncertainties remains unclear. In a study by Kuryatnikova et al. (2021), added correlations were introduced to the uncertainty ellipsoid set, aiming to narrow the boundaries of uncertainty sets. However, this approach is limited to small cases. The related literature is, therefore, still insufficient to include the uncertain ellipsoidal set with large fluctuations and apply it to significant cases.

8.2.3 Predictive and prescriptive analytics method

Considering two advantages of the intelligent learning approaches combined with prediction and optimization, namely the ability to approximate true solutions based on an infinite number of samples and tractability, a two-stage optimization model was embedded into such approaches. The two-stage optimization model has been widely used in microgrids (Liu et al., 2018), distributed energy resource systems (Li & Zhang, 2021), and even other academic fields, to further adjust the prediction error under the short-time scale. Relevant research showed that two-stage robust optimization can balance the system's economy and conservativeness through dynamic iteration. Building upon intelligent learning approaches that integrate machine learning and optimization techniques, this chapter uses a two-stage RO model to adapt to short-term deviations in prediction. It proposes a robust uncertainty set constructed by an uncertain fluctuation region for handling large volatility uncertainty using the knn + minimum volume (KMV). In detail, the contributions are as follows:

1. The intelligent learning approaches take prediction into account in the optimization and downplay the inadequacy of predictive technique with a large volume of data.
2. The two-stage RO model can adjust the output on a short timescale iteratively with certainty on a long timescale until the optimal solutions are found.
3. Containing all samples in knn, the uncertain fluctuation region in the minimum volume ellipsoid set is the most robust when comparing a box set and two ellipsoidal sets and encloses all possibilities of uncertainty.

The outstanding uncertainty region is easily available with an adjustable parameter deduced by a mathematical formula. Furthermore, all uncertain fluctuation regions in the KMV ellipsoid set follow the adjustive parameter linearly and are more robust than those in the box set. As the adjustable parameter increases, the fluctuation region becomes more extensive. This advantage facilitates the easy selection of a suitable fluctuation region in engineering practice.

The chapter construction is organized as follows. Motivation and background and corresponding challenges for optimization of IES are introduced in Section 8.1. Section 8.2 reviews relevant literature about ML and optimization technology for optimization of IES and the basics of predictive and prescriptive analytics methods. Section 8.3 overviews the proposed method framework, formulates the proposed uncertainty set using the KMV and the two-stage optimization model, and proposes an exponential formula to determine the outstanding region's adjustive parameter representing the uncertainty set budget level. Section 8.4 introduces the solution methodology to the uncertainty region and the two-stage robust optimization model. Section 8.5 performs simulation results and the superiority of the uncertain fluctuation region in the KMV ellipsoid set considering the uncertainty of wind power. Finally, Section 8.6 gives concluding remarks.

8.3 Problem formulation

The optimal multienergy flow considering uncertainty (OMEF-U) problem is crucial for ensuring the secure and stable operation as well as reliable planning of the IEM. However, the predicted error and shorter timescale fluctuations of uncertainty threaten the real-time generation-load balance in IEM. In this chapter, the optimization process combing predictive and prescriptive analytics, is designed to address problems without being significantly impacted by the predicted errors associated with uncertainty. In addition, the two-stage robust model can obtain optimal solutions quickly by formulating a reasonable uncertainty region around the prediction in the long timescale, while obtaining the short timescale adjustment results while guaranteeing safe real-time operation. In the prediction step, the knn can obtain valuable samples close to the prediction from vast historical data without complex modeling. These valuable samples are used to form a KMV ellipsoidal uncertainty set. Then the robust fluctuation region can be easily obtained from the KMV ellipsoid set by mathematical formulation. The outstanding performance of the region is verified by comparing the ones in some uncertainty sets under different settings of an adjustive parameter. In the prescriptive analytics step, the two-stage robust optimization under the proposed uncertainty set is used to make optimal solutions quickly in a better uncertainty region while reducing the conservativeness of solutions.

8.3.1 KMV ellipsoidal set

The knn can be applied to estimate from the data without the need to assume a prediction model formula or training neural network. Traditionally, a simple nonparametric regressions method can obtain the predicted value by averaging the closest k training data (the samples in knn).

$$\bar{u} = \pi_j^{knn} u_j \pi_j^{knn} = \begin{cases} 1/k, j \in N_k(\bar{u}) \\ 0, others \end{cases} \tag{8.1}$$

where $\bar{u}$ is the prediction of uncertainty, $N_k(\bar{u})$ represents all samples of uncertainty around prediction. However, all the samples in knn were used to form a minimum volume ellipsoid uncertainty set (Shunichi, 2021) to consider the predicted error.

The KMV uncertainty set is as follows:

$$\varepsilon_{knn} = \{u \big| \big| \mathbf{Pu} + \rho \big|_2 \leq 1\} \tag{8.2}$$

where $\mathbf{P}$ and ρ are optimal solutions of the following optimization model:

$$\mathrm{minlogdet}\ \mathbf{P}^{-1}\quad s.t. \big| \mathbf{Pu} + \rho \big|_2 \leq 1 \tag{8.3}$$

Once $\mathbf{P}$ and ρ are obtained, $\mathbf{R}$ and $\mathbf{V}$ are calculated:

$$\mathbf{R} = \mathbf{P}^{-1} \tag{8.4}$$

$$\mathbf{V} = -\mathbf{R}\rho \tag{8.5}$$

Then the proposed uncertainty set around $\bar{u}$ derived from knn is shown as follows:

$$\varepsilon'_{knn} = \{\mathbf{V} + \mathbf{RW} \big| |\mathbf{W}|_2 \leq 1\} \tag{8.6}$$

where $\mathbf{W}$ is a vector full of adjustive parameter W. When W is 0, there is no budget in the uncertainty set, and the lower boundary of the uncertainty region is the average of $\mathbf{V}$. The upper boundary is the average of $\mathbf{V} + \mathbf{RW}$ and increases with increasing W.

8.3.2 Robust fluctuation region selection

The appropriate fluctuation region can be found in the uncertainty set by selecting the appropriate adjustive parameter. The selection of a superior parameter follows the lemma. The lemma is derived from Appendix A.

Lemma:. The suitable W is calculated by a two-term exponential formula when k is available:

$$W = a\exp(bk) + c\exp(dk) \tag{8.7}$$

where the a-d coefficients are constrained with the optimization results of the above set model and derived from the following coefficient optimization model. The purpose of the following model is to find the suitable W to construct a robust uncertainty region containing all samples in knn:

$$\mathrm{min}\, ae^{bk} + ce^{dk} \tag{8.8}$$

$$
s.t. \begin{cases}
-\dfrac{1}{k}\sum_{j\in k}\rho_j \le a \le \dfrac{2}{k}\sum_{j\in k}V_j - \dfrac{1}{k}\sum_{j\in k}\rho_j \\[3mm]
2log\left(\dfrac{1}{k}\sum_{j\in k}V_j\right) + 0.1\dfrac{1}{k}\sum_{j\in k}\rho_j \le b \le 0.1\dfrac{1}{k}\sum_{j\in k}\rho_j \\[3mm]
-0.1\dfrac{1}{k}\sum_{j\in k}\rho_j \le c \le \dfrac{2}{k}\sum_{j\in k}V_j - 0.1\dfrac{1}{k}\sum_{j\in k}\rho_j \\[3mm]
d \le -0.1exp\left(\dfrac{1}{k}\sum_{j\in k}\rho_j\right) \\[3mm]
\dfrac{1}{k}\sum_{j\in k}V_j + \dfrac{1}{k}\sum_{j\in k}diag(\mathbf{R})_j \times (aexp(bk) + cexp(dk)) \ge Sample_{max}
\end{cases}
\tag{8.9}
$$

where ρ_j, V_j is jth element of $\boldsymbol{\rho}$, $\boldsymbol{V}$. $Sample_{max}$ represents the maximum of all samples in knn. Then the solutions (a-d coefficients) of the coefficient optimization model are the key to calculating W. Finally, when W is calculated, the robust fluctuation region can be obtained following Eq. (8.6).

8.3.3 OMEF-U formulation

The first stage is to optimize a long timescale problem, and the second stage is to optimize a short timescale problem considering the prediction uncertainty.

8.3.3.1 The first-stage optimization model
8.3.3.1.1 Objective function
The first-stage operation cost of IEM C_1^d is taken as the objective function, which mainly comprises energy purchase costs C_p^d and fuel cost C_f^d, as shown below.

$$
\min_x C_1^d = min(C_p^d + C_f^d) \tag{8.10}
$$

$$
C_p^d = p_e P_{pe} \tag{8.11}
$$

$$
C_f^d = \sum_{i=1}^{N_e}(a_{ie} + b_{ie}P_{ie}) + \sum_{i=1}^{N_h}(a_{ih} + b_{ih}P_{ih}) \tag{8.12}
$$

where P_{pe} is the purchased energy; p_e is the purchase unit price; a_{ie}, b_{ie} are cost coefficients of ith electricity; a_{ih}, b_{ih} are cost coefficients of ith heat;

N_e, N_h are the number of electricity units and heat units; P_{ie}, P_{ih} are electricity output of unit i and heat output of unit i. The first-stage decision variable $x = \{P_{pe}, P_{ie}, P_{ih}\}$.

8.3.3.1.2 Constraints

The constraints mainly comprise energy balance constraints, energy conversion constraints, energy storage constraints, and network security constraints.
Energy balance constraints:

$$P_{pe} + \sum_{i=1}^{N_e} P_{ie} + \overline{P_w} + P_{dise} = L + P_{chae} \tag{8.13}$$

$$\sum_{i=1}^{N_h} P_{ih} + P_{dish} = H + P_{chah} \tag{8.14}$$

where P_{dise}, P_{chae} are discharging and charging active power of the electricity storage system; $\overline{P_w}$ is the predicted output of wind; L is total electrical demand; P_{dish}, P_{chah} are discharging and charging power of the heat storage system; H is total heat demand.
Energy conversion constraint:

$$P_{ih} = \eta \cdot P_{ie} \tag{8.15}$$

$$P_{ie}^{\min} \leq P_{ie} \leq P_{ie}^{\max} \tag{8.16}$$

$$P_{ih}^{\min} \leq P_{ih} \leq P_{ih}^{\max} \tag{8.17}$$

where η is heat-electric efficiency of CHP; $P_{ie}^{\min}, P_{ie}^{\max}$ is lower and upper boundaries of the active power of ith CHP; $P_{ih}^{\min}, P_{ih}^{\max}$ are lower and upper boundaries of heat output of ith CHP.
Energy storage constraints:

$$\begin{cases} -P_{charq}^{\max} \leq P_q \leq P_{disq}^{\max} \\ E_q^{\min} \leq E_q \leq E_q^{\max} \end{cases} \tag{8.18}$$

$$\tau_{charq} + \tau_{disq} \leq 1 \tag{8.19}$$

where $q = e$ or h; $P_{charq}^{\max}, P_{disq}^{\max}$ are the upper active power boundaries of energy storage and release; $E_q^{\min}, E_q^{\max}$ are the lower and upper boundaries of the energy storage state; $\tau_{charq}, \tau_{disq}$ are the 0 or 1 variables of charging and discharging states.

Electricity network security constraint:

$$\begin{cases} |V_i|^{\min} \le |V_i| \le |V_i|^{\max} \\ P_i^{\min} \le P_i \le P_i^{\max} \\ P_{ij}^{\min} \le P_{ij} \le P_{ij}^{\max} \\ Q_{ij}^{\min} \le Q_{ij} \le Q_{ij}^{\max} \end{cases} \tag{8.20}$$

where $|V_i|^{\min}, |V_i|^{\max}$ are lower and upper boundaries of the voltage amplitude of bus i; $|V_i|$ is voltage amplitude of bus i; $P_i^{\min}, P_i^{\max}$ are lower and upper boundaries of the active power of bus i; P_i is injects active power of node i; $P_{ij}^{\min}, P_{ij}^{\max}$ are lower and upper boundaries of the active power of line ij; P_{ij} is the active power of line ij; $Q_{ij}^{\min}, Q_{ij}^{\max}$ are lower and upper boundaries of the reactive power of line ij; Q_{ij} is reactive power of line ij.

Heat network security constraint:

$$\begin{cases} T_i^{\min} \le T_i \le T_i^{\max} \\ 0 \le m_i \le m_i^{\max} \\ H_i^{\min} \le H_i \le H_i^{\max} \\ H_{ij}^{\min} \le H_{ij} \le H_{ij}^{\max} \end{cases} \tag{8.21}$$

where $T_i^{\min}, T_i^{\max}$ are the maximum and minimum temperature at node i; T_i is the temperature at node i; $m_i^{\max}$ is the maximum mass flow of node i; m_i is mass flow of node i; $H_i^{\min}, H_i^{\max}$ are lower and upper boundaries of heat power of node i; H_i is heat power of node i; $H_{ij}^{\min}, H_{ij}^{\max}$ are lower and upper boundaries of heat power of pipe ij; H_{ij} is heat power of pipe ij.

8.3.3.2 The second-stage optimization model
8.3.3.2.1 Objective function
The second stage aims to minizine the adjustment of CHP. The adjustment cost $(C_{chp}^U + C_{chp}^D)$ and prediction error penalty C_{err}^d are added to the objective function:

$$\min_{\gamma} C_2^d = \min(C_{chp}^U + C_{chp}^D + C_{err}^d) \tag{8.22}$$

$$C_{chp}^U + C_{chp}^D = \lambda_{chpe}^U P_{chpe}^U + \lambda_{chpe}^D P_{chpe}^D + \lambda_{chph}^U P_{chph}^U + \lambda_{chph}^D P_{chph}^D \tag{8.23}$$

$$C_{err}^d = \varpi \cdot abs(\overline{P_w} - P_w) \tag{8.24}$$

where λ_{chpe}^{U}, λ_{chpe}^{D}, λ_{chph}^{U}, λ_{chph}^{D} are the unit price of penalties adjusted up and down; P_{chpe}^{U}, P_{chpe}^{D}, P_{chph}^{U}, P_{chph}^{D} are the active power of ramping up and down and the heat power of ramping up and down; P_w is the output of wind; ϖ is the unit penalty of adjustment. The second-stage decision variable $\gamma = \{P_{chpe}^{U}, P_{chpe}^{D}, P_{chph}^{U}, P_{chph}^{D}\}$.

8.3.3.2.2 Constraints
Energy balance constraints:

$$P_{pe} + \sum_{i=1}^{N_e} P_{ie} + P_{chpe}^{U} - P_{chpe}^{D} + P_w + P_{dise} = L + P_{chae} \tag{8.25}$$

$$\sum_{i=1}^{N_h} P_{ih} + P_{dish} + P_{chph}^{U} + P_{chph}^{D} = H + P_{chah} \tag{8.26}$$

Energy conversion constraints:

$$P_{chpe}^{U\min} \leq P_{chpe}^{U} \leq P_{chpe}^{U\max} \tag{8.27}$$

$$P_{chph}^{U\min} \leq P_{chph}^{U} \leq P_{chph}^{U\max} \tag{8.28}$$

$$P_{ie}^{\min} \leq P_{ie}^{t} + P_{chpe}^{U} - P_{chpe}^{D} \leq P_{ie}^{max} \tag{8.29}$$

$$P_{ih}^{\min} \leq P_{ih}^{t} + P_{chph}^{U} - P_{chph}^{D} \leq P_{ih}^{max} \tag{8.30}$$

where $P_{chpe}^{U\min}$, $P_{chpe}^{U\max}$ are the lower and upper boundaries of ramping up of electricity units; $P_{chph}^{U\min}$, $P_{chph}^{U\max}$ are the lower and upper boundaries of ramping up of heat units.

8.4 Solution methodology

A two-stage RO model can deal with the short timescale fluctuation of uncertainty. In this chapter, the RO problem is a min-max-min problem. The first stage is to minimize operation cost under the predicted value and obtain the first-stage solution. The second stage is to minimize adjustment cost under the worst case of uncertainty in a short timescale and obtains the second-stage solution. The two-stage model can be decomposed by Column-and-Constraint Generation (C&CG) algorithm

(Ji et al., 2021). In addition, the uncertain fluctuation region in the second stage is initially shaped with machine learning and optimization techniques and then finalized by mathematical formulation.

8.4.1 The uncertainty region

The wind output P_w is uncertain and can be calculated by the auxiliary variable v. When the observation of the auxiliary variable $\bar{v}$ is available, $\overline{P_w}$ is calculated by:

$$P_w = \begin{cases} 0, v \leq v_{in}^{cut} \cup v > v_{out}^{cut} \\ \dfrac{v - v_{in}^{cut}}{v_{rated} - v_{in}^{cut}} \varphi, v_{in}^{cut} < v < v_{rated} \\ \varphi, v_{rated} < v < v_{out}^{cut} \end{cases} \qquad (8.31)$$

Then, the root-mean-square error (RMSE) with cross validation is used to obtain a heuristically optimal number k of nearest neighbors. When RMSE is lowest, the k samples around prediction are used to form the proposed uncertainty set by the KMV uncertainty set optimization model. Using the optimal result, the superior W is obtained with (8.7), where the coefficients are obtained from the coefficient optimization model with a nonlinear solver. Subsequently, the uncertainty region is formed by (8.6). Other uncertainty regions also can be acquired by setting different W.

8.4.2 Two-stage RO model for OMEF-U problem

In the optimization model, the problem can be an expression as:

$$\min_x C_1^d + \max_u \min_y C_2^d \qquad (8.32)$$

$$s.t. Ax \leq a \qquad (8.33)$$

$$By \leq b \qquad (8.34)$$

$$Cx + Dy \leq c \qquad (8.35)$$

$$Ex + Fy + Gu = d \qquad (8.36)$$

Equation (8.33) denotes all the constraints with continuous various and binary variables (8.13)−(8.21). Equation (8.34) denotes all the constraints with continuous variation of the second stage (8.27) and (8.28). Equation (8.35) contains all the constraints with continuous variation of both stages (8.29) and (8.30). Equation (8.36) summarizes the equality constraints with continuous variables and uncertainty (8.25) and (8.26).

The AC optimal power flow problem is addressed using the second order cone programming (SOCP) relaxations (Kocuk et al., 2016), while the heat network flow is analyzed according to Liu et al. (2016). Meanwhile, the C&CG algorithm decomposes the two-stage robust problem as the master problem and subproblem. The subproblem can be converted to a single-level problem with Karush−Kuhn−Tucker conditions for the internal constraints are linear, and uncertainty is at a separate level (Zeng & Zhao, 2013). The detailed C&CG algorithm can be found in Appendix B.

8.5 Case study

The case study was evaluated using JuliaPro 1.5.4 on a standard personal computer with an Intel Core i7−10700F CPU running at 2.90 GHz and 16 GB RAM. The optimization model is solved with Gurobi. The predicted output of wind turbines is 0.64 MW. The electricity load and heat load are 0.5 MW and 0.73 MW.$\eta = 0.8$. Other details of parameters of the two-stage robust model are derived from (Zhu et al., 2019).

8.5.1 Case setup

An improved industrial park testbed (Zheng et al., 2021) is used to verify the prominence of the proposed uncertainty set. As shown in Fig. 8.1, the industrial park testbed incorporates a modified distribution network with 14 buses and a heating system with 11 nodes. They are coupled through the combined heat and power (CHP) unit. Bus 1 connects the industrial park with the power grid.

Table 8.1 shows the allocation of system energy sources.

Three cases were simulated to demonstrate that the effectiveness of the proposed uncertainty set using KMV is outstanding and suitable for various uncertainty fluctuations. Case 1 includes 500 random samples with a standard deviation of 0.2, indicating a normal fluctuation. Case 2 includes 500 random samples with a standard deviation of 0.6, indicating a high fluctuation. Then 3041 large volume of historical wind data in Guangzhou is used in case 3. To provide a comprehensive understanding of the proposed uncertainty set in terms of usability, robustness, and conservativeness, a comparative analysis is conducted in Case 1 and Case 2.

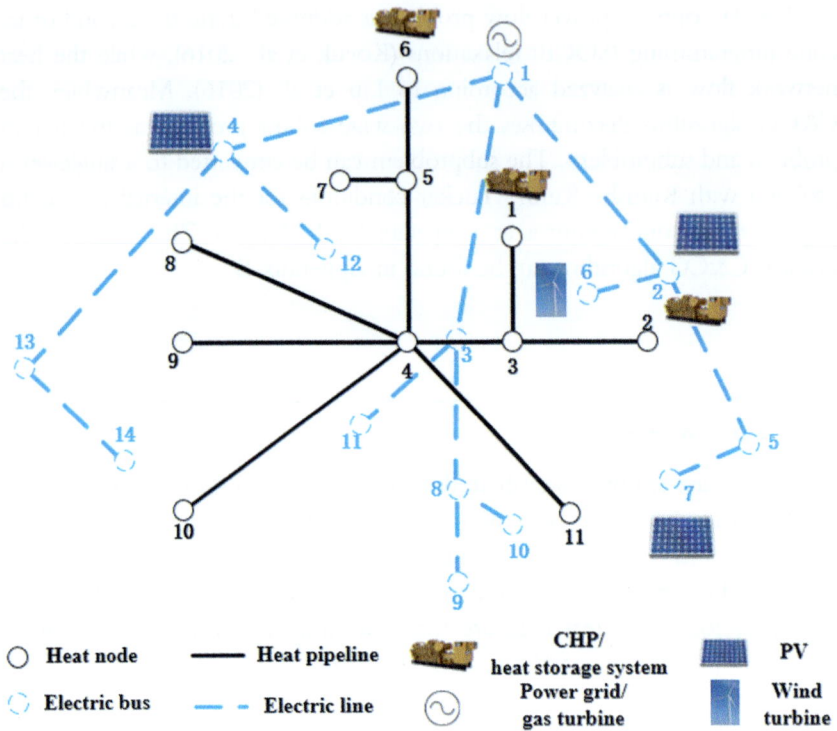

Figure 8.1 *The topology of the modified industrial park testbed.* The topology of the case.

Table 8.1 Allocation of energy sources.

Electricity bus	Heat node	Source
1	—	Power grid
2	2	CHP
6	—	Wind turbine
—	1	CHP
—	6	CHP

This analysis involves comparing the proposed uncertainty set with some state-of-the-art uncertainty sets based on fluctuation region, deviation rate of the upper boundary from the predicted value, and total operation cost, respectively. The state-of-the-art uncertainty sets are detailedly introduced in Appendix C, named as Box set, Ellipsoid Set1, and Ellipsoid Set2. During the comparison process, invalid regions are highlighted in italics.

The best fluctuation range of each uncertainty set is highlighted in bold. Finally, case 3 analyzes optimal dispatch under the proposed uncertainty set and different fluctuation regions' contribution to engineering.

8.5.2 Evaluation of the performances under case 1

RMSE indicates the optimal number k is 5, in Fig. 8.2.

Then the samples in the k-nearest neighbors: [0.63834,0.63819, 0.66186,0.63622,0.64444]. The superior W is 0.5.

8.5.2.1 Comparison of fluctuation region

Table 8.2 shows uncertain fluctuation regions with different W. When $W = 0.2$ and $W = 0.5$, the ellipsoid set2 is overly tight and does not contain the samples in the k-nearest neighbors entirely. When $W = 0.2$, the fluctuation regions of the box set and ellipsoid set1 are not robust. When $W = 0.4$, the fluctuation regions of ellipsoid set2 are not robust. Obviously, when $W = 0.5$, the fluctuation region of the last set outperforms is perfect.

In Fig. 8.3, the regional interval (upper–lower boundary length) of two ellipsoid sets changes following W irregularly, which is not convenient for selecting suitable lower and upper budgets. While the interval regions of the box set and KMV ellipsoid set are increasing linearly with

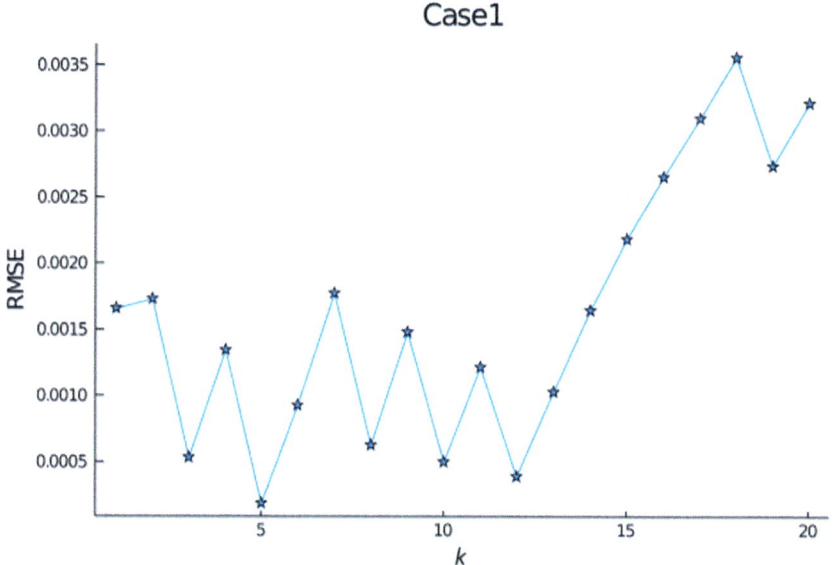

Figure 8.2 *RMSE following different* k *in case 1.* The correction between RMSE and k in case 1.

Table 8.2 Fluctuation regions under different uncertainty sets with different W in case 1.

	Box set		Ellipsoid set1		Ellipsoid set2		KMV ellipsoid set	
Adjustive parameter	Lower	Upper	Lower	Upper	Lower	Upper	Lower	Upper
$W = 0$	0.640	0.640	0.640	0.640	0.640	0.640	0.442	0.442
$W = 0.2$	0.602	0.680	0.588	0.692	*0.621*	*0.659*	0.442	0.531
$W = 0.4$	0.563	0.717	0.584	0.696	0.602	0.678	0.442	0.619
$W = 0.5$	0.544	0.736	0.405	0.875	*0.639*	*0.641*	0.442	0.663
$W = 0.6$	0.525	0.755	0.551	0.729	0.561	0.719	0.442	0.707
$W = 0.8$	0.486	0.794	0.345	0.955	0.541	0.739	0.442	0.796
$W = 1$	0.448	0.832	0.547	0.733	0.393	0.887	0.442	0.884

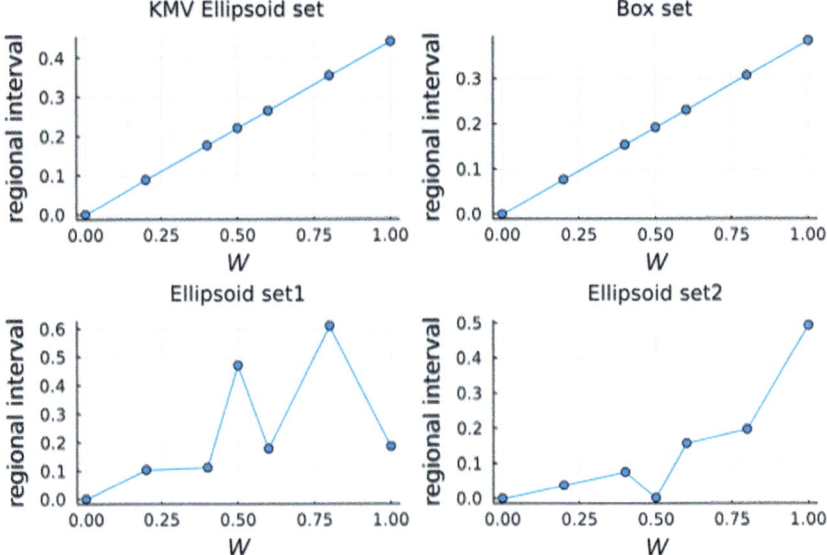

Figure 8.3 *The lower and upper budgets of different uncertainty sets.* The comparison among three uncertainty sets.

W. Based on such advantage, W was set to determine the budget of uncertainty fluctuations quickly. Unfortunately, the upper boundary of the box set is generally larger than that of the KMV ellipsoid set. The fluctuation regions of the box set are not as robust as those of the KMV ellipsoid set. In conclusion, it can be observed that the KMV ellipsoid set possesses a superior advantage over others in the uncertain fluctuation region.

8.5.3 Comparison of the deviation rate of the upper boundary from the predicted value

The deviation rate of the upper boundary from the predicted value in Table 8.3 can reveal the conservativeness of the solution under an uncertainty set. When $W = 0.5$, the fluctuation region in the KMV ellipsoid set is the most constrained when compared to others. Although the fluctuation region is perfect in the ellipsoid set1 when $W = 0.2$, its conservativeness is greater than that of the other sets. Therefore, $W = 0.5$ is verified to be the superior parameter in the KMV ellipsoid set. The fluctuation region can contain all samples in the k-nearest neighbors, and the solution is not conservative.

8.5.4 Comparison of total operation cost

Table 8.4 shows the total operation cost under different uncertainty sets with different W. It is evident that the cost under the last set is the lowest when $W = 0.5$. In the most uncertain regions, the two-stage OMEF-U is more economical than other compared sets. The total cost is reduced by 19.18% compared to the worst region in the KMV ellipsoid set. Although each set takes the promising region, the cost of this region is 2.53% less than the worst among all sets. The output of wind is contained in samples in the k-nearest neighbors. The CPU time for obtaining an optimal solution of all uncertainty sets is from 1.8 to 2.2 s. There is no distinction among all sets.

8.5.5 Evaluation of the performances under case 2

This experiment is based on higher violation simulation data. RMSE indicates the optimal number k is 9, as shown in Fig. 8.4.

Table 8.3 Deviation rate of the upper boundary from the predicted value with different W in case 1.

Adjustive parameter	Box set	Ellipsoid set1	Ellipsoid set2	KMV ellipsoid set
$W = 0$	0	0	0	0
$W = 0.2$	6.00%	8.20%	2.97%	-17.08%
$W = 0.4$	12.0%	8.81%	5.89%	-3.26%
$W = 0.5$	15.0%	36.71%	0.16%	3.64%
$W = 0.6$	18.0%	13.97%	12.36%	10.55%
$W = 0.8$	24.0%	49.27%	15.41%	24.38%
$W = 1$	30.0%	14.56%	38.59%	38.20%

Table 8.4 Total operation cost under different uncertainty sets with different *W* in Case 1.

Adjustive parameter	Box set	Ellipsoid set1	Ellipsoid set2	KMV ellipsoid set
$W = 0$	0.6055	0.6055	0.6055	0.7116
$W = 0.2$	0.6269	0.6334	0.6157	0.6639
$W = 0.4$	0.6468	0.6355	0.6259	0.6168
$W = 0.5$	0.6570	0.7315	0.6060	0.6178
$W = 0.6$	0.6671	0.6532	0.6478	0.6414
$W = 0.8$	0.6880	0.7743	0.6586	0.6891
$W = 1$	0.7084	0.6553	0.7379	0.7363

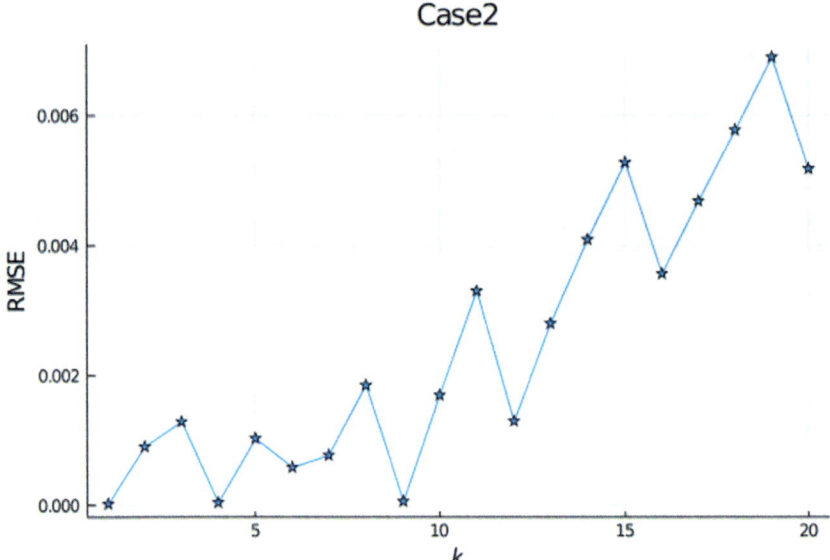

Figure 8.4 *RMSE following different* **k** *in case 2.* The correction between RMSE and *k* in case 2.

Then the samples in the k-nearest neighbors: [0.64002, 0.63817, 0.63794, 0.64369, 0.64528, 0.63138, 0.66887, 0.64939, 0.62576]. The superior *W* is 0.4.

8.5.5.1 Comparison of fluctuation region

Table 8.5 shows the fluctuation region of uncertainty with different *W*, values. the uncertainty experiences significant fluctuations, the adjustive parameter of ellipsoid set2 should be increased to 0.8 to contain all

Table 8.5 Fluctuation region under different uncertainty sets with different W in case 2.

	Box set		Ellipsoid set1		Ellipsoid set2		KMV ellipsoid set	
Adjustive parameter	Lower	Upper	Lower	Upper	Lower	Upper	Lower	Upper
$W=0$	0.64	0.64	0.64	0.64	0.64	0.64	0.4800	0.4800
$W=0.2$	0.5632	0.7168	0.5544	0.7256	*0.6239*	*0.6561*	0.4800	0.5761
$W=0.4$	0.4864	0.7936	0.5858	0.6942	0.5851	0.6949	0.4800	0.6720
$W=0.5$	0.4480	0.8320	0.5513	0.7287	*0.6392*	*0.6407*	0.4800	0.7201
$W=0.6$	0.4096	0.8704	0.4356	0.8444	*0.6399*	*0.6400*	0.4800	0.7681
$W=0.8$	0.3328	0.9472	0.5604	0.7196	0.5929	0.6871	0.4800	0.8641
$W=1$	0.2560	1.0240	0.5606	0.7193	0.5655	0.7144	0.4800	0.9601

Table 8.6 Deviation rate of the upper boundary from the predicted value with different W in case 2.

Adjustive parameter	Box set	Ellipsoid set1	Ellipsoid set2	KMV ellipsoid set
$W=0$	0	0	0	−25.00%
$W=0.2$	12.00%	13.38%	2.51%	−9.98%
$W=0.4$	24.00%	8.46%	8.58%	5.00%
$W=0.5$	30.00%	13.86%	0.00%	12.51%
$W=0.6$	36.00%	31.94%	0.00%	20.02%
$W=0.8$	48.00%	12.44%	7.36%	35.02%
$W=1$	60.00%	12.39%	11.63%	50.02%

samples in k-nearest neighbors. Furthermore, the upper boundary is smaller than the maximum of the samples in the k-nearest neighbors, and many fluctuation regions are ineffective when $W=0.2$, 0.5, 0.6. The fluctuation region of the box set is not as robust as the others in case 2. Fortunately, all fluctuation regions of the KMV ellipsoid set still work and contain all samples in the k-nearest neighbors with strong robustness.

8.5.5.2 Comparison of the deviation rate of the upper boundary from the predicted value

When $W=0.4$, the deviation rate in the KMV ellipsoid containing all samples in the k-nearest neighbors is lower than state-of-the-art uncertainty sets with any W. The box set shows the worst performance with the worst robustness in Table 8.6.

Table 8.7 Total operation cost under different uncertainty sets with different W in case 2.

Adjustive parameter	Box set	Ellipsoid set1	Ellipsoid set2	KMV ellipsoid set
$W = 0$	0.6055	0.6055	0.6055	0.6913
$W = 0.2$	0.6466	0.6514	0.6141	0.6398
$W = 0.4$	0.6878	0.6345	0.6349	0.6227
$W = 0.5$	0.7084	0.6530	0.6055	0.6484
$W = 0.6$	0.7290	0.7151	0.6055	0.6742
$W = 0.8$	0.7702	0.6482	0.6307	0.7256
$W = 1$	0.8113	0.6480	0.6454	0.7771

8.5.5.3 Comparison of total operation cost

It is obvious that the cost under the last set is the lowest in Table 8.7. When $W = 0.4$, the output of wind power is contained in samples in the k-nearest neighbors. The total cost is reduced by 24.80% compared to the worst region in the KMV ellipsoid set. In a word, the conservativeness level of uncertainty set: box set > ellipsoid set 1 > ellipsoid set2 > KMV ellipsoid set. Although each set takes the promising region, the cost of this region is 3.84% less than the worst among all sets. The CPU time obtaining the optimal solution of all uncertainty sets regions from 1.9 to 2.6 s. There is no distinction among all sets.

8.5.6 Analysis of optimal dispatch with KMV ellipsoid set

The above comparison illustrates that the fluctuation regions of the KMV ellipsoid set are outstanding for state-of-the-art uncertainty fluctuation regions. This section applies it in case 3 to analyze the advantage of different fluctuation regions with different W. When predicted value is 0.64, RMSE indicates that the optimal number k is 11, and samples in the k-nearest neighbors: [0.63935, 0.64171, 0.64184, 0.64258, 0.63738, 0.64270, 0.64273, 0.63636, 0.64436, 0.63524, 0.63460]. The superior W is 0.4.

The suitable fluctuation region is highlighted in bold in Table 8.8. Obviously, the lower and upper budgets widen as W increases. When $W = 0.4$, the fluctuation region contains all samples in the k-nearest neighbors, and the least economically feasible solution is obtained. In the most uncertain regions, the two-stage OMEF-U is more economical than the other regions. The total cost is reduced by 21.29% compared to the worst region. When W is less than 0.4, the upper boundaries are smaller than the predicted value. When W is more than 0.4, the upper boundaries are greater than the predicted value and the fluctuation regions have a large budget. Furthermore, when the real output of the wind turbine under the short

Table 8.8 Optimal dispatch in case 3 with different W.

Adjustive parameter				u_lower	u_upper	Prediction error rate			
$W = 0$	0.49	0.19	−0.14	0.49	0.49	−23.23%	0.328	0.900	2.31
$W = 0.2$	0.57	0.19	−0.06	0.49	0.58	−9.49%	0.277	0.782	2.35
$W = 0.4$	0.66	0.19	0.02	0.49	0.68	3.44%	0.271	0.742	2.38
$W = 0.5$	0.70	0.19	0.06	0.49	0.73	9.47%	0.291	0.763	2.39
$W = 0.6$	0.71	0.19	0.07	0.49	0.78	12.05%	0.300	0.771	2.40
$W = 0.8$	0.75	0.19	0.11	0.49	0.88	17.58%	0.319	0.790	2.41
$W = 1$	0.83	0.19	0.19	0.49	0.98	31.11%	0.365	0.837	2.45

timescale is larger than the prediction and energy storage capacity is abundant, the W can be set greater than 0.4 to form a wide fluctuation region. The large output of the wind turbine can be absorbed under such a fluctuation region, and the optimal solution is not conservative. When other power generations are large, such as hydropower under the short timescale is larger than the predicted value, the W can be set smaller than 0.4 to form a tight fluctuation region; therefore, proper wind power curtailment can guarantee the system's generation-load balance and safe operation.

8.6 Conclusion

This chapter explores a two-stage robust optimization model applied to OMEF-U based on intelligent learning approaches. The minimum volume ellipsoid set in the RO model containing all samples around the prediction in knn and the robustness is lower than the compared sets. The robust uncertainty region can be obtained by a two-term exponential formula from the KMV ellipsoid set. Although the ellipsoid set2 is tighter than the KMV ellipsoid set when a specific W, the fluctuation region is too narrow to cover other possibilities. Simulation results also illustrate that the ellipsoidal sets' interval regions have no regularity with the adjustive parameter, which is not convenient to select an uncertain fluctuation region quickly from their adjustive parameters. Although the interval regions of the box set and KMV ellipsoid set extend with increasing W linearly, the box set is not robust with a significant uncertainty budget. Therefore, the KMV ellipsoid set with the regular region has superiority. Finally, the KMV ellipsoid set is used to calculate two-stage OMEF-U on a large volume of real historical data. When

$W = 0.4$, the fluctuation region is robust, and the solution is not conservative. One can conclude that the superiority of the uncertain fluctuation region selected by the lemma is not affected by sharp violation and a vast quantity of uncertainty. The high-performance uncertainty region can be found for any uncertainty, and the robustness is best than other sets. The optimal result is adjustable with W and not conservative in the two-stage optimal multienergy flow model. Further work can focus on applying advanced predictive techniques and optimization techniques and considering multitimescale dispatch for stability control.

Appendix

Appendix A: Proof of Lemma

It is obvious that the upper boundary increases with W linearly in (8.6), and the knn can obtain k samples around the predicted value. Based on the above two characteristics, W can be determined with the k. In our intelligent learning approaches combined with prediction and optimization, the selection of k will affect the decision of superior W. When k is small, samples in knn are few, and W should be large to include possible uncertainty values. When k increases, there are more and more possibilities in samples around the predicted value, and W can become smaller while ensuring that the maximum value of samples is included in the uncertainty region. When k tends to a larger value, the tendency of W to become smaller will be slowed down to include all samples; therefore, W changes with k inversely and follows the law of exponential.

Equation (8.3) indicates that different k and samples in knn can emerge different optimal results of the KMV uncertainty set optimization model. Moreover, the proposed uncertainty set is indirectly affected by the selection of k and samples in knn. The optimization results also affect the robustness of the proposed uncertainty set. Therefore, the proposed uncertainty set's robust uncertainty region should be unique and calculated with an exponential formula adjusted with the machined learning and optimization results. The coefficients of the exponential formula should be constrained, and the uncertainty region should be tight while containing all samples in knn.

Appendix B: Column-and-Constraint Generation Algorithm

Table 8.9 shows the Column-and-Constraint Generation algorithm to a two-stage robust optimal multienergy flow model.

Table 8.9 Algorithm of column-and-constraint generation.

Step1: Initialization: Set upper and lower boundaries: UB = , LB = , S = , Q = 0
Step2: Solve the master problem:

$$\min_{x,\eta} C_1^d + \varsigma \tag{8.37}$$

$$s.t. Ax \leq a \tag{8.38}$$

$$\varsigma \geq C_2^d \tag{8.39}$$

$$By^q \leq b, q = 1, 2, ..., Q \tag{8.40}$$

$$Cx + Dy^q \leq c, q = 1, 2, ..., Q \tag{8.41}$$

Derive optimal solution $(x_q^*, \varsigma_q^*, y_q^*)$
Update LB
Step3: Solve the subproblem:

$$\min_{u,y,x^*} C_2^d \tag{8.42}$$

$$s.t. By \leq b : \alpha \tag{8.43}$$

$$Cx^* + Dy \leq c : \beta \tag{8.44}$$

$$Ex^* + Fy + Gu = d : \gamma \tag{8.45}$$

$$(b - By)\alpha = 0 \tag{8.46}$$

$$(c - Cx^* - Dy)\beta = 0 \tag{8.47}$$

$$(d - Fy - Gu - Ex^*)\gamma = 0 \tag{8.48}$$

$$\alpha, \beta, \gamma \geq 0 \tag{8.49}$$

Update UB
Step4: Compute gap = |UB-LB|
Step5: if gap $\leq \varepsilon$, then returns y_{q+1}^* and ends up.
Step6: else increment $q = q + 1$, and turn to Step2.

Appendix C: The Construction of the State-of-arts Uncertainty Sets

The box set has a simple structure and can quickly handle classic problems such as linear optimization, quadratic cone optimization, and semi-definite optimization; however, the robustness of the box set is worse than the ellipsoid set. We, therefore, illustrate that the region under KMV ellipsoid set is most prominent by pairing a box set and two ellipsoid sets when applied to a two-stage robust optimal multienergy flow model.

Box Set: The structure of the box set (Li et al., 2011):

$$U = \{u|e^T u = 0, \underline{\zeta} \leq u \leq \overline{\zeta}\} \tag{8.50}$$

where $\underline{\zeta}, \overline{\zeta}$ are the lower and upper boundaries of uncertainty. The uncertainty belongs to $\left[\overline{u} - W \cdot u_{err}^{max}, \overline{u} + W \cdot u_{err}^{max}\right]$

Ellipsoid Set1: the uncertain parameters u belongs to the region (Venzke et al., 2018):

$$\Omega = \{u \in R^{n_u} u \sum u \leq 1\} \tag{8.51}$$

$$\sum_1 = Diag\left(\frac{1}{W\overline{u}}\right) \tag{8.52}$$

Ellipsoid Set2: the form of an ellipsoid set Eq. (A2) is not changed, while was modified as follows (Qiu and Jiang, 2021):

$$\sum_2 = Diag\left(\frac{1}{(W\overline{u})^2}\right) \tag{8.53}$$

References

Aaslid, P., Korpas, M., Belsnes, M. M., & Fosso, O. B. (2022). Stochastic optimization of microgrid operation with renewable generation and energy storages. *IEEE Transactions on Sustainable Energy*, *13*(3), 1481−1491. Available from https://doi.org/10.1109/TSTE.2022.3156069.

Adebiyi, A. A., Adewumi, A. O., & Ayo, C. K. (2014). *Stock price prediction using the ARIMA model. Proceedings—UKSim-AMSS 16th International Conference on Computer Modelling and Simulation* (pp. 106−112). South Africa: Institute of Electrical and Electronics Engineers Inc. Available from http://doi.org/10.1109/UKSim.2014.67.

Agarwal, V., Gribok, A. V., Koschan, A., & Abidi, M. A. (2006). *Estimating illumination chromaticity via kernel regression. Proceedings—International Conference on Image Processing* (pp. 981−984). United States: ICIP. Available from http://doi.org/10.1109/ICIP.2006.312652.

Ben-Tal, A., Ghaoui, L. E., & Nemirovski, A. (2009). *Robust optimization.* Princeton University Press. Available from: http://press.princeton.edu/titles/9099.html.

Bertsimas, D., & Kallus, N. (2020). From predictive to prescriptive analytics. *Management Science*, *66*(3), 1025−1044. Available from https://doi.org/10.1287/mnsc.2018.3253.

Box, G. E. P., Jenkins, G. M., & Reinsel, G. C. (1994). *Time series analysis: Forecasting and control.* 3rd Edition, Prentice Hall PTR, Englewood Cliff, New Jersey.

Cheng, Y., Zhang, N., Lu, Z., & Kang, C. (2019). Planning multiple energy systems toward low-carbon society: A decentralized approach. *IEEE Transactions on Smart Grid*, *10*(5), 4859−4869. Available from https://doi.org/10.1109/tsg.2018.2870323.

Dragoon, K., & Milligan, P. (2003). *Assessing wind integration costs with dispatch models: A case study of PacifiCorp*.

Gey, S., & Nedelec, E. (2005). Model selection for CART regression trees. *IEEE Transactions on Information Theory*, *51*(2), 658−670. Available from https://doi.org/10.1109/TIT.2004.840903.

Hossain, M. B., Moon, J., & Chon, K. H. (2020). Estimation of arma model order via artificial neural network for modeling physiological systems. *IEEE Access*, *8*, 186813−186820. Available from https://doi.org/10.1109/ACCESS.2020.3029756.

Hua, W., Jiang, J., Sun, H., Tonello, A. M., Qadrdan, M., & Wu, J. (2022). Data-driven prosumer-centric energy scheduling using convolutional neural networks. *Applied Energy*, *308*. Available from https://doi.org/10.1016/j.apenergy.2021.118361.

Ji, Y., Xu, Q., Zhao, J., Yang, Y., & Sun, L. (2021). Day-ahead and intra-day optimization for energy and reserve scheduling under wind uncertainty and generation outages. *Electric Power Systems Research*, *195*. Available from https://doi.org/10.1016/j.epsr.2021.107133.

Jin, X., Wu, Q., Jia, H., & Hatziargyriou, N. (2021). Optimal integration of building heating loads in integrated heating/electricity community energy systems: A bi-level MPC approach. *IEEE Transactions on Sustainable Energy*, *12*(3), 1741−1754. Available from https://doi.org/10.1109/TSTE.2021.3064325.

Kocuk, B., Dey, S. S., & Andy Sun, X. (2016). Strong SOCP relaxations for the optimal power flow problem. *Operations Research*, *64*(6), 1177−1196. Available from https://doi.org/10.1287/opre.2016.1489.

Kuryatnikova, O., Ghaddar, B., & Molzahn, D. K. (2021). Adjustable robust two-stage polynomial optimization with application to AC optimal power flow. *arXiv preprint*, 2104.03107. Available from https://doi.org/10.48550/arXiv.2104.03107.

Li, D., & Zhang, S. (2021). Optimal design of distributed energy resource systems under uncertainties based on two-stage robust optimization. *Journal of Thermal Science*, *30*(1), 51−63. Available from https://doi.org/10.1007/s11630-020-1397-9.

Li, J., Fang, J., Zeng, Q., & Chen, Z. (2016). Optimal operation of the integrated electrical and heating systems to accommodate the intermittent renewable sources. *Applied Energy*, *167*, 244−254. Available from https://doi.org/10.1016/j.apenergy.2015.100.054.

Li, S., Zhou, R., Tong, X., Peng, S., Zhao, M., & Yao, L. (2011). Robust optimization with box set for maximum installed capacity of wind farm connected to grid. *Dianwang Jishu/Power System Technology*, *35*(12), 208−213.

Li, T., Sun, H., Shi, Z., Li, Q., Zhang, J., & Huang, B. (2022). *Cooperative optimal configuration of integrated energy system considering uncertainty factors of source-load*. IOP Conference Series: Earth and Environmental Science (p. 983) China: IOP Publishing Ltd. 17551315 1. Available from http://doi.org/10.1088/1755-1315/983/1/012119.

Liu, X., & Mancarella, P. (2016). Modelling, assessment and Sankey diagrams of integrated electricity-heat-gas networks in multi-vector district energy systems. *Applied Energy*, *167*, 336−352. Available from https://doi.org/10.1016/j.apenergy.2015.080.089.

Liu, X., Wu, J., Jenkins, N., & Bagdanavicius, A. (2016). Combined analysis of electricity and heat networks. *Applied Energy*, *162*, 1238−1250. Available from https://doi.org/10.1016/j.apenergy.2015.010.102.

Liu, Y., Guo, L., & Wang, C. (2018). Economic dispatch of microgrid based on two stage robust optimization. *Zhongguo Dianji Gongcheng Xuebao/Proceedings of the Chinese Society of Electrical Engineering*, *38*(14), 4013−4022. Available from https://doi.org/10.13334/j.0258-8013.pcsee.170500.

Liu, Z., Yu, H., & Liu, R. (2019). A novel energy supply and demand matching model in park integrated energy system. *Energy*, *176*, 1007−1019. Available from https://doi.org/10.1016/j.energy.2019.040.049.

Malekpour, A. R., & Pahwa, A. (2017). Stochastic networked microgrid energy management with correlated wind generators. *IEEE Transactions on Power Systems*, *32*(5), 3681−3693. Available from https://doi.org/10.1109/TPWRS.2017.2650683.

Moradzadeh, A., Mohammadi-Ivatloo, B., Abapour, M., Anvari-Moghaddam, A., & Roy, S. S. (2022). Heating and cooling loads forecasting for residential buildings based on hybrid machine learning applications: A comprehensive review and comparative analysis. *IEEE Access*, *10*, 2196−2215. Available from https://doi.org/10.1109/ACCESS.2021.3136091.

Qin, X., Sun, H., Shen, X., Guo, Y., Guo, Q., & Xia, T. (2019). A generalized quasi-dynamic model for electric-heat coupling integrated energy system with distributed energy resources. *Applied Energy*, *251*. Available from https://doi.org/10.1016/j.apenergy.2019.050.073.

Qiu, Z., & Jiang, N. (2021). An ellipsoidal Newton's iteration method of nonlinear structural systems with uncertain-but-bounded parameters. *Computer Methods in Applied Mechanics and Engineering*, *373*, 113501.1−113501.23. Available from https://doi.org/10.1016/j.cma.2020.113501.

Seo, S., Oh, S. D., & Kwak, H. Y. (2019). Wind turbine power curve modeling using maximum likelihood estimation method. *Renewable Energy*, *136*, 1164−1169. Available from https://doi.org/10.1016/j.renene.2018.090.087.

Shunichi, O. (2021). A predictive prescription using minimum volume k-nearest neighbor enclosing ellipsoid and robust optimization. *Mathematics*, *9*(2). Available from https://doi.org/10.3390/math9020119.

Sivanantham, G., & Gopalakrishnan, S. (2020). A Stackelberg game theoretical approach for demand response in smart grid. *Personal and Ubiquitous Computing*, *24*(4), 511−518. Available from https://doi.org/10.1007/s00779-019-01262-9.

Speiser, J. L., Miller, M. E., Tooze, J., & Ip, E. (2019). A comparison of random forest variable selection methods for classification prediction modeling. *Expert Systems with Applications*, *134*, 93−101. Available from https://doi.org/10.1016/j.eswa.2019.050.028.

Tan, Z., De, G., Li, M., Lin, H., Yang, S., Huang, L., & Tan, Q. (2020). Combined electricity-heat-cooling-gas load forecasting model for integrated energy system based on multi-task learning and least square support vector machine. *Journal of Cleaner Production*, *248*. Available from https://doi.org/10.1016/j.jclepro.2019.119252.

Triguero, I., García-Gil, D., Maillo, J., Luengo, J., García, S., & Herrera, F. (2019). Transforming big data into smart data: An insight on the use of the k-nearest neighbors algorithm to obtain quality data. *Wiley Interdisciplinary Reviews: Data Mining and Knowledge Discovery*, *9*(2). Available from https://doi.org/10.1002/widm.1289.

Tu, Q., Miao, S., Lin, Y., Zhang, D., Yao, F., & Han, J. (2022). Ultra-short-term interval forecasting method for regional wind farms based on dynamic R-vine Copula Model. *Gaodianya Jishu/High Voltage Engineering*, *48*(2), 456−466. Available from https://doi.org/10.13336/j.1003-6520.hve.20201711.

Vatani, B., Chowdhury, B., Dehghan, S., & Amjady, N. (2018). A critical review of robust self-scheduling for generation companies under electricity price uncertainty. *International Journal of Electrical Power and Energy Systems*, *97*, 428−439. Available from https://doi.org/10.1016/j.ijepes.2017.100.035.

Venzke, A., Halilbasic, L., Markovic, U., Hug, G., & Chatzivasileiadis, S. (2018). Convex relaxations of chance constrained AC optimal power flow. *IEEE Transactions on Power Systems*, *33*(3), 2829−2841. Available from https://doi.org/10.1109/TPWRS.2017.2760699.

Wei, Z., Mei, J., Sun, G., Zang, H., & Chen, S. (2017). Multi-period transient energy-flow simulation of integrated power and gas energy system. *Dianli Zidonghua Shebei/*

Electric Power Automation Equipment, *37*, 41−47. Available from https://doi.org/10.16081/j.issn.1006-6047.2017.060.006.

Windpower 2003 Conference Proceedings (CD-ROM), 18–21 May 2003, Austin, Texas. Proceedings Sponsored by SPS, Specialized Power Systems, Inc., Washington, DC: American Wind Energy Association; Omni Press 13 pp.; NREL Report No. CP-500-38580. http://www.nrel.gov/publications/.

Wu, W., Wang, K., Li, G., & Ge, Y. (2017). Modeling ellipsoidal uncertainty set considering conditional correlation of wind power generation. *Zhongguo Dianji Gongcheng Xuebao/Proceedings of the Chinese Society of Electrical Engineering*, *37*(9), 2500−2506. Available from https://doi.org/10.13334/j.0258-8013.pcsee.160389.

Yang, S., Tan, Z., Zhou, J., Xue, F., Gao, H., Lin, H., & Zhou, F. (2021). A two-level game optimal dispatching model for the park integrated energy system considering Stackelberg and cooperative games. *International Journal of Electrical Power and Energy Systems*, *130*. Available from https://doi.org/10.1016/j.ijepes.2021.106959.

Zeng, B., & Zhao, L. (2013). Solving two-stage robust optimization problems using a column-and- constraint generation method. *Operations Research Letters*, *41*(5), 457−461. Available from https://doi.org/10.1016/j.orl.2013.050.003.

Zeng, Q., Fang, J., Li, J., & Chen, Z. (2016). Steady-state analysis of the integrated natural gas and electric power system with bi-directional energy conversion. *Applied Energy*, *184*, 1483−1492. Available from https://doi.org/10.1016/j.apenergy.2016.050.060.

Zhao, Z., Liu, Y., Guo, L., Bai, L., & Wang, C. (2021). Locational marginal pricing mechanism for uncertainty management based on improved multi-ellipsoidal uncertainty set. *Journal of Modern Power Systems and Clean Energy*, *9*(4), 734−750. Available from https://doi.org/10.35833/MPCE.2020.000824.

Zheng, L., Li, Y., Wei, C., & Bai, X. (2021). A data-driven method for operation pattern analysis of the integrated energy microgrid. *Energy Conversion and Management: X.*, *11*. Available from https://doi.org/10.1016/j.ecmx.2021.100092.

Zhu, J., Liu, Y., Xu, L., Jiang, Z., & Ma, C. (2019). Robust day-ahead economic dispatch of microgrid with combined heat and power system considering wind power accommodation. *Dianli Xitong Zidonghua/Automation of Electric Power Systems*, *43*, 40−48. Available from https://doi.org/10.7500/AEPS20180214007.

CHAPTER NINE

Intelligent power quality disturbance detection methods in virtual power plants: state-of-the-art

Gökay Bayrak and Alper Yilmaz

Smart Grid Laboratory, Department of Electrical and Electronics Engineering, Bursa Technical University, Bursa, Turkey

9.1 Introduction

A virtual power plant (VPP) is a software and interface designed to facilitate the integration of distributed generators (DGs) and existing energy systems. It functions as an independent microgrid (Yu et al., 2019). A basic VPP structure is given in Fig. 9.1. VPPs actively participate in various aspects of energy management, such as load demand management, cost optimization, resource planning, efficient usage of DG, and emergency response for faults. Flexible loads can be managed by VPPs instead of activating or decommissioning real power plants.

A VPP is composed of multiple individual virtual generation units, which are integrated and managed as a unified system. DGs' control cannot be provided centrally due to resorting to plain DG leads instead of inefficient and costly investments in the distribution infrastructure (Thavlov & Bindner, 2015). In practice, there is often no cooperation or communication between neighboring DGs. As a result, the capacity of DG is limited to addressing only the local needs. One way to overcome these problems is to bring together many DG units and combine them into VPPs. VPPs provide an effective solution for integrating alternative energy sources into the grid, offering improved coordination and communication among DGs. Moreover, VPPs offer numerous advantages, such as alleviating the burden on the power grid, generating electricity locally, and mitigating energy losses, among others (Zamani et al., 2016). The

Intelligent Learning Approaches for Renewable and Sustainable Energy
DOI: https://doi.org/10.1016/B978-0-443-15806-3.00009-7
267

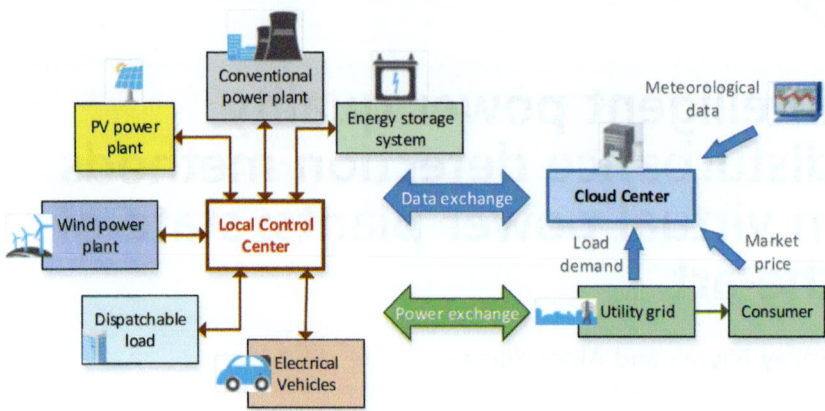

Figure 9.1 A basic VPP structure.

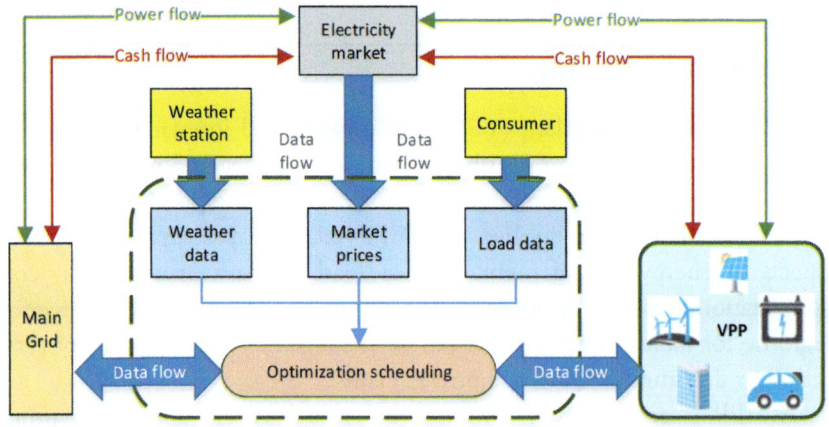

Figure 9.2 Operating structure of a general VPP system. (*PV*, Photovoltaic; *WPP*, Wind Power Plant; *ESS*, Energy Storage System; *EV*, Electric Vehicle; *Dr*, Demand Response; *CPP*, Conventional Power Plant).

general purpose of a VPP is to manage many DG units and active elements as storage and flexible loads from a standard central management system. In Fig. 9.2, the operating scheme of a general VPP system is given.

9.1.1 Relationship between distributed power systems and VPP

A VPP integrates various functional units into a centralized system. It can incorporate almost all types of energy production and storage technologies.

Distributed energy providers are established for collective management and can serve as VPPs in some grid scenarios. This approach allows the control of numerous DGs, storage elements, and flexible loads, offering flexibility without the need for commissioning or decommissioning traditional power plants. The installation of a VPP, as a reliable solution, presents itself as an alternative that can be confidently and comfortably utilized instead of traditional power plants.

Advanced solutions, such as the VPP, are essential for ensuring the efficient operation of the distribution grid by enabling real-time control and optimization of DGs. By correlating the load charts of all units, forecasting deviations, and facilitating free capacity trading during the day, a VPP can create optimized day-ahead load schedules for each balance group in the region. This system architecture is highly scalable and can accommodate a wide range of business models, from a few units to thousands of units. VPP optimizes its performance in real time by adjusting setpoints and providing balancing power while considering system limitations. The solution is capable of effectively processing vast amounts of data and signal flows, enabling optimum unit commitment and control of assets in a single step.

The load condition (critical, sensitive devices, etc.) is essential for the expected operating strategy's success in microgrid installation. The created structure must have the following properties:

- Load shedding or generation capabilities are essential in a VPP for automatic primary control functions that stabilize voltage and frequency.
- Enhancing PQ and reliability for critical and sensitive loads is of utmost importance.
- Peak load must be reduced to optimize the value of DGs.

VPP auctions refer to the trading of electricity capacity, which are considered "virtual" rather than "physical" divestments by one or more dominant players in a market. In contrast to traditional divestments where the physical plant is sold, VPP auctions involve virtual divestments of electricity capacity by one or more dominant firms. These firms retain management and control of the facility but sell its electricity capacity virtually by offering contracts to maximize its production. The contracts for electricity capacity are sold as auction items, which are considered fungible goods. The Electricité de France (EDF) auctions are globally recognized as the pioneer and most extended-running VPP auctions (Chowdhury et al., 2009). The EDF generation capacity auctions have been highly effective in maximizing electricity capacity production and ensuring grid stability.

9.1.2 Differences between microgrid and VPP

Microgrids and VPPs are two emerging technologies that are transforming the energy landscape. They share some critical features, such as DG power generation, and distribution-level storage. However, there are also differences between microgrids and VPPs, as highlighted in a study by Nosratabadi et al. (2017). The main properties of VPPs are:

1. VPPs are always connected to the grid.
2. VPPs can operate in an unexpected situation.
3. The presence or absence of archiving is possible for VPPs.
4. VPPs are highly dependent on information technology.
5. VPP is in "large geographic areas."
6. VPPs can be marketed to fill the wholesale market.
7. VPPs can be marketed on existing structures.

9.1.3 VPP structures and applications in the world

9.1.3.1 FENIX VPP (England)

The FENIX VPP project aims to enhance the integration of DGs by implementing a large-scale VPP system with a decentralized control approach (Kieny et al., 2009). FENIX box (FB) features a remote monitoring-control interface, facilitating commercial VPP, timing, and energy optimization of the DG group as well as technical VPP functionalities. It represents the production schedule and load sharing verification. In this structure, a 3 MW DG system consisting of various generators, including a 200 kW fuel cell, is used. When a market transaction occurs, a production schedule for the commercial DG is created, and this information is transferred to all DG devices. In real time, the DG monitors programs to generate/consume energy. The FENIX project seeks to improve the efficiency of DGs by incorporating them into the power grid and hierarchically managing them for optimal operation. The implementation of the FENIX VPP project is envisaged to offer a sustainable solution for ensuring future security, stability, and revenue generation of the European Union's (EU) electricity supply system, which is vulnerable to uncertainties resulting from renewable energy sources and market price fluctuations.

9.1.3.2 EDISON VPP (Denmark)

The EDISON VPP project was implemented on Bornholm Island, where the energy system connects every electric vehicle on the island via a VPP platform (Andersen et al., 2010). The VPP system comprises three main modules: a control module for each distributed generator, a data

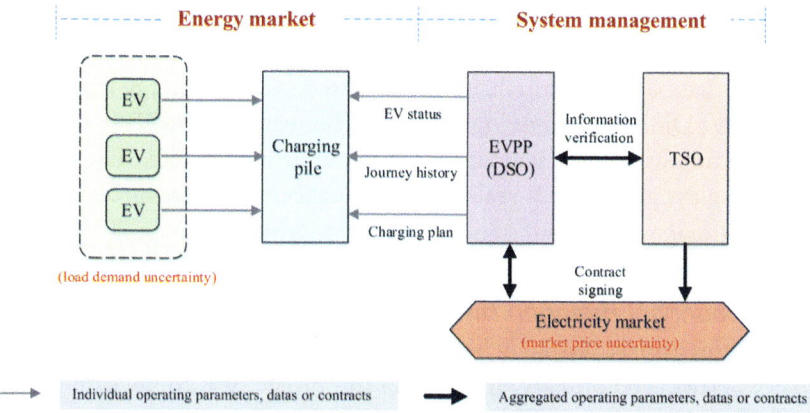

Figure 9.3 The operating structure of the EDISON project.

acquisition module, and a connectivity, collaboration, and communication module. This project involves using the energy storage capacity of electric vehicle batteries in the VPP to mitigate the output fluctuations of wind turbines and ensure grid stability (Fig. 9.3).

9.1.3.3 KONWERS VPP

The KONWERS VPP system is composed of controllable and noncontrollable units, with only the former being included in the maximum power share calculation. VPP structure includes CHP (which works with energy obtained by the combustion of biomass), wind turbines, photovoltaic (PV) power plants, and conventional power stations (Nikonowicz & Milewski, 2012). These DGs meet the central part of the electricity and heat demand, and in the case of insufficient electricity generation, extra electricity will be obtained from external sources. Therefore, the DGs must be connected to the external electricity grid to ensure a reliable energy supply.

9.1.3.4 NEMOCS VPP

The influx of renewable energy sources into electricity generation is contributing to an increase in fluctuations within the grid. A VPP can compensate for frequency variations of the grid by successfully predicting, monitoring, and controlling consumption. If the grid frequency becomes unstable, a group of flexible units step in and provide the desired power within a few seconds. Kraftwerke, an experienced energy company and operator of one of Europe's largest VPPs, has developed the NEMOCS VPP (Lehmbruck et al., 2020) platform. By combining decentralized energy sources, power generation can be adjusted depending on grid requirements, providing a stable and low-cost power supply. This system has more than 4000 decentralized energy assets with an energy capacity of 2,800 MW. Different technologies can be connected to the VPP and controlled remotely using a standard interface with the project. The control system displays and records real-time information about resources' current capacity, storage levels, and standby status (Hooshmand et al., 2018).

9.1.3.5 Shanghai Huangpu District (SHD) VPP (China)

In 2016, this project was launched in Shanghai, China (Virtual Power Plant Project, 2018). The primary objective of this project was to develop a VPP system for commercial buildings in the city, which leverages big data, smart energy, and internet technologies to enable automated, large-scale, and diverse demand responses for these buildings (Fig. 9.4). The project was located in the commercial zone, with the primary focus on the demand response of commercial buildings. However, demand response is susceptible to psychology and politics, making centralized control a challenging task. The SHD project aimed to overcome these obstacles and pave the way for efficient and effective demand response in the commercial building sector. Furthermore, uncertainty becomes more pronounced and robust following the introduction of large-scale renewable energy, rapidly leading to power system imbalances. In this project, demand and production information of DG on the consumer side are continuously monitored and evaluated by the demand side management center (dispatch center). Based on these data, critical control systems are operated in real time over the network. This enables the achievement of a balance between production and consumption, thereby realizing frequency control.

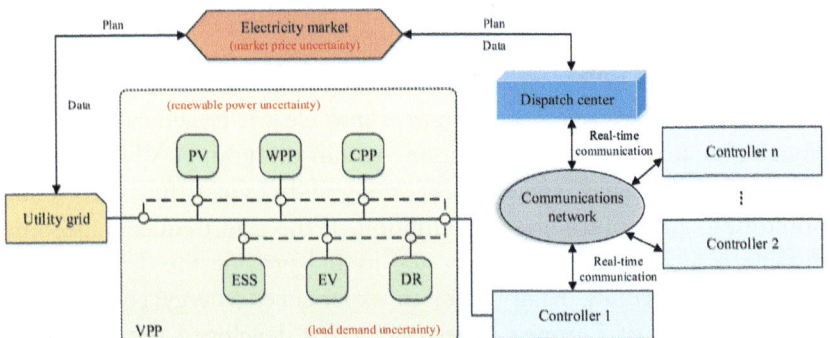

Figure 9.4 Operating structure of the SHD project.

9.2 Integration criteria and power quality

Vpps have been extensively researched and experimented with worldwide in recent years. Standards related to VPP structure are reflected by IEEE Std P1547 (Kroposki et al., 2007). The development of these standards is intended to address the gaps and insufficiencies identified in IEEE Std 1547−2003 (IEEE Std 1547, 2003).

Power quality disturbances (PQDs) such as island mode operation, harmonics, frequency changes, voltage drop/rise, flicker, notch, etc. can significantly impact the energy quality of VPPs. The IEEE Std 519 (2014) requires determining the voltage and current harmonic limits. As shown in Fig. 9.5, PQDs are affecting the VPPs' reliability.

9.2.1 Islanding operation

Islanding operation refers to the condition in which a DG continues to supply power to a portion of the electrical grid during a power outage, even though the main power grid is no longer energized. This can be a safety hazard for utility workers attempting to restore power to the affected area. Island mode operation can have various effects on power

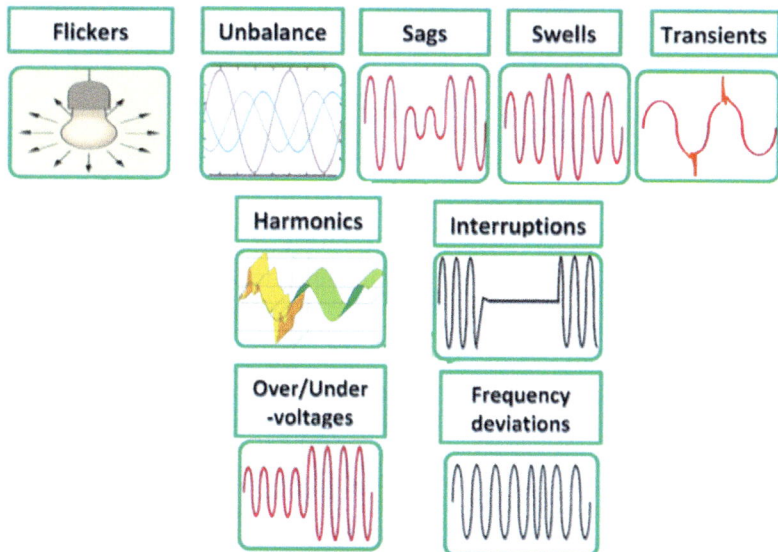

Figure 9.5 Power quality disturbances in VPPs.

system stability due to the loss of synchronization. The grid connection principles determined by IEEE 929—2008 std. are given in (Kroposki & DeBlasio, 2000). There are many studies in the literature regarding the detection of islanding conditions (Bayrak & Yılmaz, 2023; Bayrak et al., 2023; Yılmaz & Bayrak, 2022).

9.3 Power quality disturbances

9.3.1 Voltage sag, swell, and interruption

A voltage sag is defined as the decrease in voltage between 10% and 90% in a system operating at the grid frequency, with a time interval of 10 ms to 60 s. Motors cause sags with inrush currents, and other causes are overloads, failures occurring along the line, intermittent operations in the DG, etc. As per the guidelines set by IEEE 1159, a voltage swell refers to the elevation of the RMS voltage level, reaching a range between 110% and 180% of the nominal voltage. This phenomenon occurs at the power frequency and is observable for time spans ranging from half a cycle to one minute. These PQDs occur when high power loads are switched off, during malfunctions, or commissioning large capacity groups. Depending on the topological and physical location and system conditions, a fault may lead to transient voltage sag, swell, or interruption (Bayrak & Yılmaz, 2019; Yılmaz & Bayrak, 2019; Yılmaz et al., 2022, 2022).

9.3.2 Voltage fluctuations and unbalances

When the DG units are switched on and off, the voltage may drop or rise. As seen in Table 9.1, voltage variations during normal operation are limited to $\pm 3.3\%$ (IEC 61000-3-3, 2013). In addition, the nominal voltage should be within $\pm 10\%$ of the nominal value as required by the

Table 9.1 Voltage ripple limit values.

PV system operation status	Maximum value
When the PV system is connected	$\pm$ %3,3 V
When the PV system is disconnected	$\pm$ %3,3 V
As long as it is active (95%)	$\pm$ %10

standards 95% of the time it is active. Analysis should be made at the limit values that the system can withstand, especially when analyzing the voltage, such as minimum production-maximum consumption. An equal amount of DG power plant power should be distributed to each phase to prevent voltage unbalance.

9.3.3 Harmonics

The amount of total harmonic distortion in the voltage of the power system is restricted based on the IEEE 519 standards. In low-voltage grids, the total harmonic distortion limit is 8%, with each harmonic value required to be less than 5%.

9.3.4 Flicker

The flicker caused by voltage fluctuations is a power quality deterioration event caused by consumer loads independently from the distribution grid. Flicker limit values must comply with IEC 61000-2-2 and IEC 6100−3-3 standards (IEC 61000-2-2, 2002). In the Electricity Grid Regulation of the Energy Market Regulatory Authority, rapid voltage changes occurring more than 10 times in 1 hour are considered flickering. Long-term and short-term flicker limit values are given in Table 9.2.

9.3.5 Notch

Voltage waveform distortion can occur due to the number of pulses generated by a converter in a period in a grid frequency operating system. The voltage is short-term for half a period, and the notch effect creates high-frequency components that cannot be accurately measured with traditional harmonic measuring devices. The limit values for voltage waveform distortion are defined by the IEEE 519−2014 standard, which takes into account the notch area and depth, as presented in Table 9.3.

Table 9.2 Flicker limit values.

	PV plant	Consumer
P_{ST} (Short term)	≤ 1.00	≤ 1.00
P_{LT} (Long term)	≤ 0.65	≤ 0.80

Table 9.3 Notch limit values in voltage according to IEE-519−2014.

	Special applications (Hospital, Airport, etc.)	General applications	Systems using converters
Notch area (A_N)	16,400	22,800	36,500
Notch depth	%10	%20	%50

9.3.6 Noise

Noise can originate from power electronics devices, loads with switched power supplies, and high arc devices. It damages sensitive devices such as microprocessors and PLCs.

9.3.7 DC current injection

Inverters in the PV production plant connected to the LV grid can sometimes misbehave and inject DC current into the system. DC current injection drives the distribution transformer to saturation, causing the waveform to be out of nominal conditions. According to IEEE Std 929−2008 standards, this DC current supplied to the system via inverters needs to be limited to 0.5% of the nominal current.

9.3.8 Frequency deviations

The frequency deviation of a power system refers to the difference between the actual system frequency and its nominal or rated frequency. It is caused by the mismatch between the amount of load and the amount of the generated power. The frequency change's size and duration depend on the response of the control system against the load. The limit ranges of the average values of the operating frequency measured in the DG system and the grid according to the operating periods during the year characterized by the criteria outlined in the TS EN 50160 standard.

9.3.9 Inverter standards in terms of power quality

Inverters differ from the existing electrical power system due to their bidirectional nature. While the PQ in the current electricity distribution system is affected only by the load, a PV system introduces factors influenced by both the production and consumption sides. Plants generating electricity from solar energy must comply with the criteria for connection to the grid (current harmonic values, voltage fluctuation limits and allowed flickering levels, etc.) determined by IEC standards

Table 9.4 Breaker trip time limits for voltage and frequency.

IEC 61727	Lower limit	Upper limit	Trip time
Voltage	%50	%135	0,1/0,05 s
	%85	%110	2 s
Frequency	−1 Hz	+1 Hz	0,2 s

in Turkey. In addition, the PQ parameters of the energy generated in grid-connected PV systems must be within the range specified in the TS EN 50160 standard. Inverters used in PV systems are standardized in terms of electromagnetic compatibility (EMC), grid frequency monitoring, grid voltage monitoring, and grid loss. These tolerances are determined by IEC 61727. Mains voltage and frequency monitoring limits and breaker opening time, according to IEC 61727, are shown in Table 9.4.

9.4 Planning parameters in VPP systems

9.4.1 Defining the objective function and formulation

Optimal scheduling problems can be discussed through two types of formulation or modeling, namely probabilistic and deterministic. Fig. 9.6 illustrates a stochastic procedure for formulating planning and achieving optimal bidding in VPP (Nosratabadi et al., 2017). This method involves reducing all the scenarios to a lower number and performing, leading to finalizing optimum energy bids.

A microgrid aims to supply electricity at the lowest possible cost, and thus, an optimization problem can be formulated by considering various mandatory and optional constraints. The solution to this problem provides the optimal power generated by different sources in the microgrid while satisfying the constraints. The cost of the energy resources is minimized, depending on the constraints that are mandatory or optional. In contrast, VPP systems focus on maximizing profit or benefits. When a VPP reduces the load demand, it is regarded as virtual production. Energy resources may lack capacity, flexibility, and adequate controls when operating independently, hindering system management and marketing activities. These issues can be addressed by aggregating energy sources and flexible loads

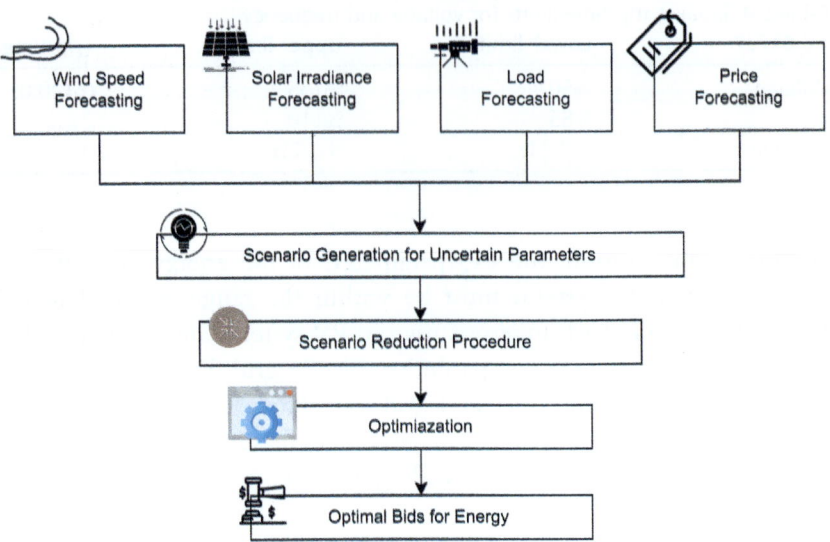

Figure 9.6 The procedure of bidding strategy in VPP.

in a VPP. The primary objective of scheduling DG resources in VPPs is to maximize profits while adhering to constraints, which can be either mandatory or optional.

9.4.2 Planning associated with the solving method

When it comes to solving scheduling problems that are defined as optimization models, there are various challenges to address. The bidding strategy in VPP involves preparing relevant input data and selecting an appropriate approach for expressing and solving the optimization model. For tackling these challenges, numerous methods have been developed for solving scheduling problems. The two types of optimization techniques used for VPPs are heuristic and mathematical approaches. In terms of mathematical optimization, VPP-related methods can be divided further into different categories, such as linear programming and nonlinear programming. Other solution methods include decision trees, event-oriented service–oriented frameworks, fuzzy logic, and net equivalent solution methods (Bayrak et al., 2023; Nosratabadi et al., 2017).

9.4.3 Uncertainties

The concept of uncertainty is often used to predict future events or explain physical measurements that lack a clear explanation. Sources of uncertainty can stem from both random and observable environments, as well as omitted factors. Uncertainty can significantly impact the planning and scheduling of microgrids and Vpps, with wind speed, solar radiation, local loads, and the market price as the critical components that may be uncertain.

9.4.4 Reactive power planning

Controlling reactive power is a crucial aspect of power system programming. Reactive power control can be considered a system-wide constraint or a result of the scheduling problem. Thus, a detailed analysis of reactive power in microgrids and VPPs is necessary.

9.4.5 Emission problem

Environmental concerns and emissions are important factors to consider in power systems due to the negative impact of conventional energy sources on the environment, particularly greenhouse gas emissions. Therefore, in scheduling renewable energy sources in both microgrid and VPP systems, minimizing emissions has been a focus in the literature.

9.4.6 Stability

Stability in VPPs refers to a power system's capacity to sustain its steady-state operating conditions and to recover from disturbances, such as faults, sudden load changes, or unexpected events. Voltage and frequency stability are two main categories of power system dynamics. The scheduling of microgrids and VPPs has been addressed in some articles regarding these stability phenomena (Liu et al., 2015; Moutis & Hatziargyriou, 2015; Unger et al., 2012).

9.5 Electrical infrastructure and physical evaluation For Gökçeada VPP

9.5.1 Gökçeada interruption details

Interruptions in Gökçeada are usually long-term period. These power interruptions can be caused by various factors, including bird interference, equipment malfunctions such as dislocated boxes or damaged transformers, falling trees, disconnections caused by damage, issues with low voltage (LV) boxes or connectors or cable damage, and maintenance requirements for medium voltage (MV) overhead lines or surge arresters. The most extended interruption recorded was 13.77 hours. The shortest interruption recorder lasted 0.01 hours. The faults occurred at a similar rate on both the LV and MV sides. Users most affected by malfunctions are LV customers both in the center and rural areas. The reasons and durations of some of the faults in Gökçeada are given in Table 9.5.

9.5.2 Gökçeada load profile, solar and wind energy potential

Within the project's scope, two wind turbines of 900 kW installed power in Gökçeada, a 210 kW solar power plant, and a 4×770 kVA diesel generator of UEDAŞ will be connected to the VPP system to be installed (*Mix Storage Approaches: Leading* Gökçeada Island (UEDAS), 2021). Approximately 100 smart counters will be installed in a transformer area. Also, an energy storage facility with a storage capacity of 100 kWh and an output power of 50 kW will be established on the island. Fig. 9.7 shows the total active power consumption profile of Gökçeada between the 2017 and 2020 years.

The load profile reveals a surge in energy consumption during the summer, particularly from June to September. Additionally, the island's overall energy consumption is correlated with the number of attendees throughout the year. With a maximum energy consumption of 5 MW on the island, renewable energy-based power plants emerge as a more suitable solution to meet the electrical energy requirements of the island.

The total active energy generation of the solar power plant (SPP) installed in Gökçeada compared to active energy consumption is shown in Fig. 9.8 Notably, the power generated by the solar power plant significantly exceeds the power consumed. This observation suggests the appropriateness of harnessing the island's solar energy potential.

In the island where the wind energy potential is quite high, it is seen that two wind power plants each with a power capacity of 900 kWp, are deemed highly suitable for the planned establishment of the Virtual Power Plant (VPP) system. Thus, the wind-based power generation is more suitable for meeting the island's electrical energy requirements. The total active energy generation of WPP-1 and WPP-2 in Gökçeada compared to active energy consumption is shown in Fig. 9.9.

Table 9.5 Interruptions.

Interruption reason	Source LV/MV	Interruption time short/long	Planned/unplanned	Interruption duration (hour)	MV customer in center	LV customer in center	MV customer in rural area	LV customer in rural area
LV box problem	LV	Long	Unplanned	0,06	0	0	0	29
LV Connector damage	LV	Long	Unplanned	0,35	0	0	0	1
LV Connector damage	LV	Long	Unplanned	0,52	0	0	0	53
LV overhead Conductor damage	LV	Long	Unplanned	0,55	0	1	0	0
LV cable damage	LV	Long	Unplanned	0,58	0	138	0	0
LV connector damage	LV	Long	Unplanned	0,64	0	0	0	81
LV box problem	LV	Long	Unplanned	0,66	0	0	0	137
LV customer cable	LV	Long	Unplanned	0,82	0	1	0	0
LV box problem	LV	Long	Unplanned	0,97	0	0	0	104
MV others	MV	Short	Unplanned	0,01	10	3794	18	1516
MV others	MV	Short	Unplanned	0,05	10	3729	17	1460
MV others	MV	Short	Unplanned	0,05	10	3758	18	1502
MV others	MV	Long	Unplanned	0,49	10	3716	17	1455
Birds touch	MV	Long	Unplanned	0,5	0	0	32	709
MV insulator damage	MV	Long	Unplanned	0,63	0	1	33	722
Transformer maintenance	MV	Long	Planned	0,63	0	374	1	193

(*Continued*)

Table 9.5 (Continued)

Interruption reason	Source LV/MV	Interruption time short/long	Planned/ unplanned	Interruption duration (hour)	MV customer in center	LV customer in center	MV customer in rural area	LV customer in rural area
Birds touch	MV	Long	Unplanned	0,64	0	0	3	206
MV insulator damage	MV	Long	Unplanned	0,65	0	1	26	721
MV cable damage	MV	Long	Unplanned	0,68	10	3716	17	1455
MV others	MV	Long	Unplanned	0,68	0	0	31	704
MV others	MV	Long	Unplanned	0,77	8	2	22	689
Birds touch	MV	Long	Unplanned	0,85	0	213	0	0
Average				0,54	2,6	883,8	10,7	533,5
Total				11,78	58	19.444	235	11.737

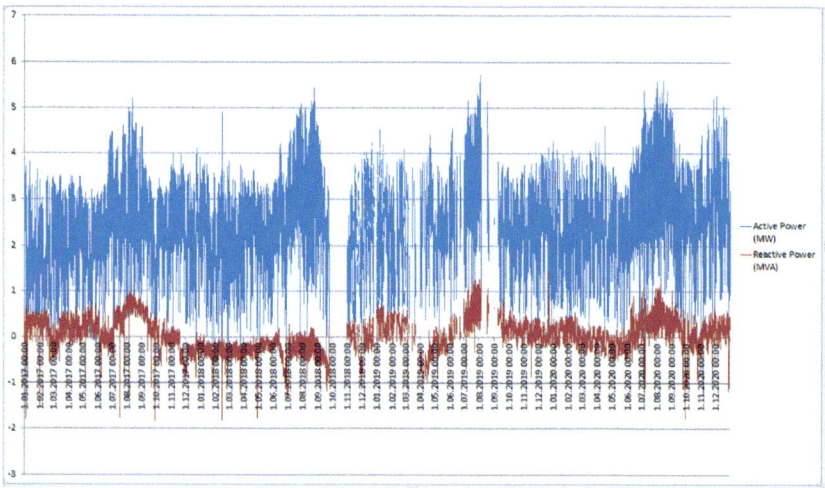

Figure 9.7 Total active power consumption profile of Gökçeada between 2017 and 2020.

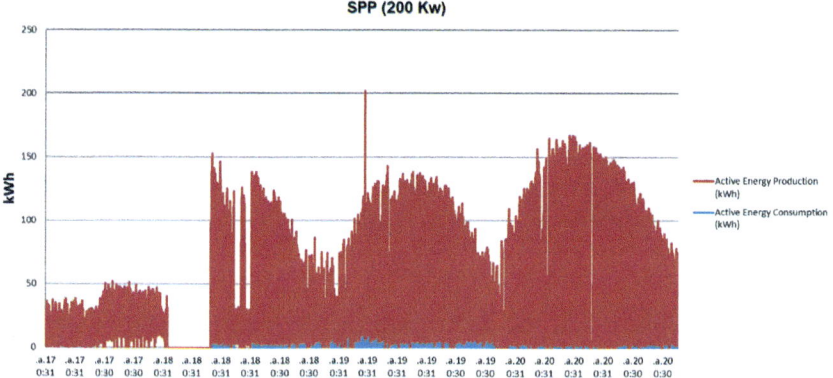

Figure 9.8 Total active energy generation of solar power plant installed in Gökçeada compared to active energy consumption.

9.5.3 Gökçeada VPP model

The generalized electrical single-line diagram of Gökçeada is shown in Fig. 9.10. To observe the short circuit power and current contribution of the integrated DGs in the power system, single line-to-ground (L-G), double line-to-ground (L-L-G), and line-to-line (L-L) asymmetric fault and three-phase symmetrical fault (L-L-L) were realized. In this part of the study, the results of the faults generated in the Santral DM (bus fault),

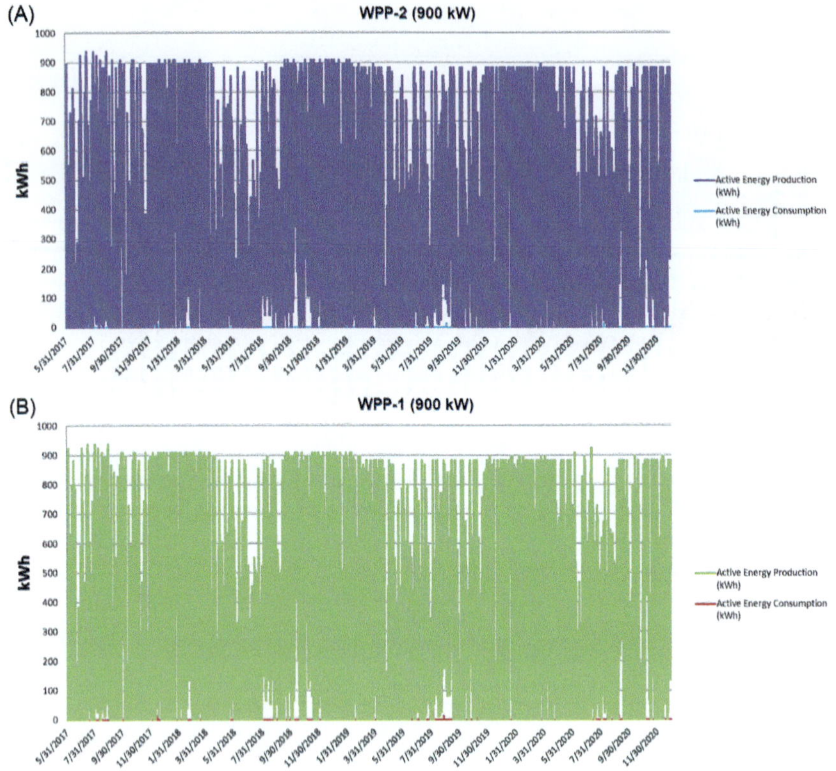

(A)

WPP-2 (900 kW)

Figure 9.9 (A, B) Total active energy generation of WPPs in Gökçeada compared to active energy consumption.

and the faults formed in the 25%th km of the Santral DM-Özlüce Kök lines (in-line fault) are presented. Fig. 9.10 depicts the simulated model in which the analyses are performed.

9.6 Results and conclusion

9.6.1 Bus fault analysis results

Table 9.6 shows the values of voltage and phase angle after the L-G fault at the Santral DM bus. In this case, the fault impedance values are R = 0.01 Ω and X = 0.07 Ω. Voltage values are given in per unit (pu).

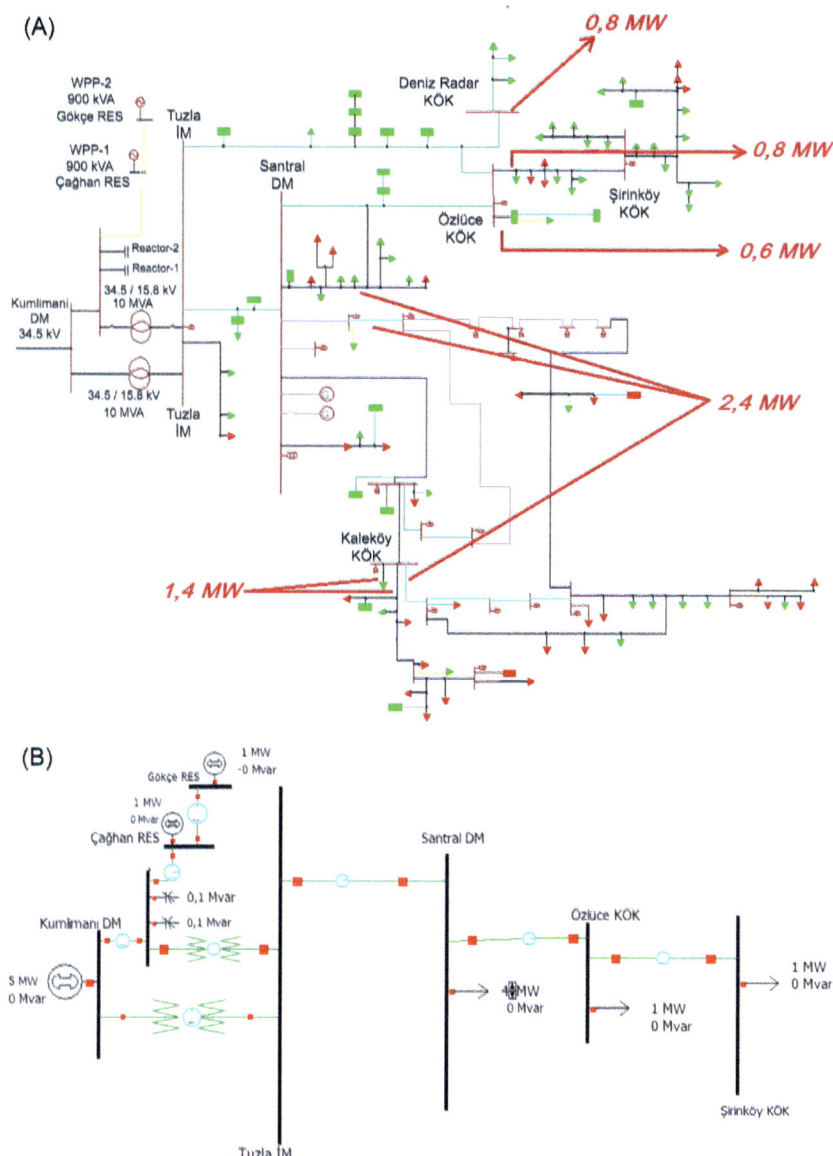

Figure 9.10 (A) Generalized power distribution in Gökçeada, (B) Simulated model.

Table 9.7 depicts the values of voltage and phase angle after L-L short circuit fault at the Santral DM bus. In this case, the fault impedance values are R = 0.01 Ω and X = 0.07 Ω. Voltage values are given in (pu).

After an L-L-L fault on the Santral DM bus, Table 9.8 shows the voltage and phase angle at system buses. The fault impedance values in this

Table 9.6 L-G short circuit fault voltage and phase angle values.

Bus name	Phase A	Phase B	Phase C	Angle A	Angle B	Angle C
Kum limanı DM	1,00000	1,00000	1,00000	−0,01	−120,01	119,99
Tuzla İM	0,01029	1,72418	1,73413	74,64	−150,46	149,45
Çağhan RES	1,00000	1,00000	1,00000	0,00	−120,00	120,00
Gökçe RES	1,00000	1,00000	1,00000	0,04	−119,96	120,04
Özlüce KÖK	0,00226	1,72739	1,72523	−108,09	−150,87	149,16
Şirinköy KÖK	0,00309	172756	1,72462	−108,35	−150,89	149,14
Santral DM	0,00000	1,72689	1,72689	−3,20	−150,79	149,21

Table 9.7 L-L short circuit fault voltage and phase angle values.

Bus name	Phase A	Phase B	Phase C	Angle A	Angle B	Angle C
Kum limanı DM	1,00000	0,61885	0,58316	−0,01	−147,43	145,13
Tuzla İM	0,99966	0,59050	0,54687	−0,22	−152,91	150,07
Çağhan RES	1,00000	0,69146	0,65940	0,00	−138,98	136,50
Gökçe RES	1,00000	0,71205	0,68204	0,04	−136,98	134,66
Özlüce KÖK	0,99635	0,50238	0,49840	−0,92	−175,54	173,67
Şirinköy KÖK	0,99610	0,50225	0,49827	−0,96	−175,59	173,62
Santral DM	0,99702	0,50272	0,49874	−0,79	−175,42	173,79

Table 9.8 L-L-L balanced short circuit fault voltage and phase angle values.

Bus name	Phase A	Phase B	Phase C	Angle A	Angle B	Angle C
Kum limanı DM	0,41742	0,41742	0,41742	−3,41	−123,41	116,59
Tuzla İM	0,34965	0,34965	0,34965	−4,92	−124,92	115,08
Çağhan RES	0,54929	0,54929	0,54929	−2,58	−122,58	117,42
Gökçe RES	0,58381	0,58381	0,58381	−2,30	−122,30	117,70
Özlüce KÖK	0,10313	0,10313	0,10313	−3,23	−123,23	116,27
Şirinköy KÖK	0,10310	0,10310	0,10310	−3,27	−123,27	116,73
Santral DM	0,10320	0,10320	0,10320	−3,10	−123,10	116,90

scenario are $R = 0.01 \ \Omega$ and $X = 0.07 \ \Omega$. The voltage values are reported in pu.

Table 9.9 shows the values of fault currents and degrees after short circuit faults at the Santral DM bus. The value given in the table shows the value of the short circuit current.

9.6.2 Line fault analysis results

Table 9.10 depicts the values of voltage and phase angle after the L-L fault at the 2.45th km (%25) of the Sentral DM-Özlüce Kök line. In this case,

Table 9.9 Simulated system short circuit current values.

Fault type	L-L		L-L-L	
Fault location	I_{kss}	Angle	I_{kss}	Angle
Santral DM Bus	4,8702	−175,10	5,3329	−84,97

Table 9.10 Voltage and phase angle values for L-L line fault.

Bus name	Phase A	Phase B	Phase C	Angle A	Angle B	Angle C
Kum limanı DM	1,00000	0,63070	0,59416	−0,01	−145,93	143,49
Tuzla İM	0,99966	0,60203	0,55751	−0,22	−151,05	148,01
Çağhan RES	1,00000	0,70219	0,66987	0,00	−138,04	135,50
Gökçe RES	1,00000	0,72228	0,69206	0,04	−136,17	133,80
Özlüce KÖK	0,99635	0,50195	0,49852	−0,92	−175,74	173,87
Fault point location	0,99685	0,50220	0,49877	−0,83	−175,65	173,96
Şirinköy KÖK	0,99610	0,50182	0,49839	−0,96	−175,79	173,82
Santral DM	0,99702	0,50909	0,50003	−0,79	−172,00	170,25

Table 9.11 Voltage and phase angle values for L-L-G line fault.

Bus name	Phase A	Phase B	Phase C	Angle A	Angle B	Angle C
Kum limanı DM	1,00000	0,61530	0,57761	−0,01	−148,13	145,75
Tuzla İM	1,59762	0,34127	0,22567	−2,81	−78,30	56,91
Çağhan RES	1,00000	0,68793	0,65437	0,00	−139,43	136,86
Gökçe RES	1,00000	0,70864	0,67726	0,04	−137,38	134,96
Özlüce KÖK	1,59484	0,11839	0,11839	−0,09	−147,64	−34,19
Fault point location	1,59527	0,11863	0,11863	−3,19	−34,58	−34,58
Şirinköy KÖK	1,59461	0,11828	0,11828	−3,28	−33,90	−33,90
Santral DM	1,59542	0,13864	0,10649	−3,17	−47,30	−18,23

the fault impedance values are $R = 0.01\ \Omega$ and $X = 0.07\ \Omega$. Voltage values are given in per unit (pu).

Table 9.11 shows the values of voltage and phase angle after L-L-G at the 2.45th km (%25) of the Sentral DM-Özlüce Kök line. In this case, the fault impedance values are $R = 0.01\ \Omega$ and $X = 0.07\ \Omega$. Voltage values are given in pu. After an L-L-L three-phase symmetrical fault in the 2.45th km (%25) of the Santral DM-Özlüce Kök line, Table 9.12 depicts the voltage and phase angle on the system buses for the same fault location.

Table 9.13 shows the values of fault currents and degrees after short circuit faults at the 2.45th km (%25) of the Sentral DM-Özlüce Kök line.

The amount of power provides the point at which they are linked, and the point at which the fault occurs in the system all influence the size

Table 9.12 Voltage and phase angle values for L-L-L symmetrical line fault.

Bus name	Phase A	Phase B	Phase C	Angle A	Angle B	Angle C
Kum limanı DM	0,43842	0,43842	0,43842	−3,36	−123,36	116,64
Tuzla İM	0,37307	0,37307	0,37307	−4,76	−124,76	115,224
Çağhan RES	0,56555	0,56555	0,56555	−2,55	−122,55	117,45
Gökçe RES	0,59882	0,59882	0,59882	−2,28	−122,28	117,72
Özlüce KÖK	0,09946	0,09946	0,09946	−2,99	−122,99	117,01
Fault point location	0,09951	0,09951	0,09951	−2,89	−122,89	117,11
Şirinköy KÖK	0,09944	0,09944	0,09944	−3,03	−123,03	116,97
Santral DM	0,13533	0,13533	0,13533	−3,62	−123,62	116,38

Table 9.13 Simulated system L-L and L-L-L short circuit current values.

Fault type	L-L		L-L-L	
Fault location	I_{kss}	Angle	I_{kss}	Angle
Santral DM Bus	4,68745	−174,87	5,14262	−84,97

of the short circuit fault current. As shown in the tables, in the event of a fault, distributed energy sources such as WT and PV systems connected to the main grid cause increased short–circuit currents. In the case where generator power is distributed by spreading to buses, the short circuit current flowing from the fault point is the lowest level except in the case of feeding on the grid. The results show that DG's contribution to short-circuit fault analysis is within acceptable bounds.

Acknowledgments

The authors are thankful to Uludag Electricity Distribution Company (UEDAS) R&D Center for providing Gökçeada electricity distribution data and also thank Supervisor Mehmet KOC, who is the supervisor of the R&D Center and Cem KIZILKAYA who is the R&D Specialist for their supports.

References

Andersen, P. B., Træholt, C., Marra, F., Poulsen, B., Binding, C., Gantenbein, D., Jansen, B., & Sundstroem, O. (2010). Electric Vehicle Fleet Integration in the Danish EDISON Project: A Virtual Power Plant on the Island of Bornholm. In *2010 IEEE Power & Energy Society General Meeting*. Available from https://doi.org/10.1109/PES.2010.5589605.

Bayrak, G., & Yılmaz, A. (2023). *A deep learning-based Islanding detection approach by considering the load demand of DGs under different grid conditions prediction techniques for renewable energy generation and load demand forecasting*. Singapore: Springer Nature Singapore. Available from https://doi.org/10.1007/978-981-19-6490-9_4.

Bayrak, G., & Yılmaz, A. (2019). Assessment of power quality disturbances for grid integration of PV power plants. *Sakarya Üniversitesi Fen Bilimleri Enstitüsü Dergisi*, *23*(1), 35–42.

Bayrak, G., Küçüker, A., & Yılmaz, A. (2023). Deep learning-based multi-model ensemble method for classification of PQDs in a hydrogen energy-based microgrid using modified weighted majority algorithm. *International Journal of Hydrogen Energy*, 48(18), 6824−6836. Available from https://doi.org/10.1016/j.ijhydene.2022.05.137.

Bayrak, G., Yılmaz, A., & Çalışır, A. (2023). A new intelligent decision-maker method determining the optimal connection point and operating conditions of hydrogen energy-based DGs to the main grid. *International Journal of Hydrogen Energy*. Available from https://doi.org/10.1016/j.ijhydene.2023.02.043.

Chowdhury, S., Chowdhury, S. P., & Crossley, P. (2009). *Microgrids and Active Distribution Networks* (pp. 1−298). South Africa: Institution of Engineering and Technology. Available from http://doi.org/10.1049/PBRN006E.

Hooshmand, R.-A., Nosratabadi, S. M., & Gholipour, E. (2018). Event-based scheduling of industrial technical virtual power plant considering wind and market prices stochastic behaviors − A case study in Iran. *Journal of Cleaner Production, 172*, 1748−1764. Available from https://doi.org/10.1016/j.jclepro.2017.12.017.

IEC 61000-2-2. (2002). Electromagnetic compatibility (EMC) − Part 2-2: Compatibility levels for low-frequency conducted disturbances and signaling in public low-voltage power supply systems.

IEC 61000-3-3. (2013). Electromagnetic compatibility (EMC) − Part 3-3: Limits − Limitation of voltage changes, voltage fluctuations and flicker in public low-voltage supply systems, for equipment with rated current $\leq$ 16 A per phase and not subject to conditional connection.

IEEE Std 519. (2014). IEEE Recommended Practice and Requirements for Harmonic Control in Electric Power Systems. Available from https://doi.org/10.1109/IEEESTD.2014.6826459.

IEEE Std 1547. (2003). IEEE Interconnecting distributed resources with electric power systems. IEEE Standard. Available from https://doi.org/10.1109/IEEESTD.2003.94285.

Kieny C. Berseneff B. Hadjsaid N. Besanger Y. Maire J. (2009, December) 2009 IEEE Power and Energy Society General Meeting, PES '09, France On the concept and the interest of Virtual Power plant: Some results from the European project FENIX. Available from https://doi.org/10.1109/PES.2009.5275526.

Kroposki, B., & DeBlasio, R. (2000). Technologies for the new millennium: photovoltaics as a distributed resource. 2000 Power Engineering Society Summer Meeting (Cat. No.00CH37134), Seattle, WA, USA, 2000, pp. 1798−1801 vol. 3. Available from https://doi.org/10.1109/PESS.2000.868807.

Kroposki B. Pink C. Basso T. DeBlasio R. (2007, December) 2007 IEEE Power Engineering Society General Meeting, PES, United States Microgrid standards and technology development. Available from https://doi.org/10.1109/PES.2007.386053.

Lehmbruck, L., Kretz, J., Aengenvoort, J., & Sioshansi, F. (2020). *Aggregation of front- and behind-the-meter: The evolving VPP business model* (pp. 211−232). Elsevier BV. Available from https://doi.org/10.1016/b978-0-12-819951-0.00010-4.

Liu, Y., Xin, H., Wang, Z., & Gan, D. (2015). Control of virtual power plant in microgrids: A coordinated approach based on photovoltaic systems and controllable loads. *IET Generation, Transmission & Distribution, 9*(10), 921−928. Available from https://doi.org/10.1049/iet-gtd.2015.0392.

Mix storage approaches: Leading Gökçeada island (UEDAS). (2021). https://vpp4islands.eu/index.php/demo-site/.

Moutis, P., & Hatziargyriou, N. D. (2015). Decision trees-aided active power reduction of a virtual power plant for power system over-frequency mitigation. *IEEE Transactions on Industrial Informatics, 11*(1), 251−261. Available from https://doi.org/10.1109/tii.2014.2371631.

Nikonowicz, è. B., & Milewski, J. (2012). Virtual power plants-general review: Structure, application, and optimization. *Journal of Power Technologies*, *92*(3), 135−149.

Nosratabadi, S. M., Hooshmand, R.-A., & Gholipour, E. (2017). A comprehensive review on microgrid and virtual power plant concepts employed for distributed energy resources scheduling in power systems. *Renewable and Sustainable Energy Reviews*, 341−363. Available from https://doi.org/10.1016/j.rser.2016.09.025.

Thavlov, A., & Bindner, H. W. (2015). Utilization of flexible demand in a virtual power plant set-up. *IEEE Transactions on Smart Grid*, *6*(2), 640−647. Available from https://doi.org/10.1109/TSG.2014.2363498.

Unger D. Spitalny L. Myrzik J.M.A. (2012, November) 2012 IEEE Energytech, Energytech 2012 Available from https://doi.org/10.1109/EnergyTech.2012.6304637 Germany Voltage control by small hydro power plants integrated into a virtual power plant.

Virtual power plant project. (2018). http://news.bjx.com.cn/html/20180403/889567.shtml/.

Yılmaz, A., & Bayrak, G. (2019). A real-time UWT-based intelligent fault detection method for PV-based microgrids. *Electric Power Systems Research*, *177*105984. Available from https://doi.org/10.1016/j.epsr.2019.105984.

Yılmaz., & Bayrak, G. (2022). A new signal processing-based islanding detection method using pyramidal algorithm with undecimated wavelet transform for distributed generators of hydrogen energy. *International Journal of Hydrogen Energy*, *47*(45), 19821−19836. Available from https://doi.org/10.1016/j.ijhydene.2022.03.114.

Yılmaz, A., Küçüker, A., Bayrak, G., Ertekin, D., Shafie-Khah, M., & Guerrero, J. M. (2022). An improved automated PQD classification method for distributed generators with hybrid SVM-based approach using un-decimated wavelet transform. *International Journal of Electrical Power & Energy Systems*, *136*107763. Available from https://doi.org/10.1016/j.ijepes.2021.107763.

Yılmaz, A., Küçüker, A., & Bayrak, G. (2022). Automated classification of power quality disturbances in a SOFC&PV-based distributed generator using a hybrid machine learning method with high noise immunity. *International Journal of Hydrogen Energy*, *47*(45), 19797−19809. Available from https://doi.org/10.1016/j.ijhydene.2022.02.033.

Yu, S., Fang, F., Liu, Y., & Liu, J. (2019). Uncertainties of virtual power plant: Problems and countermeasures. *Applied Energy*, *239*, 454−470. Available from https://doi.org/10.1016/j.apenergy.2019.01.224.

Zamani, A. G., Zakariazadeh, A., & Jadid, S. (2016). Day-ahead resource scheduling of a renewable energy based virtual power plant. *Applied Energy*, *169*, 324−340. Available from https://doi.org/10.1016/j.apenergy.2016.02.011.

Index

Note: Page numbers followed by "*f*" and "*t*" refer to figures and tables, respectively.

CPI Antony Rowe
Eastbourne, UK
April 22, 2024